The State of the Nation's Ecosystems

We all rely on a familiar set of indicators—interest rates, unemployment, inflation, the Dow Jones index, and GDP, for example—to gauge the performance of the national economy. No such measures are currently available to describe the environment.

The State of the Nation's Ecosystems lays out a blueprint for periodic reporting on the condition and use of ecosystems in the United States. Developed by experts from businesses, environmental organizations, universities, and federal, state, and local government agencies, it is designed to provide policymakers and the general public with a succinct and comprehensive—yet scientifically sound and nonpartisan—view of "how we are doing." Participants and contributors represent a wide array of political perspectives and include experts from the fields of ecology, chemistry and toxicology, hydrology, oceanography, limnology, use of satellite remote sensing, forestry, farming, range management, and many others. The book identifies the major characteristics of ecosystems that should be tracked through time to provide this view, and where possible, provides information on both current conditions and historic trends. The book also highlights key gaps—situations where data do not exist or have not been assembled to support national reporting. Separate chapters report on coasts and oceans, farmlands, forests, fresh waters, grasslands and shrublands, and urban and suburban areas. These ecosystem-specific indicators are complemented by "core national indicators" that provide a highly aggregated view of overall conditions.

Who should be interested in reading this book?

- Decision makers in natural resource management and environmental policy in federal, state, and local government agencies, as well as in environmental organizations, businesses, and trade associations
- Academics with a research or teaching interest in environmental quality and ecosystem condition
- Interested laypersons
- Instructors for environmental studies and ecosystems courses, who may also wish to use the volume as a main or supplementary textbook for students to illustrate key aspects of ecosystems in the United States

The State of the Nation's Ecosystems provides a prescription for "taking the pulse" of America's lands and waters. It identifies what should be measured, counted, and reported so that decision makers and the public can understand the changes that are occurring in the American landscape.

Coasts and Oceans

Farmlands

Forests

Fresh Waters

Grasslands and Shrublands

Urban and Suburban Areas

The State of The Nation's Ecosystems

Measuring the Lands, Waters, and Living Resources of the United States

THE H. JOHN HEINZ III CENTER FOR SCIENCE, ECONOMICS AND THE ENVIRONMENT

About the Heinz Center

Established in December 1995 in honor of Senator John Heinz, The H. John Heinz III Center for Science, Economics and the Environment is a nonprofit, nonpartisan institution dedicated to improving the scientific and economic foundation for environmental policy through multisectoral collaboration. Focusing on issues that are likely to confront policymakers within two to five years, the Center fosters collaboration among industry, environmental organizations, academia, and government in each of its program areas and projects. It uses the best scientific and economic analyses to develop viable options to solving problems, and its findings and recommendations are widely disseminated to public and private sector decision makers, the scientific community, and the public.

PUBLISHED BY THE PRESS SYNDICATE OF THE UNIVERSITY OF CAMBRIDGE
The Pitt Building, Trumpington Street, Cambridge, United Kingdom

CAMBRIDGE UNIVERSITY PRESS
The Edinburgh Building, Cambridge CB2 2RU, UK
40 West 20th Street, New York, NY 10011-4211, USA
477 Williamstown Road, Port Melbourne, VIC 3207, Australia
Ruiz de Alarcón 13, 28014 Madrid, Spain
Dock House, The Waterfront, Cape Town 8001, South Africa

http://www.cambridge.org

First published 2002

Printed in the United States of America

A catalog record for this book is available from the British Library.

Library of Congress Cataloging-in-Publication Data
The state of the nation's ecosystems : measuring the lands, waters, and living resources of the United States / The H. John Heinz III Center for Science, Economics, and the Environment
p. cm.
Includes bibliographical references.
ISBN 0-521-52572-1 (pb.)
1. Environmental monitoring—United States—Methodology. 2. Ecosystem health—United States. I. H. John Heinz III Center for Science, Economics and the Environment.
QH104 .S73 2002
333.7'2—dc21 2002073890

Designed by Janin/Cliff Design, Inc., Washington, D.C.
Printed on recycled paper

The State of the Nation's Ecosystems is also available on line at www.heinzctr.org/ecosystems.

The H. John Heinz III Center for Science, Economics and the Environment
1001 Pennsylvania Avenue, NW, Suite 735 South
Washington, DC 20004
Tel: (202) 737-6307 Fax: (202) 737-6410 e-mail: info@heinzctr.org

Contents

Foreword

The State of the Nation's Ecosystems initiates a series of periodic reports on the lands, waters, and living resources of the United States.

The report has been prepared *for* decision makers, opinion leaders, and informed citizens who seek an authoritative, comprehensive, and succinct overview of what the nation most needs to know about the changing state of its ecosystems.

The report has been prepared *by* experts from government, the private sector, environmental organizations, and academia through an intense five-year collaborative process. This involved hundreds of contributors and reviewers from all four sectors, publication of a prototype to solicit public commentary, and feedback on several drafts from a wide array of interested groups and experts.

The report emerging from this process presents a unique system of indicators that is simultaneously *relevant* to contemporary policy and decision making, *balanced* and *unbiased* in what it chooses to report on, and *scientifically credible* in the data it presents. We hope and believe that *The State of the Nation's Ecosystems* and its planned successors will help to strengthen the empirical foundation for American environmental policymaking in the same way that the emergence of solid data about changes in GDP, employment, and inflation helped to strengthen the country's economic policymaking in the last half-century.

The completion of this first report on *The State of the Nation's Ecosystems* shows that a sustained, multisector collaborative approach to environmental reporting can make inroads on many of the problems of parochialism, perceived bias, and variable quality that have plagued previous efforts. We believe that the articulation of a coherent framework for reporting, a clear-eyed assessment of the strengths and weaknesses of available data, and the identification of data gaps are important advances. Its strengths notwithstanding, however, we are well aware that this report is at best an early step on a long path toward realization of the comprehensive, mature, and well-grounded system of ecosystem and environmental reporting that the nation deserves.

A number of specific steps are needed over the next five years in preparation for a second full edition of *The State of the Nation's Ecosystems.* First, the Heinz Center will actively solicit feedback on this report, continuing the practice—begun with the 1999 prototype report—of using each completed step as the basis for future improvements. Second, we believe that a multisector effort is needed to address key gaps identified in this report. For almost half the indicators identified in this report as necessary to characterize the state of the nation's ecosystems, gaps in scientific understanding, operational monitoring, or data coordination have made it impossible to produce useful national data. Finally, we hope to foster a broad and inclusive dialogue on where and how a permanent effort to produce a continuing series of high-quality reports on the state of the nation's ecosystems could best be housed, administered, and funded. We pledge our own commitment to working with government at all levels, the private sector, environmental groups, and academia in ensuring that these issues are addressed in a timely and serious manner.

It is our pleasure to thank the extraordinary group of individuals and organizations that have worked together to realize this first report on *The State of the Nation's Ecosystems.*

The foundations of this effort are the countless professionals and supporting organizations involved in the exacting work of ecosystem monitoring. Without them, there would simply be nothing of quality to report. The sources of data drawn on in this report—sources from government, the private sector, environmental groups, and academia—are cited on the individual indicator pages and in the technical notes.

Despite the substantial donations of time and talent from the groups already described, this report would not have been possible without substantial financial support from more than twenty federal, private, and philanthropic sources. These funders are named—and thanked—on p. xvii, and we would like to add our grateful appreciation for their support.

The work of defining an overall indicator system, applying it to specific ecosystems, and identifying and evaluating candidate data sets fell largely on the backs of the Design Committee and Work Groups convened by the Heinz Center. These individuals—nearly 150 in all—are listed on pp. x–xvi. To a person, they took part with enthusiasm, openness, creativity, and dedication.

Oversight and review of the work of the Design Committee and Working Groups were provided at two levels. Strategically, the balance and relevance of the overall reporting effort was reviewed periodically by a small group of senior advisors (see p. x) and the Heinz Center Board of Trustees (see p. xvi). Quality assurance on more specific aspects of the report was provided through a rigorous process of peer review, involving nearly 100 experts from all four sectors (these reviewers are listed on the Heinz Center's Web site, www.heinzctr.org/ecosystems).

At the Heinz Center itself, our first thanks go to the first president of the Center, Bill Merrell. Bill's leadership of the Center in its formative years made a reality of the multisector, nonpartisan, science-based principles on which it was founded. He was instrumental in seeing "environmental reporting" as a key area for enhancing the contribution of science and economics to policy, in conceptualizing the present effort, in recruiting those who have led it over the past 5 years, and in putting together the broadly based funding package that has supported it.

On the Heinz Center staff, a wonderfully creative, adaptable, and dedicated group of professionals herded the multiple cats of the *State of the Nation's Ecosystems* project to produce an integrated product. Robert M. Friedman, the Center's Vice President for Research, guided the overall effort with a light hand and a keen, insightful mind. Kent Cavender-Bares, Research Associate and analyst par excellence, served as the project's nerve center for data analysis and presentation and contributed in countless ways to every aspect of the report. Jeannette L. Aspden, the Center's Research Editor, exhibited true flexibility and creativity in ensuring that the final product was of excellent quality and consistency, despite having been written in literally hundreds of separate pieces over several years. And Elissette Rivera, Kate Wing, and Heather Blough, Research Assistants, provided technical, logistical, and administrative support for the project, without which the data needed to produce this report would not have been obtained or analyzed, the meetings needed to reach agreement on what indicators were appropriate would not have been held, and the myriad other necessary details would not have been attended to. These individuals were aided in their work by the frequent and cheerful efforts of—at one point or another—every member of the Heinz Center staff, all of whom pitched in at critical points to lighten the load.

Finally, however, we must single out for thanks Robin O'Malley, the Project Manager of the *State of the Nation's Ecosystems* project. He has been a consummate project manager, keeping an immensely complicated and dynamic process running on time with a reasonable degree of synchrony; alternately prodding, chiding, and soothing multiple contesting egos; writing not only the text that he promised, but also the text that others promised but forgot to complete; and delivering an uncounted number of ever-better briefings. Beyond these impressive managerial accomplishments, however, Robin has also played a central role in shaping the structure and content of this report, coming up with original analytical approaches, prescient criticism and comments, and original syntheses. He has, in fact, emerged as one of the nation's foremost experts on the state of the nation's ecosystems. It has been an honor and a pleasure to work with him in creating this report.

William C. Clark
Harvey Brooks Professor
John F. Kennedy School of Government
Harvard University

Thomas Jorling
Vice President
Environmental Affairs
International Paper

Thomas E. Lovejoy
President
The Heinz Center

Participants

Senior Advisors

Thomas C. Jorling *(Chair)* (1997–2002)
Vice President
Environmental Affairs
International Paper

William Clark (1997–2002)
Chair, Design Committee (1997–2002)
Harvey Brooks Professor of International Science, Public Policy and Human Development
John F. Kennedy School of Government
Harvard University

Jared Cohon (2000–2002)
President
Carnegie Mellon University

John Flicker (1997–1999)
President
National Audubon Society

Jack Hunt (2000–2002)
President
The King Ranch

Theodore Hullar (1997–1999)
Director
Cornell Center for the Environment
Cornell University

Charles Johnson (1997–1999)
Chairman, President and CEO
Pioneer Hi-Bred International

Fred Krupp (2000–2002)
Executive Director
Environmental Defense

The Honorable James Saxton (1997–2002)
Congressman
New Jersey 3rd District

John Sawhill *(deceased)* (1997–1999)
President and CEO
The Nature Conservancy

Design Committee

William C. Clark *(Chair)* (1997–2002)
Harvey Brooks Professor of International Science, Public Policy and Human Development
John F. Kennedy School of Government
Harvard University

Lee Alverson (2000–2002)
Chairman of the Board
Natural Resources Consultants

Tom Bancroft (1997–2002)
Vice President, Ecology and Economics Research Department
The Wilderness Society

Rosina Bierbaum (1997–2002)
Professor and Dean, School of Natural Resources and Environment
University of Michigan
(Formerly, Associate Director for Environment, Office of Science and Technology Policy)

Carrol Bolen (1997–1999)
Vice President of Legal and Government Affairs
Pioneer Hi-Bred International, Inc.

Mike Burton (2000–2002)
Executive Officer
Portland Metro

Lawrence Clark (1999–2000)
Deputy Chief, Science and Technology
Natural Resources Conservation Service, USDA

Terry Davies (1997–1999)
Senior Fellow
Resources for the Future

Steve Daugherty (1997–2002)
Director, Government Affairs
Pioneer Hi-Bred International, Inc

David E. Ervin (1997–1999)
Professor
Environmental Sciences and Resources Program
Portland State University, and
Senior Policy Analyst
Wallace Center for Agricultural and Environmental Policy, Winrock International

Terry Garcia (1997–1999)
Executive Vice President
National Geographic Society
(Formerly, Assistant Secretary for Oceans and Atmosphere, National Oceanic and Atmospheric Administration)

Mike Goodman (2000–2002)
Chair, Croplands Work Group (2000–2001)
Grand Prairie Area Director
The Nature Conservancy

Richard Guldin (1997–2002)
Director of Science Policy, Planning, Inventory and Information
USDA Forest Service

James R. Hendricks (2000–2002)
Vice President
Corporate Environment, Health and Safety
Duke Energy Corporation

William Robert Irvin (2000–2002)
Director, U.S. Ecoregional Conservation
World Wildlife Fund

Deborah Jensen (1997–1999)
Vice President for Conservation Science
The Nature Conservancy

John L. Knott, Jr. (2000–2002)
Chair, Urban/Suburban Work Group (2000–2002)
CEO / Managing Director
Dewees Island

Alan Lucier (1997–2002)
Chair, Forest Work Group (1997–1999)
Senior Vice President
National Council for Air and Stream Improvement, Inc.

Gail Mallard (2000–2002)
Chair, Freshwater Work Group (2000–2002)
Senior Advisor for Water Resources
U.S. Geological Survey

Thomas Malone (2000–2002)
Co-chair, Coasts and Oceans Work Group (2000–2002)
Director, Horn Point Laboratory
University of Maryland Center for Environmental Science

Suzanne Iudicello Martley (1997–2002)
Co-chair, Coasts and Oceans Work Group (2000–2002)
Independent Marine Conservation Writer
(Formerly, Special Counsel for Fisheries, Center for Marine Conservation)

Maurice Mausbach (1997–1999)
Deputy Chief, Soil Survey and Resource Assessment, Natural Resources Conservation Service, USDA

Jerry Melillo (1997–2002)
Co-director and Senior Scientist
Ecosystem Center
Marine Biological Laboratory

Reed Noss (2000–2002)
Chair, Forest Work Group (2000–2002)
Principal Scientist
Conservation Science Inc.

Mike Nussman (1997–1999)
Vice President
American Sportfishing Association

Gordon Orians (2000–2002)
Professor Emeritus
Department of Zoology
University of Washington

Duncan Patten (2000–2002)
Chair, Grasslands and Shrublands Work Group (2000–2002)
Professor Emeritus
Arizona State University, and
Research Professor
Montana State University

Fran Pierce (1997–1999)
Chair, Farmlands Work Group (1997–1999)
Director, Center for Precision Agricultural Systems
Washington State University
(Formerly, Professor, Crop and Soil Sciences Department, Michigan State University)

Louis F. Pitelka (1997–2002)
Director, Appalachian Laboratory
University of Maryland Center for Environmental Science

William Riley (2000–2002)
General Manager, Environmental Affairs
Bethlehem Steel Corporation

Paul Sabatier (1997–1999)
Department of Environmental Sciences and Policy
University of California, Davis

Don Scavia (1997–2002)
Chair, Coasts and Oceans Work Group (1997–1999)
Chief Scientist
National Ocean Service
National Oceanic and Atmospheric Administration

Mark Schaefer (1997–2000)
President
NatureServe
(Formerly, Deputy Assistant Secretary for Science, U.S. Department of the Interior)

Bruce Stein (2000–2002)
Vice President for Programs
NatureServe

Bud Ward (1997–1999)
Executive Director
Environmental Health Center

Nancy Wheatley (2000–2002)
Vice President
Royal Caribbean Cruise Lines

Douglas Wheeler (1997–2002)
Partner
Hogan and Hartson, LLP
(Formerly, Secretary, Resources Agency of California)

Agency Liaison

Tom Muir (2000–2002)
Office of Science and Technology Policy/ U.S. Geological Survey

Coasts and Oceans Work Group

Thomas Malone *(Co-chair 2000–2002, member 1997–2002)*
Director, Horn Point Environmental Laboratory
University of Maryland Center for Environmental Science

Suzanne Iudicello Martley *(Co-chair 2000–2002, member 1997–2002)*
Independent Marine Conservation Writer

Bob Bailey (2000–2002)
Ocean Program Administrator
State of Oregon

Brian Baird (1997–1999)
Ocean Program Manager
California Resources Agency

Gilbert T. Bergquist (1997–1999)
Associate Director
Florida Center for Public Management Florida State University

Brock Bernstein (2000–2002)
Independent Consultant

Denise Breitburg (1997–1999)
Estuarine Research Center
The Academy of Natural Sciences

David J. Detlor (2000–2002)
Oceanographer
National Marine Fisheries Service

Carlos Fetterolf (1997–1999)
National Sea Grant Review Panel

William S. Fisher (2000–2002)
Senior Research Biologist
Office of Research & Development, NHEERL/Gulf Ecology Division
Environmental Protection Agency

John Gauvin (2000–2002)
Director
The Groundfish Forum

John Hoey (1997–1999)
National Marine Fisheries Service
(Formerly, National Fisheries Insititute)

Dale Keifer (2000–2002)
Professor
Department of Biological Sciences
University of Southern California

Lynn Kutner (1997–1999)
NatureServe

Douglas Lipton (1997–1999)
Coordinator, Sea Grant Extension Program
Department of Agricultural and Resource Economics
University of Maryland

Michael Mac (1997–1999)
Director
Columbia Environmental Research Center
U.S. Geological Survey

Patrick O'Brien (1997–2002)
Team Leader, Ecological Services
ChevronTexaco Energy Research and Technology Company

Andrew Robertson (2000–2002)
Director, Center for Coastal Monitoring and Assessment
National Ocean Service
National Oceanic and Atmospheric Administration

Don Scavia *(Chair 1997–1999)*
Chief Scientist
National Ocean Service
National Oceanic and Atmospheric Administration

Jack A. Sobel (1997–1999)
Director, Ecosystem Programs
The Ocean Conservancy *(formerly the Center for Marine Conservation)*

Kevin Summers (1997–2002)
Associate Director for Science
Office of Research & Development, NHEERL/ Gulf Ecology Division
Environmental Protection Agency

Steven B. Weisberg (2000–2002)
Executive Director
Southern California Coastal Water Research Project Authority

David Wilmot (2000–002)
Executive Director, Living Oceans Program
National Audubon Society

Farmlands Work Group

Mike Goodman, *Chair* (2000–2001)
Grand Prairie Area Director
The Nature Conservancy

Jeri Berc (1997–1999)
Special Assistant to the Deputy Chief for Soil Survey and Resource Assessment
Natural Resources Conservation Service, USDA

Chet Boruff (2000–2002)
Farmers National Marketing Group

Frank Casey (2000–2002)
Natural Resources Economist
Defenders of Wildlife

Craig Cox (2000–2002)
Executive Director
Soil and Water Conservation Society

Daniel Dooley (1997–1999)
Partner
Dooley, Herr & Williams

George Hallberg (1997–2002)
Principal
Cadmus Group

Jerry L. Hatfield (2000–2002)
Director
National Soil Tilth Laboratory
Agricultural Research Service, USDA

Maureen Kuwano Hinkle (1997–1999)
Director, Agricultural Policy *(Retired)*
National Audubon Society

Jim Hrubovcak (1997–1999)
Resource Economics Division
Economic Research Service, USDA

Richard R. Johnson *(deceased)* (1997–1999)
Deere & Co. Technical Center

James A. LaGro, Jr. (2000–2002)
Professor
Department of Urban & Regional Planning
University of Wisconsin-Madison

Daryl Lund (2000–2002)
Program Analyst
Resources Inventory Division, Natural Resources Conservation Service, USDA

Deborah Neher (2000–2002)
Department of Earth, Ecological and Environmental Sciences
The University of Toledo

Thomas E. Nickson (2000–2002)
Ecological Technology Center
Monsanto Company

Egide Nizeyimana (1997–1999)
Senior Research Associate, Office for Remote Sensing of Earth Resources
Pennsylvania State University

Gary W. Petersen (1997–1999)
Co-director, Office for Remote Sensing of Earth Resources
Pennsylvania State University

Fran Pierce *(Chair 1997–1999)*
Director, Center for Precision Agricultural Systems
Washington State University
(Formerly, Professor, Crop & Soil Sciences Department, Michigan State University)

Terry Z. Riley (2000–2002)
Director of Conservation
Wildlife Management Institute

David S. Shriner (1997–1999)
Assistant Director for Research
North Central Research Station
USDA Forest Service

B. A. Stewart (1997–2002)
Director, Dryland Agriculture Institute
West Texas A&M University

William G. Wilber (2000–2002)
National Synthesis Coordinator
National Water Quality Assessment Program,
U.S. Geological Survey

Fresh Waters Work Group

Gail Mallard, *Chair* (2000–2002)
Senior Advisor for Water Resources
U.S. Geological Survey

Patrick Brezonik (2000–2002)
Director
Water Resources Research Center
University of Minnesota

Scott Dyer (2000–2002)
Principal Scientist, Miami Valley Labs
Procter & Gamble

Leo M. Eisel (2000–2002)
Principal Engineer
Brown and Caldwell

Otto Gutenson (2000–2002)
Ecologist
Office of Wetlands, Oceans and Watersheds
Environmental Protection Agency

Daniel Markowitz (2000–2002)
Senior Scientist
Malcolm Pirnie, Inc.

Lawrence L. Master (2000–2002)
Chief Zoologist
NatureServe

Judy L. Meyer (2000–2002)
Science and Technical Advisor
American Rivers, and
Institute of Ecology
University of Georgia

Ralph Tiner (2000–2002)
National Wetlands Inventory
U.S. Fish and Wildlife Service

Terry F. Young (2000–2002)
Senior Consulting Scientist
Environmental Defense

Agency Liaisons

Paul Bertram (2000–2002)
Great Lakes National Program Office
Environmental Protection Agency

Kim Devonald *(deceased)* (2000–2002)
Office of Research and Development
Environmental Protection Agency

Charles Spooner (2000–2002)
Office of Wetlands, Oceans and Watersheds
Environmental Protection Agency

Forest Work Group

Reed Noss, *Chair* (2000–2002)
Principal Scientist
Conservation Science, Inc.

Marc D. Abrams (2000–2002)
Steimer Professor
School of Forest Resources
Pennsylvania State University

Tom Bancroft (1997–2002)
Vice President, Ecology and Economics Research Department
The Wilderness Society

Jimmy Bullock (2000–2002)
Manager
Sustainable Forestry & Wildlife
International Paper

Arthur W. Cooper (1997–1999)
Professor
Department of Forestry
North Carolina State University

David Darr (1997–1999)
Staff Assistant
Resource Valuation and Use Research
USDA Forest Service

Dominick A. DellaSala (1997–1999)
World Wildlife Fund

Peter Farnum (1997–1999)
Director, Timberlands Forestry Research
Weyerhaeuser Company

Thomas R. Fox (2000–2002)
Associate Professor
Virginia Polytechnic Institute and State University *(formerly with Rayonier, Inc.)*

Dennis Grossman (2000–2002)
Vice President for Science
NatureServe

Rich Guldin (1997–2002)
Director of Science Policy, Planning, Inventory and Information
USDA Forest Service

Deborah Jensen (1997–1999)
Vice President for Conservation Science
The Nature Conservancy

Dennis H. Knight (2000–2002)
Professor Emeritus
Botany Department
University of Wyoming

Alan Lucier *(Chair 1997–1999, member 1997–2002)*
Senior Vice President
National Council for Air and Stream Improvement, Inc.

Robert Mangold (1997–1999)
National Manager, Forest Health Monitoring Program
USDA Forest Service

Helga Van Miegroet (1997–1999)
Associate Professor
Wildland Soils and Biogeochemistry
Utah State University

Eric Palola (2000–2002)
National Wildlife Federation

Brad Smith (2000–2002)
National Program Leader
Forest Inventory and Analysis
USDA Forest Service

Grasslands and Shrublands Work Group

Duncan Patten, *Chair* (2000–2002)
Research Professor
Montana State University, and
Professor Emeritus
Arizona State University

Hugh Barrett (2000–2002)
Rangeland Management Specialist
Bureau of Land Management

Carl E. Bock (2000–2002)
Professor
Department of Environmental, Population, and Organismic Biology
University of Colorado

Jane Bock (2000–2002)
Professor Emeritus
Department of Biology
University of Colorado

Bob Budd (2000–2002)
Director of Science, Stewardship and Planning
The Nature Conservancy of Wyoming

Len Carpenter (2000–2002)
Southwest Field Representative
The Wildlife Management Institute

Leonard Jolley (2000–2002)
State Range Ecologist–California
Natural Resources Conservation Service, USDA

Bill Miller (2000–2002)
Rancher and President
Malpai Borderlands Project

John E. Mitchell (2000–2002)
Rangeland Scientist
Rocky Mountain Research Station
USDA Forest Service

Timothy Seastedt (2000–2002)
Arctic & Alpine Research Institute
University of Colorado

Gregg Simonds (2000–2002)
President, Agricultural Enterprises
The Ensign Group

Thomas J. Stohlgren (2000–2002)
USGS/Biological Resources Division
Colorado State University

Stephen C. Torbit (2000–2002)
Senior Scientist
National Wildlife Federation

Agency Liaisons

Alison Hill (2000–2002)
National Program Leader
Grassland Ecology Research
USDA/Forest Service

Richard Mayberry (2000–2002)
Rangeland Management Specialist
Bureau of Land Management

Dennis W. Thompson (2000–2002)
Natural Resources Conservation Service, USDA

Urban and Suburban Work Group

John L. Knott, Jr. *Chair* (2000–2002)
CEO/Managing Director
Dewees Island

William S. Alverson (2000–2002)
Conservation Ecologist
Environmental and Conservation Programs
The Field Museum

Roger Bannerman (2000–2002)
Environmental Specialist
Wisconsin Department of Natural Resources

Margaret Carreiro (2000–2002)
Associate Professor
Biology Department
University of Louisville

Caren E. Glotfelty (2000–2002)
Director, Environment Program
The Heinz Endowments

Michael W. Klemens (2000–2002)
Director, Metropolitan Conservation Alliance
Wildlife Conservation Society

Michael Pawlukiewics (2000–2002)
Director, Environmental Research
The Urban Land Institute

Pete Sandrock (2000–2002)
Chief Operating Officer
Portland Metro

Sara Vickerman (2000–2002)
Director, West Coast Office
Defenders of Wildlife

Jianguo Wu (2000–2002)
Associate Professor
Department of Plant Biology
Arizona State University

Wayne Zipperer (2000–2002)
Research Forester
USDA Forest Service
SUNY College of Environmental Sciences

Heinz Center Project Staff

Robin O'Malley (1997–2002)
Fellow and Program Manager

Robert M. Friedman (1997–2002)
Senior Fellow and Vice President for Research

Jeannette Aspden (2001–2002)
Fellow and Research Editor

Kent Cavender-Bares (2000–2002)
Fellow and Research Associate

Elissette Rivera (2000–2002)
Research Assistant

Kate Wing (1998–2000)
Research Assistant

Heather Blough (1997–1998)
Research Assistant

Heinz Center Board of Trustees

Acknowledgments

This report would not have been possible without the support, encouragement, and patience of many, many individuals and institutions.

We would like to name—and thank—the many individuals who have assisted in some way, but we cannot. Besides those who participated as members of one of our many committees (and who are named on pages x–xv), there are literally hundreds who have contributed time and expertise, including many who work within our partner organizations, but with whom we have not worked directly. We are also deeply grateful to the nearly 100 reviewers who commented on the draft of this report; your insightful contributions helped us immensely, and your names are listed on our Web site (www.heinzctr.org/ecosystems). Any list would inevitably miss someone, so, rather than provide an incomplete list, we offer our sincere thanks to each and every one of you—you know who you are!

Funding for this project was also provided by a diverse array of sources. We would like to thank our corporate and foundation funders: Chevron Company, Cleveland Foundation Special Fund No. 6, Foundation for Environmental Research, John Deere & Company, Electric Power Research Institute, Georgia-Pacific Corporation, Vira I. Heinz Foundation, Andrew W. Mellon Foundation, Richard King Mellon Foundation, Charles Stewart Mott Foundation, David and Lucile Packard Foundation, Procter & Gamble Company, and Royal Caribbean Cruise Lines.

We would also like to thank the federal agencies that provided financial support: the Department of Agriculture, the Department of Defense, the Department of Energy, the Department of the Interior, the Environmental Protection Agency, the Federal Emergency Management Agency, the National Aeronautics and Space Administration, the National Oceanic and Atmospheric Administration, and the National Science Foundation. The Office of Naval Research also provided grants administration support, for which we are grateful.

We worked with a host of public and private institutions in preparing this report. They provided data, enabled their staff to serve on committees, assisted with reviews, and made any number of other valuable contributions. We are deeply grateful for their help and support. There were, however, several institutions whose assistance went well beyond the norm: the U.S. Geological Survey, the USDA Forest Service, and NatureServe (which was a part of the Nature Conservancy when this project began) all provided insight, information, and expertise in amounts beyond what we could have asked for.

We are also grateful to Cynthia Cliff and James Durham of Janin/Cliff Design, Inc., for their skill, insight and patience—this volume would not be what it is without them.

Finally, we thank the members of The Heinz Center Board of Trustees, without whose support this project would not have come to fruition, and all of our present and former colleagues at The Heinz Center, whose patience and pitching in made it all possible.

Part I:

The State of the Nation's Ecosystems: Philosophy, Framework, and Findings

This volume is intended as the first in a series of periodic reports on the extent, condition, and use of the lands, waters, and living resources of the United States. The "ecosystem indicators" that form the heart of the *State of the Nation's Ecosystems* have been selected through a nonpartisan collaboration among government, environmental organizations, the private sector, and the academic community. The indicators thus represent a unique consensus on how the nation's ecosystems can be described—and their status tracked over time—in a fair and balanced way. They characterize what is most important to know about the nation's ecosystems, rather than merely reflecting what happens to have been measured. Finally, the data presented for each indicator are based on solid science—on verified measurements, not opinion—that have been reviewed by experts from all sectors of society.

For all these reasons, this report will be a valuable tool for environmental decision makers at all levels and in all sectors of society. It should also provide Americans with a new way of looking at and talking about ecosystems that will help them evaluate the potential, and actual, effects of both public and private management decisions.

Part I lays out the fundamental principles on which this report is based, describes the nature of the indicators and defines the ecosystems on which we report, and summarizes the highlights of our findings.

The State of the Nation's Ecosystems

- Is designed as a blueprint for periodic reporting
- Is written for decision makers and the public, by scientists and other experts
- Presents a succinct set of indicators chosen by representatives from business, environmental organizations, academia, and federal, state, and local government
- Describes conditions without saying whether they are "good" or "bad" or recommending policies or actions
- Reports on the state or condition of ecosystems, not on pollution or other stresses, or on government or private programs and actions
- Describes a balanced range of ecosystem conditions and goods and services that benefit society
- Includes trends or other comparative information where available
- Highlights key information gaps

Chapter 1:

Reporting on the State of the Nation's Ecosystems

A Clear Need

Americans' support for sound environmental policy is strong, nonpartisan, and consistent,[1] reflecting recognition of the high cost—both monetary and otherwise—of a damaged environment. But the costs of ensuring a clean, safe, and healthy environment are also significant. In 1994, the last year for which government estimates are available, the United States spent more than $120 billion on pollution abatement and control—nearly 2% of the nation's gross domestic product[2]—and this amount is only a part of the total cost of ensuring a clean, healthy, and vibrant environment.[3]

Each year, the federal government alone spends more than $600 million collecting environmental data and, through regulatory requirements, imposes additional costs on the private sector, for monitoring of emissions and effluents.[4] State and local government and environmental organizations also devote considerable resources to environmental monitoring, as does the private sector, above and beyond what is required for simple compliance. These efforts, reported in a host of individual documents and Web sites, provide crucial information without which this project would not have been possible. They do not, however, provide the high-level, comprehensive account on the state of the nation's ecosystems that is the goal of this project.

Given the importance and cost of environmental protection, it is hardly surprising that the need for a periodic report on "how we are doing" in our environmental management efforts has been recognized for at least three decades. In 1970, the Council on Environmental Quality noted in its first annual report to Congress that the efforts of that time did "not provide the type of information or coverage necessary to evaluate the condition of the Nation's environment or to chart changes in its quality and trace their causes."[5] Since then, virtually every comprehensive study of national environmental protection has called for more coherent and comprehensive information on the state of our environment. The National Academy of Sciences and the National Academy of Public Administration are among the many organizations that have recognized this need.[6]

But despite some excellent syntheses of data on specific problems and places, there is no periodic, comprehensive, and reliable compilation of essential information about the overall state of the nation's environment.[7] As a result, policymakers and other stakeholders are swamped by increasing volumes of data that nonetheless seem to neglect important issues. Society all too often ends up arguing not about the issues, but about the relevance and validity of the data on which the prospects for a substantive policy debate depend.

For a nation deeply committed to protecting the environment, this is an unacceptable state of affairs. It is as though we would seek to develop sound economic policy without having reliable measures of the nation's GDP, unemployment, or inflation rate, relying instead on idiosyncratic reports from individual firms, sectors, unions, and local chambers of commerce. We cannot know whether our current environmental policies and practices are sound, and we cannot make new policy with confidence, without a similar set of generally accepted measures of fundamental properties of the environment.

Origin, Principles, and Process

Late in 1995, as part of its review of federal environmental monitoring efforts, the White House Office of Science and Technology Policy (OSTP) asked The Heinz Center to create a nonpartisan, scientifically grounded report on the state of the nation's environment. Acknowledging the relatively sophisticated reporting that already existed on many physical and chemical components of the environment (e.g., air quality, stream flows), OSTP proposed that The Heinz Center focus on ecosystems—that is, on the nation's living resources and the landscapes and waters they inhabit.

In undertaking this effort, The Heinz Center and its collaborators were guided by a fundamental conviction that, to be useful, *The State of the Nation's Ecosystems* must

- **Be scientifically credible.** Too many earlier efforts were disregarded because they were perceived as willing to accept any data available, or because their conclusions were not based in sound science. The report's content must benefit from input and review from a wide range of scientific and technical experts.
- **Be nonpartisan, both in content and in process.** Too many previous reporting efforts failed because they were perceived to be politicized or because they seemed to promote the perspectives of particular interests. Any hope for greater success requires that this effort be seen as fair and unbiased by a broad cross section of political interests.
- **Engage the expertise and experience of the nation's environmental monitoring programs and professionals.** Any attempt to characterize the state of the nation's ecosystems will fail without the cooperation of those who are engaged full time in the exacting and important profession of ecosystem monitoring and reporting.
- **Benefit from experimentation and learning.** No effort as ambitious as this could be expected to get everything right the first time around. Any hope for success depends on the ability to learn from the inevitable mistakes and to incorporate new data and understanding as they become available.

To implement these principles, the Center developed a funding strategy that depended upon joint support from government, industry, and private foundations. It assembled a small in-house staff and a large team of part-time collaborators drawn from government, the private sector, environmental organizations, and academia. A Design Committee, with members drawn from all of these groups, oversaw the entire project and made crucial decisions regarding approach, indicator selection, content, tone, and format of the overall document. Technical Work Groups, also representing a cross section of societal perspectives, were assembled to provide expertise in particular ecosystems. Their members identified the indicators included in this report, selected and assessed the data sources we relied upon, and drafted much of the descriptions and technical materials. Finally, a group of senior advisors and the Center's own Board of Trustees reviewed the project's strategic directions, with special attention to ensuring broad and balanced representation. Overall, nearly 150 individuals have participated in the project as committee and group members, with many more involved as contributors, reviewers, and advisors. (See p. x for a listing of committee members and p. xvii for additional acknowledgments.)

The Heinz Center established its working committees and began working intensively in late 1997. The project reached a key milestone with the release in late 1999 of a prototype report for public comment, covering three ecosystems: forests, farmlands, and coasts and oceans. This prototype was revised significantly in response to comments, and three additional ecosystems (fresh waters, grasslands and shrublands, and urban and suburban areas) were added. The process concluded with an extensive external review of a draft version of the present text in late 2001. Nearly 100 sets of comments were received from reviewers in business, environmental, government, and academic institutions. The end result of these steps—the first full report on *The State of the Nation's Ecosystems*—is presented in the chapters that follow.

Next Steps

This first edition of the *State of the Nation's Ecosystems* is issued simultaneously in a print version, published by Cambridge University Press, and in a Web version available at www.heinzctr.org. Subsequent reports in this series will incorporate new data and understanding, as well as comments, criticism, and suggestions from users of this initial edition. The Heinz Center actively solicits feedback, either by mail or through the *State of the Nation's Ecosystems* Web site.

This report is the first in what is intended to be a regular series of reports on the state of the nation's ecosystems. A variety of activities will be needed to produce the next edition. These include filling data gaps and improving the consistency of both data and indicators, consulting with key scientific communities in order to refine and clarify certain indicators, working with public and private agencies to regularize the provision of data in the form needed for national reporting, and strengthening the linkages between this project and others concerned with ecosystem reporting. The Heinz Center plans to undertake such activities following publication of this first report and is currently seeking the resources to do so, in anticipation of publishing the next report in the series in 2007.

One of the needs for the immediate future is to create the mechanisms for producing and updating the report on a regular basis. New editions will be issued in print and on the Web every five years; these will incorporate new understanding of the performance of ecosystems and of the most appropriate indicators and monitoring techniques to track that performance. Between these major new editions, substantial revisions—for example, to incorporate new data sets that become available—will be issued in an annual update to the Web version, with minor updates and corrections published on the Web as necessary.

Regular production of the report will require both long-term stable funding and an appropriate institutional "home." While no decision has been made about whether The Heinz Center should continue to host the effort after the 2007 edition, what is clear is that the institutional and funding arrangements that support the project must ensure its continued independence and scientific credibility. Finally, besides what is required to produce the next report, it is likely that additional resources will need to be marshaled in order to fill some of the data gaps identified here.

Meeting the Need

This document responds to a clearly defined need—periodic information, worthy of trust, about the condition of our nation's lands, waters, and living resources. Where it is possible to do so, the extent, condition, and use of these precious assets are described. Where it is not possible, we have provided a road map to guide future efforts. These are valuable steps, but the true and lasting value of this project will be realized only if the effort is repeated regularly and is accompanied by significant enhancement of the base of scientific understanding and by continuation and improvement of high-quality monitoring programs.

The Structure of this Report

The remainder of Part I summarizes the findings of this project. Chapter 2 describes the reporting framework developed by the Design Committee for characterizing the state of the nation's ecosystems. Chapter 3 summarizes the overall findings of the report, including both what can be reported now and those gaps in data and understanding that will have to be filled before a fully comprehensive account of the state of the nation's ecosystems is possible.

Part II presents the indicators that characterize the state of the nation's ecosystems. Chapter 4 presents the core national indicators, which cut across the six ecosystems, and chapters 5–10 present the indicators that describe the state of the individual ecosystems that the project identified—Coasts and Oceans, Farmlands, Forests, Fresh Waters, Grasslands and Shrublands, and Urban and Suburban Areas.

An appendix describes in greater detail the data gaps identified in this document (see page 199).

Finally, the extensive technical notes (pp. 207–270) provide the technical foundations for the indicators. They include not only information on data sources and access, but also discussions of how the data have been manipulated and comments on their quality.

Notes and References

1. S.P. Hays. 1989. Beauty, Health, and Permanence: Environmental Politics in the United States, 1955–1985. Cambridge, UK: Cambridge University Press.

   The following sources provide polling data that demonstrate the nature and depth of public opinion on the environment:

   The Polling Report, Inc. Multiple polls, including Gallup, ABC News, Newsweek, Harris. Accessed at www.pollingreport.com/enviro.htm on April 26, 2002.

   League of Conservation Voters Education Fund. Multiple polls from 1999 and 2000. Accessed at http://www.voteenvironment.org/media_debunking_env_myths_data.html on November 21, 2001.

2. Christine Vogan. Pollution Abatement and Control Expenditures, 1973–94. Survey of Current Business, Bureau of Economic Analysis, U.S. Department of Commerce. http://www.bea.doc.gov/bea/an/0996eed/maintext.htm; accessed August 8, 2001.

3. Note that costs of compliance with wetlands, endangered species, and similar regulatory programs, plus voluntary actions by the public and private sector, are probably also significant as well, but are not included.

4. Executive Office of the President, National Science and Technology Council, Committee on Environment and Natural Resources. National Environmental Monitoring and Research Workshop Proceedings. February 25, 1997.

5. Council on Environmental Quality. 1970. Environmental Quality: The First Annual Report of the Council on Environmental Quality, p. 237. Washington, DC: U.S. Government Printing Office.

6. National Academy of Public Administration. 1995. Setting Priorities, Getting Results: A New Direction for the Environmental Protection Agency. Washington, DC: National Academy of Public Administration.

   National Research Council, Committee to Evaluate Indicators for Monitoring Aquatic and Terrestrial Environments. 2000. Ecological Indicators for the Nation. Washington, DC: National Academy Press. http://www.nap.edu/catalog/9720.html.

7. This situation exists in spite of several efforts to prepare and sustain periodic reporting on indicators and trends. For example, the Council on Environmental Quality published two major reports on environmental indicators and trends, one in 1981 and one in 1989, and the Conservation Foundation, a nonprofit organization, prepared three major reports (in 1982, 1984, and 1987) on environmental indicators and trends:

   Conservation Foundation. 1982. State of the Environment 1982: A Report from the Conservation Foundation. Washington, DC.

   Conservation Foundation. 1984. State of the Environment: An Assessment at Mid-decade. Washington, DC.

   Conservation Foundation. 1987. State of the Environment: A View toward the Nineties. Washington, DC.

   Council on Environmental Quality. 1981. Environmental trends. Executive Office of the President, Washington, DC.

   Council on Environmental Quality. 1989. Environmental trends. Cosponsored by the Interagency Advisory Committee on Environmental Trends, Executive Office of the President. Washington, DC.

Chapter 2:
The Reporting Framework

This chapter describes the basic framework developed by the Design Committee to characterize the state of the nation's ecosystems. It discusses the strategic guidelines that shaped the report, defines both the major ecosystem types and the major categories of indicators described in this report, and concludes with an overview of the nature of the data included in the report.

Goals

In developing a framework for reporting on the state of the nation's ecosystems, the Design Committee reviewed a wide range of previous reporting efforts, consulted broadly with relevant stakeholders, users of environmental information, and experts, and incorporated feedback from the 1999 prototype of the present report. In addition, it built on three seminal documents: the proceedings of a National Environmental Monitoring and Research Workshop held at the Smithsonian Institution in 1996[1]; the National Science and Technology Council's *Integrating the Nation's Environmental Monitoring and Research Networks and Programs: A Proposed Framework*,[2] published in 1997; and the National Research Council's study *Ecological Indicators for the Nation*,[3] published in 2000. Recruitment of key contributors to each of these documents as members of this report's Design Committee ensured continuity and cumulative learning across the several efforts. The Design Committee developed and refined the goals for this report:

- **The report is written for decision makers and opinion leaders concerned about the "big picture" of the nation's ecosystems.** Its goal is to identify what the nation most needs to know about its ecosystems in order to conduct enlightened policy debate; we also summarize what is known—and what is not known—about those key characteristics. More generally, the report seeks to educate a broader audience by highlighting important aspects of the nation's ecosystems and by characterizing patterns of change in those conditions.
- **The report identifies a succinct set of strategic indicators to characterize the nation's ecosystems.** It does not characterize every aspect of the environment or the ecosystems of particular regions. Rather, it identifies strategic indicators that can serve as meaningful reference points for broad-ranging policy discussions.[4] In doing so, we seek to complement, not replace, existing reporting frameworks developed for particular management, regulatory, or scientific needs. Such programs provide data on many characteristics of ecosystems that we do not describe, and they can highlight changes that may not appear large at a national scale but are nonetheless quite important at a local scale.
- **The report provides scientific information on which decisions can be based, while avoiding value judgments and policy recommendations.** It thus seeks to be policy relevant while avoiding bias or advocacy. Rather than imposing our judgments of whether conditions are "good" and "bad," the report assists readers in interpreting its content by including time trends and maps from which regional comparisons can be made. When possibles, the report characterizes conditions in terms of departures from generally accepted standards (e.g., safe drinking water standards), while recognizing that there are judgments involved in setting such standards.
- **The report focuses on the *state* (or condition) of the nation's ecosystems.** It leaves to others the task of identifying the stresses (pressures) that might be changing ecosystems, and of analyzing the effects

of actions taken by governments, private individuals, or businesses to reduce those stresses. Information on pressures and societal responses is clearly important, and it has been incorporated in widely used environmental reporting frameworks.[5] For this project, however, we chose to focus on *state* for two reasons. First, there is a strong need to complement existing reporting about environmental pressures and responses with information about society's ultimate concern: the state of the nation's ecosystems. Second, the difficulties of determining "cause and effect" can influence perceptions of the scientific credibility and political neutrality of both data and reporting efforts. Experience with other national reporting efforts (particularly those concerned with the nation's economy) suggests that a broadly accepted characterization of system state can make an enormous contribution to policy development and understanding, even when disagreements persist on the causes of and appropriate policy responses to that state.

- **The indicators selected for this report reflect both key properties relating to ecosystem condition and the goods and services derived from ecosystems.** Ecosystems are incredibly complex, and reporting on them necessarily involves focusing on some characteristics and excluding many others. In addition, the values held by different people can lead them to place greater importance on some aspects of ecosystems than on others; some people place primary emphasis on the goods and services ecosystems produce, while others focus on their condition. The question is not *whether* to select, but only who does the selecting, and *how* it is done. The indicators included here were extensively discussed and negotiated by the members of our Design Committee and technical Work Groups, which included a balanced array of representatives from the private sector, environmental organizations, government, and academia. Although the selection of the indicators was inevitably a value-driven process, we took great care to make it fair and inclusive. The specific numbers assigned to those indicators were determined through a peer-reviewed scientific process, which we took great care to make transparent and credible.
- **The report identifies critical gaps in data and in monitoring programs that must be filled in order to fully, and in a balanced way, characterize the state of the nation's ecosystems.** It leaves to the future, however, any discussion of how to fill those gaps. In preparing this report, we first identified ecosystem characteristics most important for a balanced national report. We then made extensive and good faith efforts to locate sufficiently high-quality and extensive data to report on those characteristics. Where such data are not available, the report calls attention to the gaps. In implementing this strategy, we have resisted the temptation to focus only on what happens to be illuminated by the lamp-posts of existing monitoring and reporting programs. Instead, the report identifies where lamps need to be posted in order to provide the kind of illumination of ecosystems that the nation most needs.

Defining Ecosystems

At the heart of this report are a set of six ecosystem types (coasts and oceans, forests, farmlands, fresh waters, grasslands and shrublands, and urban and suburban areas) and the indicators that, taken together, describe the state of these ecosystems and of the nation as a whole. It is reasonable to think about—and to seek indicators for characterizing—the ecosystem of a small watershed, or of the planet as a whole, or of places at any scale in between. However, like the recent National Research Council study on *Ecological Indicators for the Nation*, this report focuses on indicators that can support policy debate and decision making at the national scale.[6]

Ecosystems, Land Cover, and Geography

The word "ecosystem" is used in a number of ways, and there are two common organizational approaches we might have taken—land cover and geographic. The land cover approach defines ecosystem types based on their dominant vegetation or other physical characteristics. Thus, one would speak of a "forest ecosystem," a "cropland ecosystem," or a "freshwater ecosystem." The geographic approach considers all living and nonliving things in a region to be an ecosystem[7] regardless of vegetation type. In this approach, boundaries can be defined in many ways: watersheds and ecoregions[8] are common examples.

We have chosen the land cover approach and we use the terms "land cover types," "ecosystems," and "ecosystem types" more or less interchangeably. However, we also use a more geographic approach in some cases, such as when we define a farmland landscape that includes both croplands and interspersed natural areas.

We have chosen the land cover approach in large part because many natural resource management decisions are differentiated by land type. Forests, grasslands and shrublands, farmlands, and so on produce different products, respond to different management approaches, are owned for different reasons, and are, in plain terms, different. Significant government and private activities are aligned with these land cover distinctions, and we believed that a report reflecting this structure would be most useful at this time.

Nevertheless, a growing number of "place-based" efforts are working to implement management strategies that consider all of the interactions within a watershed, ecosystem, or region. These efforts are supported by monitoring and information systems that help decision makers and the public see their region as an integrated whole, rather than as distinct elements to be managed separately.[9] We strongly support the development of such reporting and information systems, and we have had preliminary discussions on the application of the reporting framework presented in this report to smaller geographic regions.[10]

Ecosystem Types

This report uses six major ecosystem types as its basic reporting units.

- Coasts and Oceans
- Farmlands
- Forests
- Fresh Waters
- Grasslands and Shrublands
- Urban and Suburban Areas

This scheme is intended to cover all the lands and waters of the United States, including the ocean out to the limit of U.S. national jurisdiction. Obviously, these broad ecosystem or land cover types are neither homogeneous nor mutually exclusive. For example, the grasslands and shrublands ecosystem includes bare-rock desert and tundra, as well as the prairies and shrubland its name evokes. Freshwater wetlands are described along with lakes, streams, and so on, but are also tallied within the acreage of forests, farmlands, and other land covers. We describe each ecosystem type, including overlaps with other types, in greater detail in the opening section of each ecosystem chapter.

Map 4.2 (p. 40) shows where these ecosystems occur.

Coasts and Oceans. This ecosystem consists primarily of estuaries and ocean waters under U.S. jurisdiction. Estuaries are partially enclosed bodies of water (this term includes bays, sounds, lagoons, and fjords); they are generally considered to begin at the upper end of tidal or saltwater influence and end where they meet the ocean. By definition, U.S. waters extend to the boundaries of the U.S. Exclusive Economic Zone (EEZ), which extends 200 miles from the U.S. coast, but not all indicators report on this entire area. In addition, several indicators characterize shorelines along both estuaries and oceanfront areas. In these cases, we focus on the margin between land and water, not on uplands or watersheds that may influence coastal conditions.

Farmlands. We focus both on *croplands*—lands used for production of annual and perennial crops and livestock—and on a larger *farmland landscape*, which includes field borders and windbreaks, small woodlots, grassland or shrubland areas, wetlands, farmsteads, small villages and other built-up areas, and similar areas within and adjacent to croplands. Some indicators focus on croplands only, while some describe the entire farmland landscape.

Forests. We generally rely on the USDA Forest Service definition of forest: lands at least 10% covered by trees of any size, at least one acre in extent. This includes areas in which trees are intermingled with other cover, such as chaparral and pinyon–juniper areas in the Southwest, and both naturally regenerating forests and areas planted for future harvest (plantations or "tree farms").

Fresh Waters. Our freshwater ecosystems include

- Rivers and streams, including those that flow only part of the year
- Lakes, ponds, and reservoirs, from small farm ponds to the Great Lakes
- Groundwater, which is often directly connected to rivers, streams, lakes, and wetlands
- Freshwater wetlands, including forested, shrub, and emergent wetlands (marshes), and open water ponds
- Riparian areas—the usually vegetated margins of streams and rivers (although this term can also apply to lake margins)

Obviously, there are overlaps and gradations among these systems. Wetlands often occur at the margins of streams and rivers, in what is also considered the riparian area. Some ponds are shallow and thus may also be classified as wetlands. In some rivers, dams create reservoirs, and these may be classified as rivers, reservoirs, or both.

Grasslands and Shrublands. The title of this system (which many people call *rangelands*) is quite descriptive: lands in which the dominant vegetation is grasses and other nonwoody vegetation, or where shrubs (with or without scattered trees) are the norm. Bare-rock deserts, alpine meadows, and arctic tundra are included in this system as well. We also include pastures and haylands, which represent an overlap with the farmland system; less-managed pastures and haylands fit well within the grassland/shrubland system, while more heavily managed ones fit well as part of the farmlands system. Most monitoring programs do not distinguish between the levels of management for pastures, however.

Urban and Suburban Areas. This system consists of those places where the land is primarily devoted to buildings, houses, roads, concrete, grassy lawns, and other elements of human use and construction. Urban and suburban areas, in which about three-fourths of all Americans live, span a range of density, from the unmistakable city center, characterized by high-rise buildings, concrete, and relatively little green space, to the suburban fringe—where development thins to an obviously rural landscape. This definition does not include all developed lands. It includes areas that we believe are large enough and built-up enough to qualify as "urban and suburban." Many areas—small residential zones, the area of rural interstate highways, farmsteads, and the like—are "developed" but would not be considered "urban or suburban."

Indicator Categories

This report identifies ten major characteristics of ecosystem condition and use that together provide a broad, balanced description of any ecosystem type. These ten characteristics cover the physical dimensions of the systems, their chemical and physical conditions, the status of their biological components, and the amounts of goods and services people receive from them.

These ten major characteristics are described for each of the six major ecosystem types, using between fourteen and eighteen indicators to cover all ten characteristics. As a general rule, for each of the six ecosystem types, there is at least one indicator describing each of the ten major ecosystem characteristics.

We have also identified ten "core national indicators" that provide a very broad and succinct view of national ecosystem condition and use.

Table 2.1 lists the ecosystem characteristics and briefly describes the related indicators. The table on pp. 28–29 lists all indicators in the report by ecosystem type and ecosystem characteristic.

Table 2.1

Ecosystem Characteristic	Indicator Description
SYSTEM DIMENSIONS	
Extent	Area of an ecosystem or land cover type and its major components
Fragmentation and Landscape Pattern	Shapes and sizes of patches of an ecosystem type, and their relation to one another
CHEMICAL AND PHYSICAL CONDITIONS	
Nutrients, Carbon, Oxygen	Amounts and concentrations of key plant nutrients (nitrogen and phosphorus) and key ecosystem elements (oxygen and carbon)
Chemical Contaminants	Numbers of selected contaminants found in ecosystems, and how often these chemicals exceed regulatory or advisory thresholds
Physical Conditions	Condition of key aspects of the physical makeup of an ecosystem, such as erosion or water temperature
BIOLOGICAL COMPONENTS	
Plants and Animals	Status of native and non-native plant and animal species
Biological Communities	Condition of the plant and animal communities that make up an ecosystem
Ecological Productivity	Plant growth on land and in the water
HUMAN USE	
Food, Fiber, and Water	Amounts and values of key products for human use
Other Services, Including Recreation	Tangible and intangible services provided by ecosystems

Indicators of System Dimension

Extent. The extent of an ecosystem and its various components—measured either as area or as linear distance, as for rivers—is one of the most basic aspects of its condition, and provides background and context for other indicators. Indicators in this category generally describe the overall dimensions of the system—in absolute size and as a fraction of total U.S. land area. Some indicators also provide information on the composition of the system (e.g., acreage of major forest types) or on land use characteristics (e.g., area of grassland and shrublands used for livestock raising).

Fragmentation and Landscape Pattern. The size and shape of patches of forest, farmland, or other ecosystem types, and how patches of different ecosystem types are intermingled, help determine the quality and quantity of some ecosystem benefits or services. Examples of services that are believed to be strongly affected by landscape pattern include wildlife habitat, the ability to filter sediment and other contaminants from runoff, and the value for solitude and recreation.

Indicators of Chemical and Physical Condition

Nutrients, Carbon, Oxygen. These are key ecosystem building blocks. We report the amount of nitrogen and phosphorus in water, the amount of carbon in soil and other ecosystem components, and, for aquatic systems, the amount of oxygen in the water. Nitrogen and phosphorus are key plant nutrients; in excess, however, they can contribute to water quality degradation. Most animals need oxygen to survive, and carbon is a critical component of living tissue. Moreover, increased carbon storage by ecosystems can offset emissions of carbon dioxide, of concern because of climate change.

Chemical Contaminants. Chemical contaminants can harm people and impair ecosystem functioning through their effects on plants and animals. We report on two key characteristics of this phenomenon: how many chemicals are found in water, sediments, and soil, and how often their concentrations exceed standards and guidelines set to protect human health and ecosystem condition. Indicators report on selected contaminants in stream water, groundwater, stream and estuary sediments, fish tissue, and soil.

Physical Conditions. Features such as the degree of erosion of farmland soils and the timing and size of low and high flows in streams have a strong influence on the plants, animals, and microorganisms that inhabit ecosystems and on the goods and services ecosystems produce. The specific physical features that are most important differ greatly among ecosystems, so there is less consistency among these indicators than among indicators describing other major characteristics.

Indicators of Biological Condition

Plants and Animals. Plants and animals are fundamental components of ecosystems, their condition can reflect broader ecosystem conditions, and many people care deeply about their status. Indicators generally focus on the relative risk of extinction of specific groups of species, the number and extent of non-native species, and unusual mortality events.

Biological Communities. Species do not exist in isolation; rather, they occur in characteristic groupings, adapted to a particular location and climate. These communities—each with its own characteristic set of species—form the biological "neighborhood" within which individual species exist.

Ecological Productivity: Plants, including algae, capture the sun's energy, which is the basis for almost all life on earth. The amount of plant growth in various ecosystems is a fundamental indicator of their condition.

Indicators of the Microscopic World

A number of indicators in this report touch upon the microscopic world, which exists in all ecosystems. For many people, terms like "plants and animals" and "biological communities" may bring to mind trees, flowers, fish, mammals, birds, and the like, along with their communal groupings. However, microscopic plants—algae—capture the sun's energy and thus support much of life in the oceans; they also produce much of the oxygen necessary for animal life. Bacteria, which are neither plants nor animals, perform a host of chemical transformations in soil and water, without which these systems simply would not function. See Coastal Areas with Depleted Oxygen (p. 71); Harmful Algal Blooms (p. 78); Coastal Chlorophyll Concentrations (p. 80); Soil Biological Condition (p. 102); and Forest Disturbance: Fire, Insects, and Disease (p. 127).

Indicators of Human Use

Food, Fiber, and Water. The major commodity goods produced by ecosystems meet human needs and are important to the national economy. For each ecosystem, except for urban and suburban areas, we report on major commodity or commodity-like products: fish landings, timber harvest, agricultural production, fresh water withdrawals, and range-fed cattle. We report basic quantities of the commodity, often accompanied by information that relates to the long-term stability of production: factors such as agricultural yield, status of fish stocks, and the ratio between timber harvest and annual growth.

Other Services, Including Recreation. Ecosystems produce an enormous variety of "services"—from opportunities for recreation to the building of soil, reduction in flooding, and pollination of crops. This is an area of intense scientific interest, but the methods for quantifying these services are not well developed. In several instances, we highlight the importance of the underlying services but also the lack of developed indicators.

Data: Quality, Coverage, and Context

The final major element of our reporting framework involves how we selected and reported data. As noted above, we selected *indicators* on the basis of what is needed to fairly characterize the state of the nation's ecosystems rather than because the *data* happened to be available. We then had to decide on criteria for including data from particular sources, on what to do when adequate data were not available, and on how to give meaning to the measurements we report. We summarize our design decisions below.

Quality and Coverage

For each indicator, we reviewed available data sources, using both the knowledge of individuals on our various working groups and input from a large number of collaborators and reviewers. Data included in this report had to meet three key criteria:

- Data had to be of sufficiently high quality to provide a scientifically credible description of actual ecosystem conditions
- Data had to have adequate geographic coverage to represent the state of the *nation's* ecosystems
- Data had to be collected through an established monitoring program that offers a reasonable likelihood of future data availability

Data quality—meaning that the data provide a reasonably accurate representation of actual conditions and do not include any substantial known sources of bias or distortion—was the key criterion for selection of data sources. Quality was assessed using the expert knowledge of the participants in the project, supplemented by information provided by the managers of certain data sets; we also commissioned analyses of data sets specifically for this project.

A data set must also provide enough information on the resource or issue in question. This criterion is met by data sets with complete coverage (such as those based on satellite measurements) and those based on representative samples from which reasonably accurate estimates of overall conditions can be made. In practice, this led to the selection of data sets that covered most states or a significant fraction of the ecosystem in question. Obviously, there are large amounts of high-quality data that do not meet this criterion. For example, states and research institutions collect many potentially relevant data, but unless they are aggregated and reviewed to determine whether the collection methods are compatible, the data are not available in practice for national reporting, and so are not included in this report.

Third, we decided that data must be from ongoing programs, with a reasonable chance of the measurements being repeated at regular intervals in the future. Although all monitoring and reporting programs are subject to changes in funding and priorities, established programs are clearly different from one-time studies. One-time efforts can be quite valuable, since they often break new ground scientifically and may serve as baselines against which to compare future conditions. But until and unless they are performed regularly, they do not advance the goal of periodic national reporting.

Inadequate Data and Indicators Requiring Development

Applying the data selection criteria noted above, we identified a number of high-quality, nationally representative data sets with good prospects for future continuity. Inevitably, however, adequate data sets did not exist for all indicators.

Confronted with this dilemma, we tried to be pragmatic. Where small changes in the definition of an indicator would enable us to use existing data, we considered revising the indicator—provided, of course, that the modification would not compromise the indicator's basic purpose. We also avoided indicators that seemed likely to require extraordinary technical advancement beyond current monitoring methods, or extraordinary human or fiscal resources.

Nevertheless, for a substantial number of the indicators selected for this report, adequate data could not be assembled. We identify such cases in the text, to highlight where future data monitoring work is needed. (These gaps are discussed in more detail in Chapter 3, p. 17, and in the Appendix, p. 199.)

Data Not Adequate for National Reporting. There are several causes for these data shortfalls, each with a distinctly different remedy. In some cases, the data needed for reporting are available, but additional processing or analysis—requiring either more money or more time than was available for this project—was needed. For example, there are several indicators of fragmentation and landscape pattern for which the appropriate remote sensing data are available, but which would require additional processing to calculate the relevant measures (see, for example, pp. 93 and 94). These cases represent relatively simple, low-cost opportunities for filling gaps identified in this report. Table A.2 (p. 205) lists the indicators in this category; the table also lists several indicators for which data are currently being collected, thus requiring no new action to fill a gap.

Second—and by far the largest category of indicators with missing data—are those cases where many data probably exist, but they are not available in a form that we could use. Most commonly, relevant data are collected, but by different entities (e.g., states, local governments, research institutions), potentially using different methods. For example, data on groundwater levels in major aquifers are collected by a wide variety of entities to help them understand their water supply situation (see p. 151). However, no group has gathered these data and assessed whether the monitoring methods are comparable. Filling these data gaps might simply require aggregation of existing data, or it might require development and adoption of consistent methods by data collectors.

Third, there are situations where data are not widely collected, but could be if an adequately funded program were in place. The condition of microscopic animals in cropland soils (p. 102) and the contamination of bottom sediments in ocean waters (p. 72) are two examples. The challenge here is operational rather than conceptual.

Indicator Development Needed. Finally, in several cases, we could not select a specific measure, and thus could not evaluate whether data are available. For some indicators, there are multiple competing approaches to measuring a particular

What does "Data Not Adequate for National Reporting" mean?

Data selected for this report had to

- Be scientifically credible and high quality
- Cover most of the United States
- Have a reasonable likelihood of being available for future reporting

We use the phrase "data not adequate for national reporting" to indicate that we were not able to identify a data set meeting these criteria.

In many cases, some indicator data are available. However, these data may cover only a limited geographic area, may never have been assembled from the states, local governments, or research institutions that collected them, or may have been the result of one-time studies. Many of these data sets are excellent examples of the kind of monitoring necessary, and they may serve as the basis for future national reporting.

phenomenon, and progress could be made rapidly if a single method could be selected (see, for example, the stream habitat index, p. 105). Other indicators require conceptual development before data availability can be assessed (see the suburban/rural land use change indicator, p. 182).

Trends and Other Context-Setting Information

Data without context are apt to have little meaning. In order to provide context, and instead of providing "grades" for particular indicator values, we have, wherever possible, provided one or more of the following:

- Information on how the indicator value has changed over time (trends). We tried to find data for the period from 1950 through the present, although this was possible in relatively few cases. In addition, where appropriate, we also provide information on long-term historical comparisons (to presettlement conditions, for example). Presettlement comparisons are meant to give context, not to represent "ideal" conditions.
- Information on regional differences. Frequently, we display data on a regional basis to allow users of the report to compare values in one part of the country with those in another.
- Comparisons with widely accepted reference points. Where they exist, we compare data to regulatory and related standards and guidelines that have become widely used and accepted national reference points, while recognizing that there are judgments inherent in setting such benchmarks. Such standards, guidelines, and related reference points are available primarily for indicators related to nutrients and chemical contaminants. In several cases, indicators are based on comparison to relatively undisturbed "reference sites."

A Note About Regions

We have generally relied on the regional groupings used by the agency providing the data. So, for example, we report on many forest indicators using USDA Forest Service regions and on several coastal indicators using National Oceanic and Atmospheric Administration regions. In a few cases, we used a set of regions developed by The Heinz Center that considered climate, topography, and vegetation.

Since no two agencies share the same regional boundaries, the regions used in this report vary considerably. While it may be desirable to report all indicators on a common geographic basis, in practice this is not currently possible. We are certainly not the first to make this observation, and there are many efforts under way within federal agencies and elsewhere to address this issue. Ideally, data on ecosystem conditions, as reported here, should be gathered and managed so as to enable reporting on any geographic basis; this would allow comparison and aggregation of information collected by different agencies and programs.

Notes and References

1. Executive Office of the President, National Science and Technology Council, Committee on Natural Resources. 1997. National Environmental Monitoring and Research Workshop: Proceedings. February 25, 1997. http://www.epa.gov/cludygxb/Pubs/nemrwork.pdf.
2. Executive Office of the President, National Science and Technology Council, Committee on Natural Resources, Environmental Monitoring Team. 1997. Integrating The Nation's Environmental Monitoring and Research Networks and Programs: A Proposed Framework. http://www.epa.gov/cludygxb/Pubs/framewrk.pdf.
3. National Research Council (NRC), Committee to Evaluate Indicators for Monitoring Aquatic and Terrestrial Environments. 2000. Ecological Indicators for the Nation. Washington, DC: National Academy Press. http://www.nap.edu/catalog/9720.html.
4. The scientific feasibility of such a strategic approach to ecosystem characterization has recently been endorsed by the National Academy of Sciences. See NRC op. cit.

5. The Organization for Economic Cooperation and Development developed what is widely known as the "pressure-state-response" framework for reporting on environmental conditions. See Organization for Economic Cooperation and Development. 1991. The State of the Environment (Paris).
6. Local- and regional-scale ecosystem indicators clearly are needed to guide many types of public and private decisions. The need for global-scale indicators to support international environmental agreements is increasingly recognized, and has called forth large scale efforts such as the UN Environment Programme's *Global Environmental Outlook* reports and the international Millennium Ecosystem Assessment. National ecosystem indicators are nonetheless also needed, not only to support sound policymaking by nations but also to provide context for domestic regional efforts and input to global reporting efforts. NRC, op. cit.

   Millennium Ecosystem Assessment (http://www.ma-secretariat.org/en/index.htm).

   United Nations Environment Programme. 2000. *Global Environmental Outlook*—2000. http://www.unep.org/Geo2000/ov-e/index.htm.
7. For example, Odum (1971) defines an ecosystem as "Any unit that includes all of the organisms (i.e., the "community") in a given area interacting with the physical environment so that a flow of energy leads to a clearly defined trophic structure, biotic diversity, and material cycles (i.e., exchange of materials between living and nonliving parts) within the system is an ecological system or *ecosystem*." E.P. Odum. 1971. Fundamentals of Ecology. Philadelphia: Saunders.
8. An ecoregion is "a relatively large area of land or water that contains a geographically distinct assemblage of natural communities". R.A. Abell et al. 2000. Freshwater Ecoregions of North America: A Conservation Assessment. Washington, DC: Island Press.
9. L.H. Gunderson, C. S. Holling, S. S. Light (eds.). 1994. Barriers and Bridges to the Renewal of Ecosystems and Institutions. New York: Columbia University Press.
10. In May 2000, a 1½-day meeting was held in Bozeman, Montana, under the joint sponsorship of The Heinz Center and the Department of the Interior. This meeting was with representatives of a range of public and private interests in the Greater Yellowstone area (GYA). While the meeting was not intended to be conclusive, there was general agreement that the basic framework of indicator categories used in this report was applicable in the GYA, and perhaps in other regional/ecosystem contexts as well.

Chapter 3:

The State of the Nation's Ecosystems: What We Know and What We Don't Know

Choosing Indicators and Data

This report is the collective effort of close to 150 researchers, organized into seven committees, working over nearly five years. A multidisciplinary "Design Committee," with members drawn from industry, environmental groups, government, and universities, identified ten key characteristics of ecosystem condition that are valued by Americans and that, in our judgment, need to be addressed in any credible, balanced and useful report. These ten characteristics describe the physical dimensions of the systems, their chemical and physical conditions, the status of their biological components, and the amounts of goods and services people receive from them (see Figure 3.1). We also decided to report on these indicators for the nation as a whole and for six major ecosystem types that have long been the focus of policy debate, research, management, and monitoring—coasts and oceans, farmlands, forests, freshwaters, grasslands and shrublands, and urban and suburban areas.

Six ecosystem-specific work groups, each with representation from business, environmental, academic, and government institutions, identified between 15 and 20 specific indicators for each system, as well as a set of "core national indicators." The indicators were selected based on their importance; no indicator was ruled out simply because the data to report on it is not currently available. Each of the ecosystem-specific work groups then carefully examined potential data sources for reporting on each indicator. We used data only if it met high professional standards for integrity and overall quality and allowed us to report on most of the United States, and if there was a reasonable likelihood that the underlying measurements would be repeated over time. Key data gaps became apparent and are identified throughout the report.

Finally, we obtained the required data from the government agencies and private organizations that collect and maintain them. Our primary focus was to present current conditions and to lay the groundwork for future reporting, but wherever possible we sought datasets with records long enough to

Figure 3.1. The State of the Nation's Ecosystems: Characteristics and Indicators

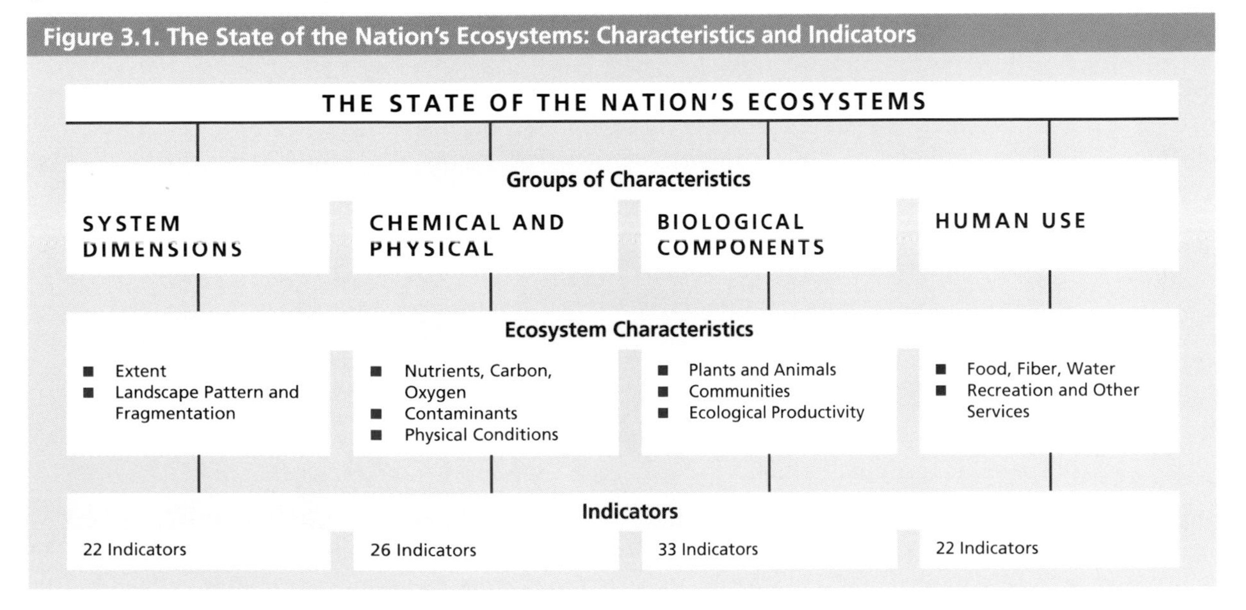

reveal trends. When they were available, we compared data on current conditions with widely accepted reference points, primarily regulatory and related standards and guidelines, while recognizing that there are judgments involved in setting such standards. In many cases, we also provided data on a regional basis, allowing comparisons between regions.

The State of the Data for Reporting on the Nation's Ecosystems

In seeking data, we found a classic case of a glass that is both half empty and half full. In applying the selection criteria outlined above, we found adequate data for more than half of the selected indicators, with trends or other context information on many of these, allowing us to report meaningfully on many aspects of ecosystem condition. However, substantial gaps remain, and until and unless these gaps are filled, Americans will not have access to a complete picture of the "state of the nation's ecosystems." Even with these gaps, however, consistent tracking and reporting of those indicators for which we found adequate data would produce a much more useful picture of the state of the nation's ecosystems than has ever been available.

Our full analysis of data availability and gaps is presented in the Appendix, p. 199. Highlights are summarized in Figure 3.2 and described below.

Figure 3.2. The State of the Data

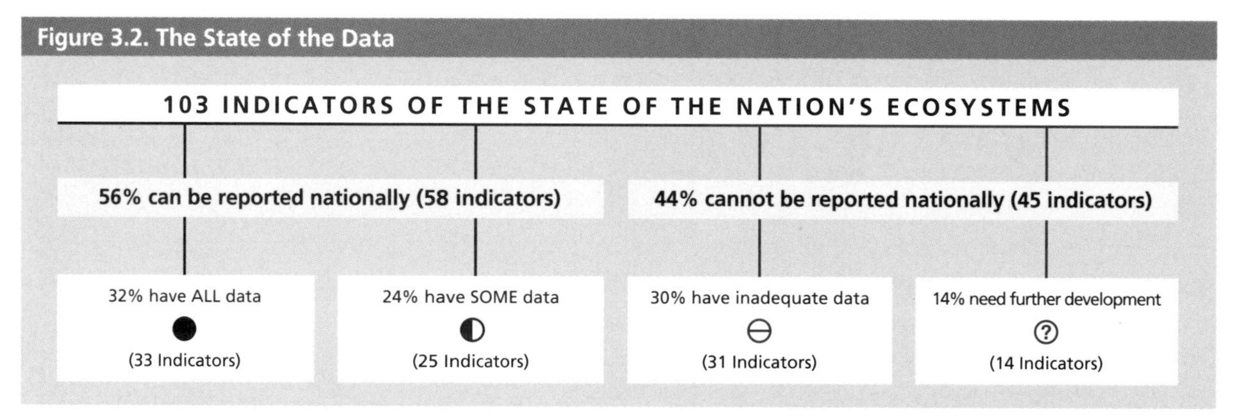

- This report presents 103 indicators. Data are adequate to support national reporting for 58. Of these, we have all the desired data for 33 indicators (●). Important gaps remain for the other 25 indicators (◐).
- Of the 58 indicators with data, we present trends for 31. For 11 other indicators, we provide comparisons against widely accepted standards, or against undisturbed or "reference" conditions. For the remaining 16 indicators, neither trends nor appropriate reference points were available.
- We provide no data for 45 indicators. For 31 of these, data availability is the only impediment to national reporting. These indicators are clearly marked with a "data not adequate for national reporting" label and with this icon: ⊖.
- For the other 14 indicators for which no data are reported, the problem is more fundamental: a lack of agreement on how the relevant ecosystem characteristic can be measured most meaningfully and effectively. For these indicators, additional work is required in the appropriate scientific communities to build a consensus on the specific measurements that should be reported. Indicators in this category are marked with an "indicator development needed" label and with this icon: ⓘ.
- Data availability varies by ecosystem: about three-fourths of forest indicators have some or all data, contrasting with grasslands and shrublands and urban and suburban areas, where only about 40% have data. Data availability also varies by ecosystem characteristic: more than 80% of the indicators of ecosystem extent, chemical contamination, and the quantities of food, fiber, and water produced in ecosystems have some or all data, while for several characteristics (landscape pattern and fragmentation, biological communities, and recreation and other services), fewer than a third of the selected indicators have adequate data for national reporting.

The State of the Nation's Ecosystems

What follows is a very brief overview of the findings of this report. In it we present highlights of both actual ecosystem conditions and the availability of data and indicators. We have organized this summary according to the ten major aspects of ecosystem condition that form a key part of our reporting framework. These characteristics are identified in Figure 3.1, and are discussed in detail in Chapter 2. The table on pages 28 and 29 shows all indicators included in this report.

While the summary below does not generally provide highlights of indicators for which adequate data are not available, such indicators are just as important as those for which data *are* presented. It is what should be measured that is important, not whether it has been measured yet.

System Dimensions: Extent

The acreage of a particular ecosystem type (or for features like shorelines and rivers, their length) is a basic characteristic of their condition. Gains or losses in the area devoted to different ecosystem types, or in the acreage devoted to particular uses of land, such as wilderness areas or livestock grazing, change the landscape in important ways. Gains or losses within an ecosystem type—for example, conversion from one forest type to another—are also important.

We present 15 indicators of ecosystem extent. For 13 of these indicators, we located either full or partial data. Historical trend data are available for eight of these.

Systems Dimensions: Extent

Core National
◐ Ecosystem Extent

Coasts and Oceans
◐ Coastal Living Habitats
◐ Shoreline Types

Farmlands
● Total Cropland
● The Farmland Landscape

Forests
● Forest Area and Ownership
● Forest Types
● Forest Management Categories

Fresh Waters
◐ Extent of Freshwater Ecosystems
◐ Altered Freshwater Ecosystems

Grasslands/Shrublands
● Area of Grasslands and Shrublands
◐ Land Use

Urban/Suburban
● Area of Urban/Suburban Lands
⊖ Total Impervious Area
? Stream Bank Vegetation

● *Complete data available*
◐ *Partial data available*
⊖ *Data not adequate for national reporting*
? *Indicator development needed*

Highlights: Ecosystem Extent

- Forests and grasslands and shrublands each occupy about a third of the land area of the lower 48 states, and croplands about a quarter; wetlands and urban and suburban areas each occupy a few percent of the total area. See Table 3.1.
- Since European settlement, the area of both forest and grasslands and shrublands has declined by about a third. Each had initially occupied about half of the land area of the lower 48 states.
- Since the 1950s, the area of forests has declined by about 1%, and the area of croplands by about 5%. Nonfederal grassland/shrubland area

Table 3.1. Core National Extent Measurements (lower 48 states)

Ecosystem	Core National Extent Measurements	Area in Millions of Acres	Percent of Land Area[a]	Estimated Presettlement Area (as % of Total Land Area)	Changes from 1950s, Millions of Acres (%)[a]
Grasslands and Shrublands	Total area (not including pastures)	683	36%	52%	Declining, amount and rate unknown
Forests	Total area	618	33%	48%	-9 (-1.1%)
Farmlands	Area of croplands	455	24%	—	-23 (-4.8%)
Freshwater	Area of freshwater wetlands	94	5%	11%	-11 (-10%)
Urban and Suburban areas	Urban and suburban lands	32	1.7%	—	Increasing, amount and rate unknown
Coasts and Oceans	Coastal brackish water	Unknown	—	Unknown	Unknown

[a] *This table does not include 100% of lands in the United States. For example, urban and suburban areas, as defined in this report, do not include all developed areas (some developed areas are too small to be considered "suburban" or "urban"). Thus, declines in the area of forests, grasslands and shrublands, croplands, and freshwater wetlands are not—nor should they necessarily be—offset by corresponding gains in urban and suburban lands. In addition, the area of wetlands and portions of urban and suburban areas may also be counted as croplands, forests, or grasslands and shrublands. For these reasons, the figures in this table should not be added to obtain an overall estimate of U.S. land area.*

has decreased since the 1980s by about 3%. The area of urban and suburban lands, although comparatively small, has increased considerably.[1]

- The acreage of wetlands has declined by more than half since European settlement, with both freshwater wetlands and coastal wetlands declining (by 10% and 8%, respectively) since the 1950s, although the rate of loss has slowed in recent decades. There are inadequate data to report on coastal wetlands on the West Coast.
- The acreage of forests that are replanted for future harvest and those in wilderness areas and national parks has increased over the past 50 years. Information on land use in grasslands and shrublands is not available.

System Dimensions: Fragmentation and Landscape Pattern

Scientists agree that the pattern of ecosystems on the landscape affects their condition. For example, whether forests are found in large patches or small, and how these patches intermingle with other ecosystem types within a region, affects their value as habitat for different species and the quantities of other goods and services they provide. However, there remain considerable gaps in scientific understanding about which aspects of the size, shape, and proximity of patches of an ecosystem type matter most in different ecosystems and to different species.

We identified seven indicators of fragmentation and landscape pattern. Data are available for only two of these, and those data are for a single point in time only (i.e., no trends). There is no consensus on what should be measured as a national-level indicator of fragmentation and landscape pattern.

System Dimensions: Fragmentation and Landscape Pattern

Core National
⑦ Fragmentation and Landscape Pattern

Farmlands
⊖ Fragmentation of Farmland Landscapes by Development
⊖ Shape of "Natural" Patches in the Farmland Landscape

Forests
◐ Forest Pattern & Fragmentation

Grasslands/Shrublands
⊖ Area and Size of Grassland/Shrubland Patches

Urban/Suburban
● Patches of Forest, Grasslands/Shrublands, and Wetlands
⑦ Suburban/Rural Land Use Change

● Complete data available
◐ Partial data available
⊖ Data not adequate for national reporting
⑦ Indicator development needed

Highlights: Fragmentation and Landscape Patterns

- Landscape pattern and fragmentation are important, but they can be measured in many different ways. No single method has appeared that is "best" for all ecosystems.
- About two-thirds of all points in both eastern and western forests are surrounded by an "immediate neighborhood" (roughly 250-foot radius) that is mostly forested (90% or greater forest cover). About a quarter of all forest points are surrounded by larger (roughly 2½-mile radius) neighborhoods that are mostly forest.
- About half of all natural lands (forests, grasslands and shrublands, wetlands) in urban and suburban areas are in patches smaller than 10 acres.

Chemical and Physical: Nutrients, Carbon, Oxygen—Chemical Building Blocks of Life

Four elements—carbon, oxygen, nitrogen, and phosphorus—play key roles in ecosystems. Nitrogen and phosphorus are important plant nutrients, but human activities sometimes increase their levels to a degree that causes water quality problems. Carbon stored as organic matter in soil improves soil fertility. Moreover, increased storage of carbon in ecosystems can offset emissions of carbon dioxide, of concern because of climate change. Finally, water must have sufficient oxygen if aquatic animals are to survive.

We selected 12 indicators related to nitrogen, phosphorus, carbon, and oxygen. Full or partial data are available for eight of these. For six of these indicators with data, we provide comparisons to regulatory standards or similar benchmarks. For two, we present trend data.

Highlights: Nutrients, Carbon, Oxygen

- The amount of nitrogen carried by major U.S. rivers has increased over recent decades. The amount carried by the Mississippi River, which drains 40% of the lower 48 states, has tripled since the 1950s.
- Farmland streams and groundwater have higher levels of nitrate than those in forests or urban and suburban areas.
- About 20% of groundwater wells and 10% of streams tested in farmland areas exceeded the federal drinking water standard for nitrate.
- Farmland and urban/suburban streams have similar phosphorus levels; both are higher than forest streams.
- At least half of larger rivers in the United States, three-fourths of streams in farmland areas, and two-thirds of urban/suburban streams had phosphorus levels at or exceeding the limit recommended by the Environmental Protection Agency (EPA) for avoiding excess algae growth.
- From the 1950s to the 1990s, carbon stored in trees increased by 80% in the East and remained constant in the West.
- There are inadequate data for national reporting on areas with depleted oxygen in coastal waters.

Chemical and Physical Conditions: Nutrients, Carbon, Oxygen

Core National
● Movement of Nitrogen

Coasts and Oceans
⊖ Areas with Depleted Oxygen

Farmlands
● Nitrate in Farmland Streams and Groundwater
● Phosphorus in Farmland Streams
⊖ Soil Organic Matter

Forests
● Nitrate in Forest Streams
◐ Carbon Storage

Fresh Waters
◐ Phosphorus in Lakes, Reservoirs, and Large Rivers

Grasslands/Shrublands
⊖ Nitrate in Groundwater
⊖ Carbon Storage

Urban/Suburban
● Nitrate in Urban/Suburban Streams
● Phosphorus in Urban/Suburban Streams

● *Complete data available*
◐ *Partial data available*
⊖ *Data not adequate for national reporting*
⓪ *Indicator development needed*

Chemical and Physical: Chemical Contaminants

Our indicators of chemical contamination generally present two aspects of this issue. First, we report the number of contaminants detected in streams, groundwater, sediments, or fish tissue, which provides a perspective on how widespread such chemicals are. However, because the presence of contaminants does not necessarily mean that levels are high enough to cause problems, we also report on how frequently regulatory and other guidelines or standards are exceeded. See Table 3.2 for a summary of findings.

Table 3.2. Summary of Findings of Contaminants Indicators

	Percent with One or More Contaminants Detected	Percent with One or More Contaminants Exceeding Aquatic Life Guidelines[a]	Percent with One or More Contaminants Exceeding Human Health Guidelines
Streams			
All	100%	77%	13%
Farmlands (pesticides only)	100%	84%	4%
Urban/Suburban	100%	100%	5%
Groundwater			
All	90%	Not applicable	26%
Farmlands (pesticides only)	61%	Not applicable	Less than 1%
Stream Sediments	99%	48%	Not applicable
Freshwater Fish	94%	50%	Data Not Available
Coastal Sediments (estuary data only)	100%	60%	N/A
Coastal Fish	Data Not Available	Data Not Available	Data Not Available

[a] For fish, guidelines used refer to fish-eating wildlife, such as eagles and other predatory birds. For coastal sediments, the figure presented here (60%) includes sediments with concentrations exceeding guidelines for possible harmful effects (19% with 1-4 such contaminants; 39% with 5 or more such contaminants), as well as those whose contaminant levels exceed guidelines for probable effects (2%).

Note: The data presented here reflect testing for different chemicals in different environmental media—some compounds typically are found in stream water, for example, but not in sediments. Tested contaminants include many pesticides, selected degradation products, polychlorinated biphenyls (PCBs), polyaromatic hydrocarbons (PAHs), volatile organic compounds, other industrial contaminants, trace elements, nitrate, and ammonium. *See the technical note for the national contaminants indicator, p. 210, for details.*

Chemical and Physical Conditions: Contaminants

Core National
◐ Chemical Contaminants

Coasts and Oceans
◐ Contamination in Bottom Sediments

Farmlands
● Pesticides in Farmland Streams and Groundwater

Fresh Waters
See the core national, farmlands, and urban/suburban indicators.

Urban/Suburban
● Air Quality
◐ Chemical Contamination

● *Complete data available*
◐ *Partial data available*
⊖ *Data not adequate for national reporting*
? *Indicator development needed*

We present five indicators of chemical contamination. All have at least partial data, and all include some comparison to regulatory standards or similar benchmarks. Trend data are available for only one indicator.

Highlights: Chemical Contaminants

- All or almost all streams, groundwater, sediments (stream and estuarine), and freshwater fish sampled have at least one contaminant at detectable levels.
- Thirteen percent of streams and 26% of groundwater tested had at least one contaminant at a concentration that exceeded human health standards. (Farmland streams and groundwater show fewer exceedances, but these data cover only pesticides.)
- Guidelines for protection of aquatic life are exceeded more often than are human health standards. Half or more of the streams, freshwater fish, and coastal sediments had at least one contaminant that exceeded aquatic life guidelines.

Chemical and Physical: Physical Conditions

The physical makeup and condition of an ecosystem is critical to its functioning. For example, ocean temperature determines what kind of fish and other aquatic animals will live or thrive in an area, the depth to groundwater influences the ability of plants to survive, and the degree of erosion affects both soil quality in farmlands and the degree of off-farm impacts from sedimentation. Because these physical conditions are different for different ecosystems, we include a wide variety of indicators of key physical conditions.

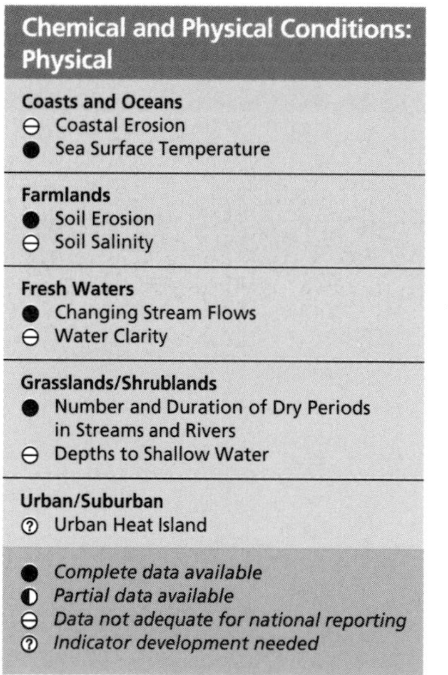
Chemical and Physical Conditions: Physical

Coasts and Oceans
⊖ Coastal Erosion
● Sea Surface Temperature

Farmlands
● Soil Erosion
⊖ Soil Salinity

Fresh Waters
● Changing Stream Flows
⊖ Water Clarity

Grasslands/Shrublands
● Number and Duration of Dry Periods in Streams and Rivers
⊖ Depths to Shallow Water

Urban/Suburban
? Urban Heat Island

● *Complete data available*
◐ *Partial data available*
⊖ *Data not adequate for national reporting*
? *Indicator development needed*

We selected nine indicators of physical conditions. Adequate data for national reporting, including time trends, were available for four of these.

Highlights: Physical Conditions

- Since 1982, the area of cropland with high potential for wind erosion decreased by one-third (to 63 million acres, or 15% of croplands); the area with high potential for water erosion also dropped by a third, to 89 million acres (22% of croplands).
- The number of streams or rivers with major changes in flow compared to a 1930–1949 reference period increased slightly from the 1970s to 1990, to 60%. Streams with high flows well above the 1930–1949 reference period increased markedly after the 1980s, to about 30% of streams. Changes in low flows were more modest.
- Compared to the 1950s and 1960s, fewer grassland/shrubland streams have at least one day with no flow (about 15% in the 1990s), and when no-flow periods occur, they are generally shorter.
- While data are available for sea surface temperature, no trends (either warming or cooling) are evident.

Biological Components: Plants and Animals

Individual species of plants and animals are fundamental building blocks of ecosystems. Species-oriented indicators in this report include those focusing on the percentage of species in particular areas or ecosystems that are at risk of extinction; the degree to which non-native species are gaining a foothold and spreading; and the frequency of unusual mortality events among selected groups of species.

Sixteen indicators relate to plant and animal species, and complete or partial data are available for eight. Three indicators have sufficient data to report trends.

Highlights: Plants and Animals

- About 19% of native animal species and 15% of native plants species in the U.S. are ranked as "imperiled" or "critically imperiled"; such species are typically found in 20 or fewer places, may have experienced steep or very steep declines, or display other risk factors. In addition, about 4% of animals and 1% of plants are, or are believed to be, extinct. However, because the number of at-risk species is affected both by the number of naturally rare species and by a variety of human activities, it is difficult to interpret these data without information on trends in the number of at-risk species. Trend information is not currently available.
- When species ranked as "vulnerable" are included, about a third of all plant and animal species are "at risk." The degree of risk for "at risk" species varies considerably, from those species at relatively low risk, to those that are in imminent danger of extinction.
- About 20% of native freshwater animal species are ranked as "imperiled," as are 9% of forest and grassland/shrubland animals. An ecosystem with a larger percentage of at-risk species does not necessarily have a larger percentage of species that are declining, because some ecosystems have more naturally rare species. Again, it is difficult to interpret these numbers without information on trends, which is not available.
- The only national data on non-native species are for birds and freshwater fish. Only 1% of the 350 major watersheds in the U.S. have no non-native fish; almost two-thirds have between 1 and 10 non-native fish, and the rest have more. In grassland and shrubland areas, populations of invasive and native, non-invasive bird species were changing in about the same proportion for most of the past 35 years.
- About 20% fewer incidents of unusual waterfowl mortality occurred in 1990–1995 than in the previous two 5-year periods. Particularly large mortality events for marine mammals occurred in 1992 (more than 2500 sea lions) and 1999 (215 harbor porpoises and 270 gray whales).

Biological Components: Plants & Animals

Core National
◐ At-Risk Native Species

Coasts and Oceans
⊖ At-Risk Marine Species
? Non-native Species
◐ Unusual Marine Mortalities

Farmlands
? Status of Animal Species in Farmlands Areas
? Native Vegetation in Areas Dominated by Croplands

Forests
◐ At-Risk Native Species
⊖ Area Covered by Non-native Plants

Fresh Waters
◐ At-Risk Native Species
◐ Non-native Species
◐ Animal Deaths and Deformities

Grasslands/Shrublands
◐ At-Risk Native Species
⊖ Non-native Plant Cover
● Population Trends in Invasive and Non-invasive Birds

Urban/Suburban
⊖ Species Status
⊖ Disruptive Species

● *Complete data available*
◐ *Partial data available*
⊖ *Data not adequate for national reporting*
? *Indicator development needed*

Biological Components: Communities

Biological communities are the more-or-less stable groupings of plants and animals found in particular habitats. These interacting communities form the biological "neighborhood" within which individual species exist, and their condition reflects a broad array of influences on an ecosystem. As with the indicators of physical condition, indicators of biological community condition differ greatly among ecosystems.

Fifteen indicators describe the condition of biological communities. All or partial data are available for only four of the 15 indicators, and trends for only one. Six indicators lack adequate national data, and five of the indicators require additional development.

Biological Components: Communities

Core National
⑦ Condition of Plant and Animal Communities

Coasts and Oceans
⑦ Harmful Algal Blooms
◐ Condition of Bottom-Dwelling Animals

Farmlands
⊖ Soil Biological Condition
⑦ Stream Habitat Quality

Forests
◐ Forest Age
● Forest Disturbance: Fire, Insects, and Disease
⊖ Fire Frequency
⊖ Forest Community Types with Significantly Reduced Area

Fresh Waters
⊖ Status of Freshwater Animal Communities
◐ At-Risk Freshwater Plant Communities
⑦ Stream Habitat Quality

Grasslands/Shrublands
⊖ Fire Frequency
⑦ Riparian Condition

Urban/Suburban
⊖ Status of Animal Communities in Urban/Suburban Streams

● *Complete data available*
◐ *Partial data available*
⊖ *Data not adequate for national reporting*
⑦ *Indicator development needed*

Biological Components: Ecological Productivity

Core National
● Plant Growth Index

Coasts and Oceans
◐ Chlorophyll Concentrations

● *Complete data available*
◐ *Partial data available*
⊖ *Data not adequate for national reporting*
⑦ *Indicator development needed*

Highlights: Communities

- At least half of the estuary area in the Mid-Atlantic, South Atlantic, and Gulf of Mexico regions has bottom-dwelling animal communities that are "undegraded," compared to an undisturbed site; about one-fifth has "degraded" bottom-dwelling animal communities. Data are not available for other regions.
- About 65% of eastern timberlands and 30% of western timberlands are less than 60 years old. About 5% of eastern timberlands and 35% of those in the West are 100 or more years old. (Data for the roughly one-third of forests that are not classified as "timberlands" are not yet available.)
- Since 1980, wildfires in both forests and grasslands and shrublands have affected between 2 and 7 million acres per year, down from a high of 52 million acres in 1930. (While national data do not show an overall increase in acreage burned over the past 20 years, data from national forests, which are mostly in the West, do show an increase.) Insect damage in forests affected between 8 and 46 million acres per year over the past 20 years; the overall trend is downward.
- About 12% of freshwater wetland plant community types are considered at very high risk of being eliminated, and a total of 60% are considered to be at risk of elimination.

Biological Components: Ecological Productivity

The amount of plant growth in an ecosystem is a direct measure of the amount of energy (from the sun) entering the ecosystem and thus of the amount of energy available to all organisms in the system.

This report includes two related indicators: one measures the solar energy captured by plants across the United States, which is closely related to the amount of plant growth, while the other reports on the concentration of chlorophyll in coastal waters, a measure of growth of algae. Data are available for both.

Highlights: Ecological Productivity

- For plant growth nationwide, no overall upward or downward trends are apparent over the 11-year period for which data are available. However, there is large year-to-year variation, both regionally and by ecosystem type.
- Data on coastal chlorophyll concentrations are available for only three years, which is too short to determine trends.

Human Uses: Production of Food and Fiber and Use of Water

Ecosystems produce goods that meet a variety of societal demands. In this report, we include 13 indicators of major ecosystem-related commodities. Most of these indicators describe the goods society derives from ecosystems; several also provide information on the ability of the system to continue producing those goods. Data, including trends, are available for ten of these indicators.

Highlights: Production of Food and Fiber and Use of Water

- Agricultural production has increased by about 85% since the 1950s, although there were noticeable fluctuations within the overall increasing trend.
- Per-acre yields of the major crops grown in the United States have increased dramatically over the past 50 years. For all five major crops (corn, wheat, soybeans, hay, and cotton), the increase in yield was close to, or greater than, 100%, with corn yields increasing almost fourfold. The amount of key inputs required to produce a unit of farm output—with the exception of pesticides—has decreased; pesticide inputs have leveled off since 1980.
- Timber harvest is about 40% higher than it was during the 1950s, but it is lower now than at its peak in the 1980s.
- Annual timber growth in both the East and West regions exceeds harvest on both public and private timberlands. This has been largely true for the past 50 years. Private lands account for almost 90% of total harvest.
- Freshwater withdrawals for various human uses increased nearly 60% from 1960 to 1980, when they dropped sharply, followed by a gradual increase.
- The number of human disease outbreaks attributable to contaminated drinking water has declined significantly overall since the mid-1970s; during the same period, the number of outbreaks associated with recreational contact increased significantly. Since 1990, there have been fewer than 20 outbreaks per year in each category.
- Marine fish landings grew by about 10% from the mid-1970s, when reliable data became available, to the mid-1990s. Recent declines mean that current levels are about equal to those of the late 1970s.
- Nationally, from 1981 to the present, about 40% of fish stocks with known population status had decreasing population trends, while about 20% had increasing trends. Population trends are not known for about three-quarters of commercially important stocks.
- The number of range-fed cattle decreased slightly during the 1990s, to about 93 million animals.

Human Uses: Food, Fiber, & Water

Core National
● Production of Food and Fiber and Water Withdrawals

Coasts and Oceans
● Commercial Fish and Shellfish Landings
◐ Status of Commercially Important Fish Stocks
⊖ Selected Contaminants in Fish and Shellfish

Farmlands
● Major Crop Yields
● Agricultural Inputs and Outputs
● Monetary Value of Agricultural Production

Forests
● Timber Harvest
● Timber Growth and Harvest

Fresh Waters
● Water Withdrawals
⊖ Groundwater Levels
● Waterborne Human Disease Outbreaks

Grasslands/Shrublands
● Production of Cattle

● *Complete data available*
◐ *Partial data available*
⊖ *Data not adequate for national reporting*
⑦ *Indicator development needed*

Human Uses: Recreation and Other Services

Ecosystems provide "services" to people, such as soil building, plant pollination, natural flood control, and the like, as well as outdoor recreation. We defined nine indicators in this category, seven of which deal with either the number of days of recreational activity or the quality or availability of recreational resources. Data are inadequate for national reporting on all but one of these indicators; partial data, with no trends, are available for one indicator, and two require further development before data availability can be assessed.

While there have been efforts to characterize and measure ecosystem services, there is currently little agreement on how such characteristics measures should be defined, and no national data on conditions or trends. We therefore identify, in two instances, the need for indicators of ecosystem services, but recognize that these indicators require additional development.

Human Uses: Recreation and other Services

Core National
◐ Outdoor Recreation
⑦ Natural Ecosystem Services

Coasts and Oceans
⊖ Recreational Water Quality

Farmlands
⊖ Recreation on Farmlands

Forests
⊖ Recreation in Forests

Fresh Waters
⊖ Freshwater Recreation Activities

Grasslands/Shrublands
⊖ Recreation on Grasslands and Shrublands

Urban/Suburban
⊖ Publicly Accessible Open Space per Resident
⑦ Natural Ecosystem Services

● *Complete data available*
◐ *Partial data available*
⊖ *Data not adequate for national reporting*
⑦ *Indicator development needed*

Highlights: Recreation and Other Services

- "Fitness activities," such as walking and biking, and nature viewing—each with more than 10 billion "recreation days" per year—are by far the most common outdoor recreation activity for which information is available. Swimming and beachgoing, which together account for about 5 billion recreation days, is the next most popular activity.
- It is not possible to report on the amount of recreation taking place in specific ecosystem types, like forest or grasslands/shrublands. For most recreational activities, it is not possible to distinguish freshwater activities from saltwater.
- Indicators of ecosystem services, such as soil building and pollination, require additional development.

Notes

1. We estimated urban/suburban land area using a satellite-based method that does not allow for comparison with previous estimates. However, data from the Economic Research Service (see the core national extent indicator, p. 40) indicate that the area of urban lands has grown by more than 300% since the 1950s. Also, as noted below, the USDA Natural Resources Inventory showed substantial increases in nonfederal developed lands from 1982 to 1997.

The State of the Nation's Ecosystems: The Indicators at a Glance

	Core National Indicators	Coasts and Oceans	Farmlands
SYSTEM DIMENSIONS			
Extent	◐ Ecosystem Extent	◐ Coastal Living Habitats ◐ Shoreline Types	● Total Cropland ● The Farmland Landscape
Fragmentation and Landscape Pattern	(?) Fragmentation and Landscape Pattern		⊖ Fragmentation of Farmland Landscapes by Development ⊖ Shape of "Natural" Patches in the Farmland Landscape
CHEMICAL AND PHYSICAL CONDITIONS			
Nutrients, Carbon, and Oxygen	● Movement of Nitrogen	⊖ Areas with Depleted Oxygen	● Nitrate in Farmland Streams and Groundwater ● Phosphorus in Farmland Streams ⊖ Soil Organic Matter
Contaminants	◐ Chemical Contaminants	◐ Contamination in Bottom Sediments	● Pesticides in Farmland Streams and Groundwater
Physical		⊖ Coastal Erosion ● Sea Surface Temperature	● Soil Erosion ⊖ Soil Salinity
BIOLOGICAL COMPONENTS			
Plants and Animals	◐ At-Risk Native Species	⊖ At- Risk Marine Species (?) Non-native Species ◐ Unusual Marine Mortalities	(?) Status of Animal Species in Farmland Areas (?) Native Vegetation in Areas Dominated by Croplands
Communities	(?) Condition of Plant and Animal Communities	(?) Harmful Algal Blooms ◐ Condition of Bottom Dwelling Animals	⊖ Soil Biological Condition (?) Stream Habitat Quality
Ecological Productivity	● Plant Growth Index	◐ Chlorophyll Concentrations	
HUMAN USES			
Food, Fiber, and Water	● Production of Food and Fiber and Water Withdrawals	● Commercial Fish and Shellfish Landings ◐ Status of Commercially Important Fish Stocks ⊖ Selected Contaminants in Fish and Shellfish	● Major Crop Yields ● Agricultural Inputs and Outputs ● Monetary Value of Agricultural Production
Recreation and Other Services	◐ Outdoor Recreation (?) Natural Ecosystem Services	⊖ Recreational Water Quality	⊖ Recreation on Farmlands

	Forests	Fresh Waters	Grasslands and Shrublands	Urban and Suburban Areas
	● Forest Area and Ownership ● Forest Types ● Forest Management Categories	◐ Extent of Freshwater Ecosystems ◐ Altered Freshwater Ecosystems	● Area of Grasslands and Shrublands ◐ Land Use	● Area of Urban/ Suburban Lands ⊖ Total Impervious Area ? Stream Bank Vegetation
	◐ Forest Pattern and Fragmentation		⊖ Area and Size of Grassland/Shrubland Patches	? Suburban/Rural Land Use Change ● Patches of Forest, Grasslands/Shrublands, and Wetlands
	● Nitrate in Forest Streams ◐ Carbon Storage	◐ Phosphorus in Lakes, Reservoirs, and Large Rivers Also see Core National, Farmlands, Forest, Grasslands/Shrublands, and Urban/Suburban Indicators	⊖ Nitrate in Groundwater ⊖ Carbon Storage	● Nitrate in Urban/ Suburban Streams ● Phosphorus in Urban/ Suburban Streams
		Also see Core National, Farmlands, and Urban/Suburban Indicators		● Air Quality ◐ Chemical Contamination
		● Changing Stream Flows ⊖ Water Clarity	● Number and Duration of Dry Periods in Streams and Rivers ⊖ Depth to Shallow Groundwater	? Urban Heat Island
	◐ At-Risk Native Species ⊖ Area Covered by Non-native Plants	◐ At-Risk Native Species ◐ Non-Native Species ◐ Animal Deaths and Deformities	◐ At-Risk Native Species ⊖ Non-native Plant Cover ● Population Trends in Invasive and Non-invasive Birds	⊖ Species Status ⊖ Disruptive Species
	◐ Forest Age ● Forest Disturbance: Fire, Insects, and Disease ⊖ Fire Frequency ⊖ Forest Community Types with Significantly Reduced Area	⊖ Status of Freshwater Animal Communities ◐ At-Risk Freshwater Plant Communities ? Stream Habitat Quality	⊖ Fire Frequency ? Riparian Condition	⊖ Status of Animal Communities in Urban/Suburban Streams
	● Timber Harvest ● Timber Growth and Harvest	● Water Withdrawals ⊖ Groundwater Levels ● Waterborne Human Disease Outbreaks	● Production of Cattle	
	⊖ Recreation in Forests	⊖ Freshwater Recreation Activities	⊖ Recreation on Grasslands and Shrublands	⊖ Publicly Accessible Open Space Per Resident ? Natural Ecosystem Services

● All Necessary Data Available ◐ Partial Data Available ⊖ Data Not Adequate for National Reporting ? Indicator Development Needed

Part II: The Indicators

In this part of *The State of the Nation's Ecosystems*, we move from background and principles to the indicators themselves. In each of the following seven chapters, we present a suite of indicators, describing, for each one, its significance, current conditions, and historic trends when data are available and, when they are not, why not.

We begin with the core national indicators, which provide a succinct description of ten key aspects of the condition and use of ecosystems in the United States. These core national indicators are followed by chapters that present the indicators for coastal waters, farmlands, forests, fresh waters, grasslands and shrublands, and urban and suburban areas. Each of these ecosystems is described using 14 to 18 indicators.

Each chapter in this part begins with a summary table that briefly describes the indicators, including whether data adequate for national reporting are available or not and, if so, whether there are trends or other useful reference points against which to compare the data. This overview table is followed by a summary of the highlights of each indicator, as well as information on the definition of the ecosystem (e.g., "what do we mean by grasslands and shrublands?"). Finally, since data are presented using a variety of regional schemes, we define these for each ecosystem.

The heart of each chapter is the indicators themselves, which are generally presented in a single page (given their broad scope, the core national indicators are accorded two pages or more, as are several more-complex indicators throughout the report). Each indicator is linked to technical notes, which provide detail on the indicators and the data sources used to report on them; these technical notes begin on page 207.

Core National Indicators

	What are the Core National Indicators?		Can we report trends? Are there other useful reference points?
SYSTEM DIMENSIONS			
◐	Ecosystem Extent	What is the area of the six major ecosystem types?	Some trends
ⓘ?	Fragmentation and Landscape Patterns	How fragmented are natural lands into smaller, more isolated patches? How are developed lands intermingled within the natural landscape?	No data reported
CHEMICAL AND PHYSICAL CONDITIONS			
●	Movement of Nitrogen	How much nitrogen leaves watersheds across the country, and how much is delivered to coastal waters?	Trends
◐	Chemical Contamination	How frequently are chemical contaminants found in ecosystems, and how often do they exceed standards and guidelines for the protection of human health and aquatic life?	Current data only, federal standards and guidelines
BIOLOGICAL COMPONENTS			
◐	At-Risk Native Species	How many native species are at different levels of risk of extinction?	Current data only
?	Condition of Plant and Animal Communities	What fraction of U.S. lands and waters are highly managed or highly altered, and what levels of disturbance are found on natural/semi-natural lands?	No data reported
●	Plant Growth Index	What are the trends in plant growth in different regions and different ecosystems?	Trends
HUMAN USES			
●	Production of Food and Fiber and Water Withdrawals	How are the quantities of key ecosystem-related commodity goods changing over time?	Trends
◐	Outdoor Recreation	How often do people take part in outdoor recreation activities, and which kinds?	Current data only
?	Natural Ecosystem Services	What other services, such as soil building and flood protection, are provided by natural ecosystems?	No data reported

● All Necessary Data Available ◐ Partial Data Available ⊖ Data Not Adequate for National Reporting ? Indicator Development Needed

Chapter 4:
Core National Indicators

America's ecosystems are enormous, and enormously diverse. They range from deep ocean trenches to wide grassy plains, from above the Arctic Circle to the tip of Florida. In this chapter, rather than focusing on specific ecosystems, as we do in succeeding chapters, we present ten indicators that describe key characteristics of the entire array of America's ecosystems.

These ten core national indicators provide a broad, yet succinct, description of the condition and use of ecosystems in the United States. They describe and track changes in key aspects of the area and configuration of ecosystems, significant chemical and physical conditions, biological components, and the goods and services that people derive from these systems. In doing so, they parallel the indicators presented in the six following chapters, each of which focuses on a single ecosystem. These chapters also cover ecosystem area and configuration, chemical and physical properties, biological components, and human uses, but they do so using a larger number of indicators that focus on a subset of the nation's lands and waters.

What can we say about the condition and use of U.S. ecosystems, based on these core national indicators?

Partial or complete data are available for seven of the ten core national indicators. Four of the seven have data from a long enough period to judge trends, and one uses federal benchmarks to help readers judge the significance of ecosystem conditions. The three indicators for which data are not presented require further development.

After the following brief summaries of the findings and data availability for each indicator, the remainder of this chapter consists of the indicators themselves. Each indicator presentation offers a graphic representation of the available data, defines the indicator and explains why it is important, and describes either the available data or the gaps in those data.

System Dimensions

The national indicators include two measures of extent and pattern. The first is the most basic description of the state of our nation's lands and waters, the area of each of the component systems and how they change through time. The second measure, not yet developed, will describe the intermingling of the various system types across the national landscape.

- **What is the area of the six major ecosystem types?** Grasslands and shrublands and forests each occupy about a third of the area of the lower 48 states, and farmlands about a quarter. The area of forest and grasslands and shrublands has declined since European settlement, as has the area of freshwater wetlands, and the extent of cropland and urban and suburban areas has grown. More recent trends show decline in forest, croplands, grassland and shrublands, and freshwater wetlands, and increases in urban and suburban areas (Table 4.1).

  The area of ecosystems is a very basic characteristic but, for various reasons, is complex to report. The main reason is that the area of different ecosystems is often tallied by different agencies, using different methods and definitions of the systems. Satellite remote sensing, which can provide an integrated view, is available at the appropriate scale for only one time period (1992) and thus cannot provide information on changes in the area of different ecosystems.

Table 4.1. Core National Extent Measurements (lower 48 states)

Ecosystem	Core National Extent Measurements	Area in Millions of Acres	Percent of Land Area[a]	Estimated Presettlement Area (as % of Total Land Area)	Changes from 1950s, Millions of Acres (%)[a]
Grasslands and Shrublands	Total area (not including pastures)	683	36%	52%	Declining, amount and rate unknown
Forests	Total area	618	33%	48%	–9 (–1.1%)
Farmlands	Area of croplands	455	24%	—	–23 (–4.8%)
Freshwater	Area of Freshwater Wetlands	94	5%	11%	–11 (–10%)
Urban and Suburban areas	Urban and suburban lands	32	1.7%	—	Increasing, amount and rate unknown
Coasts and Oceans	Coastal brackish water	Unknown	—	Unknown	Unknown

Source: Grasslands and shrublands and urban and suburban areas: Multi-Resolution Land Characterization Consortium and the U.S. Geological Survey; data are for 1992. Forests: USDA Forest Service Forest Inventory and Analysis program; data are for 1997. Croplands: USDA Economic Research Service (see p. 91 for estimates from other agencies); data are for 1997. Freshwater and coastal wetlands: U.S. Fish and Wildlife Service; data are for mid-1990s. Coverage: lower 48 states.

[a] This table does not include 100% of lands in the United States. For example, urban and suburban areas, as defined in this report, do not include all developed areas (some developed areas are too small to be considered "suburban" or "urban"). Thus, declines in the area of forests, grasslands and shrublands, croplands, and freshwater wetlands are not—nor should they necessarily be—offset by corresponding gains in urban and suburban lands. In addition, the area of wetlands and portions of urban and suburban areas may also be counted as croplands, forests, or grasslands and shrublands. For these reasons, the figures in this table should not be added to obtain an overall estimate of U.S. land area.

Table 4.1 presents estimates from multiple sources, which means that care must be taken in comparing and adding data about different ecosystems and in tracking gains and losses from one system to another.

- **How fragmented are natural lands into smaller, more isolated patches? How are developed lands intermingled within the natural landscape?** This indicator requires further development. There is widespread recognition that these patterns are important ecologically and that they can affect people's quality of life, but there is less agreement among scientists on the most appropriate indicators to use in measuring such patterns and the most appropriate geographic scales on which to apply them. This report includes several indicators of fragmentation and landscape pattern—for forests (p. 120), farmlands (pp. 93 and 94), grasslands/shrublands (p. 163), and urban/suburban areas (pp. 182 and 183)—but these indicators focus on different aspects of fragmentation and pattern from system to system. This indicator requires further development.

Chemical and Physical Conditions

Out of the many important indicators of chemical and physical condition, we have identified two as national indicators. Nitrogen is a vital plant nutrient, but if present in excess it can cause ecological problems, especially in coastal waters. One indicator tracks the amount of nitrogen that leaves the land and is delivered to coastal waters. The second is a multipart indicator that tracks such contaminants as pesticides, PCBs, and heavy metals in streams, sediment, groundwater, and fish.

- **How much nitrogen leaves watersheds across the country, and how much is delivered to coastal waters?** Delivery of nitrogen from rivers and streams to coastal waters can cause excess algae growth, which reduces recreational and aesthetic values and can contribute to low-oxygen conditions. Watersheds in the upper Midwest and Northeast contribute the most nitrogen per square mile to rivers and streams. The amount of nitrate carried by the four largest rivers in the United States increased over the past few decades, with the amount carried by the Mississippi River—which drains more than 40% of the area of the lower 48 states—tripling since the 1950s.

- **How often are chemical contaminants found in ecosystems, and how often do they exceed standards and guidelines for the protection of human health and aquatic life?** This indicator describes the numbers of contaminants that can be detected and the frequency with which concentrations exceed applicable standards or guidelines. The indicator covers stream water, streambed and coastal sediments, freshwater and saltwater fish, and groundwater. Numbers of contaminants found, and the frequency with which they exceed applicable standards or guidelines, vary by ecosystem. For example, all streams tested averaged one or more contaminants at detectable levels throughout the year, as did 94% of freshwater fish samples and about 90% of groundwater wells tested. About three-quarters of stream samples and half of stream sediments tested had one or more compounds that exceeded guidelines for the protection of aquatic life, and about 60% of estuary sediments exceeded levels that indicate probable negative effects on aquatic life. About 15% of stream sites and one-quarter of groundwater wells had concentrations of contaminants that exceeded standards or guidelines for the protection of human health. No trend data are available for this indicator.

Biological Components

Three indicators describe biological conditions. The first tracks how many plant and animal species are at risk of extinction, because plants and animals are important as components of ecosystems and because people value them for many reasons. A second indicator, not yet developed, will measure how much of U.S. lands and waters are altered, to varying extents, from natural conditions. A third indicator tracks trends in annual plant growth, the energy that drives and sustains ecosystems.

- **How many native plant and animal species are at different levels of risk of extinction?** About 19% of native animal species and 15% of native plants species in the U.S. are ranked as "imperiled" or "critically imperiled"; such species are typically found in 20 or fewer places, may have experienced steep or very steep declines, or display other risk factors. In addition, about 4% of animals and 1% of plants are, or are believed to be, extinct. When species ranked as "vulnerable" are included, about one-third of all plant and animal species are "at risk." The degree of risk for any particular plant or animal species varies considerably, from those species at relatively low risk, to those that are in imminent danger of extinction. Hawaii has a much higher percentage of at-risk plants and animals than any other region, followed by the Pacific Coast. The Midwest and Northeast/Mid-Atlantic have the lowest percentages.

  Interpreting these figures is complicated, however, because the rankings are influenced by differences in the number of naturally rare species among regions and species groups, as well as by different types and levels of human activities that can cause species declines.
- **What fraction of U.S. lands and waters are highly managed or highly altered, and what levels of disturbance are found on natural/semi-natural lands?** How highly managed or altered an area is affects the type of species the area can support, and this directly influences the goods and services available from the area. This indicator requires further development. It is possible to identify areas that are physically altered (that is, they have a high percentage of asphalt, concrete, etc.) or highly managed (that is, they are farms, forest plantations, golf courses, etc.). However, it is not now possible to distinguish among the different levels of disturbance in natural/semi-natural lands.

- **What are the trends in plant growth in different regions and different ecosystems?** The plant growth index utilizes satellite data to estimate the amount of energy (sunlight) that is captured by plants. Changes in this index, over large regions, could signal changes in ecosystem functioning that may affect crop yields, timber growth, or other ecosystem outputs. No overall trend in plant growth can be seen for the 11-year period for which data are available, either nationally or within any region or ecosystem type. Year-to-year variation is quite high, and this variation is similar among regions and ecosystems. During 2000, plant growth nationwide was less than the 11-year average. Growth was about average in the Pacific states and the Midwest and lower than the 11-year average in the other four regions. Plant growth was farthest below the 11-year average in the Southwest.

Human Use

People rely—in many ways—on the goods and services that ecosystems provide. We distill these ways into three core national indicators. The first focuses on the commodities we get from natural ecosystems: the crops, livestock, fish, timber, water, and other goods that are sold on the market. The second tracks another major use, outdoor recreation. A third indicator, not yet developed, will focus on other services provided by ecosystems, such as flood protection and purification of air and water.

- **How are the quantities of key ecosystem-related commodity goods changing over time?** Over the past half-century or so, agricultural and forest production and freshwater withdrawals have all increased. Agricultural production grew the fastest; its growth has generally been at a higher rate than that of the U.S. population. Forest production has generally tracked population growth; in the late 1970s and early 1980s, production increased to record levels, but it has fallen somewhat in more recent years. Withdrawals of freshwater increased faster than population through 1980, declined by about 10% by the mid 1980s, and has grown slowly since then. Marine fish landings grew slowly from the late 1970s, when reliable statistics became available, through the mid-1990s, but have declined since then. Most of the regional patterns of food and fiber production and water withdrawals match the national patterns above.
- **How often do people take part in outdoor recreation activities, and which kinds?** "Fitness activities" such as walking and biking are by far the most common outdoor recreation activity for which information is available. Nature viewing and swimming and beachgoing are next in terms of overall popularity, followed by outdoor social activities like picnics and family gatherings. Altogether, people camped and hiked about as much as they went to picnics and family gatherings, and more than they hunted and fished. For many water-related activities (e.g., swimming, boating), it is not possible to distinguish whether the activity took place in fresh or salt water.
- **What other services, such as soil building and flood protection, are provided by natural ecosystems?** This indicator requires further development. There is widespread recognition that such services are important to society, but measuring them is quite challenging.

A Note about Regions

Data for three of these core national indicators (at-risk native species, p. 52; plant growth index, p. 56; and production of food and fiber and water withdrawals, p. 58) are presented on a regional basis, while a fourth (movement of nitrogen, p. 46) is presented in mapped form. The regional scheme, developed specially for this project, is also used to report the at-risk species indicators in the forest, grasslands and shrublands, and freshwater chapters. See Map 4.1.

Map 4.1. Regions Used for Reporting Selected Core National Indicators

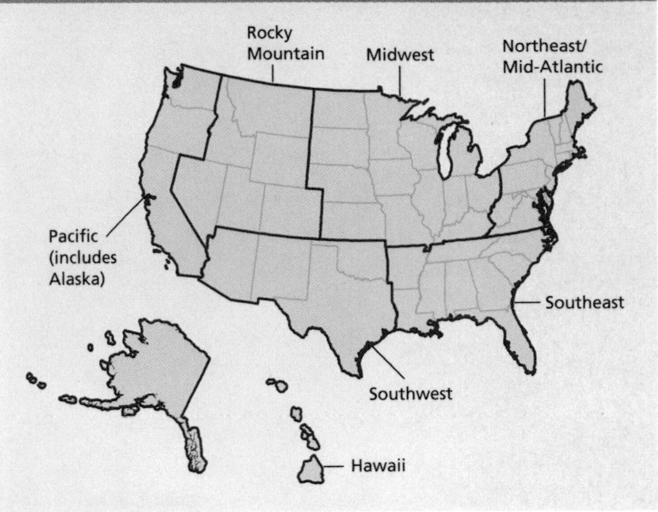

SYSTEM DIMENSIONS	CHEMICAL AND PHYSICAL	BIOLOGICAL COMPONENTS	HUMAN USES
Extent	Nutrients, Carbon, Oxygen	Plants and Animals	Food, Fiber, and Water
Pattern	Contaminants	Communities	Recreation and Other Services
	Physical	Ecological Productivity	

◐ Ecosystem Extent

Map 4.2. U.S. Land Cover and Ocean Depth

This map uses satellite remote sensing information to show the distribution of the ecosystems described in this report. It covers forests, croplands (including pastures and haylands), grasslands and shrublands, urban and suburban areas, most wetlands, and rivers with flows that exceed 1000 cubic feet per second. The map also includes information on the depth of coastal waters, which will be replaced by data on the extent of brackish coastal waters, when such data become available.

Data Source: lower 48 states: Multi-Resolution Land Characterization (MRLC) Consortium; Alaska: Flemming (1996); Hawaii: NOAA; Bathymetry data: NOAA; anaylsis by USGS EROS Data center.

◐ Ecosystem Extent *(continued)*

What Is This Indicator, and Why Is It Important? This indicator presents the area of the four major land-based ecosystem types covered in this report (forests, farmlands, grasslands and shrublands, and urban and suburban areas) as a percentage of the total U.S. land area, for the most recent 50-year period and compared to presettlement estimates. It also reports on a key component of freshwater ecosystems (freshwater wetlands) and will report on the area of brackish water, a key component of coastal and ocean ecosystems when data become available. The change in area since 1955 is also shown for each ecosystem type.

The area occupied by an ecosystem is one of the most basic elements of its condition. The area devoted to different ecosystem types directly influences the character of the American landscape and largely determines the ecosystem goods and services that are derived from it. Conversion from one ecosystem to another means that the ecosystem goods and services that can be derived from the original ecosystem are no longer available, replaced by the goods and services provided by the new system.

Even though ecosystem area is a basic ecosystem characteristic, reporting on it is not simple. The area of different ecosystem types is tallied by different agencies, using different methods and definitions of the ecosystems. These estimates provide important trend data and are generally well regarded. However, because they use different methods and definitions, data from these different sources cannot be compared or pieced together for a full national picture. Satellite remote sensing can provide such an overall, integrated view (see Map 4.2). However, it is only available at the appropriate scale for one time period (1992) and thus cannot provide information on changes in ecosystem area. In this report, we have generally used the estimates provided by the various agencies as the basis for reporting on ecosystem extent. We present the satellite data for comparison purposes and because, if repeated, it can provide frequent, consistent, and non-overlapping estimates of changes in ecosystem extent.

What Do the Data Show? Before European settlement, the land that was to become the United States was dominated by forests and grasslands and shrublands. Researchers have estimated that, before European settlement, there were about 920 million acres of forests

Ecosystem Area: Long-Term Changes and Recent Trends

Data Not Adequate for National Reporting on

- **Extent of brackish coastal waters**

Partial Indicator Data: Long-term Changes for Forests, Croplands, Grasslands/Shrublands, Urban/Suburban

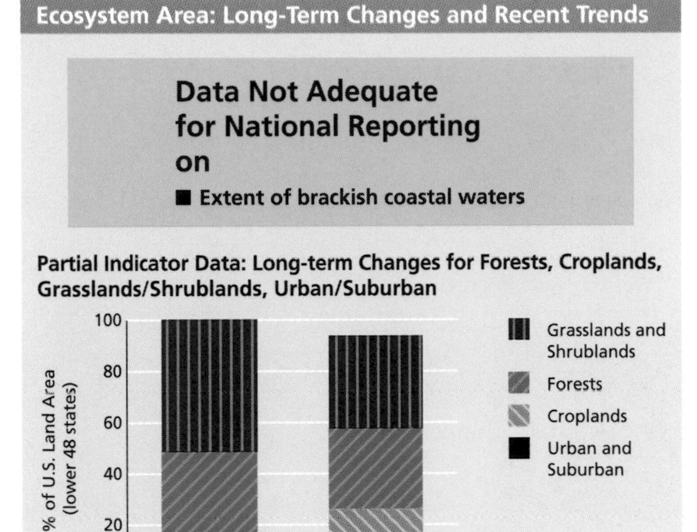

Partial Indicator Data: Recent Trends for Forests, Croplands, Grasslands/Shrublands, Urban/Suburban, Freshwater Wetlands

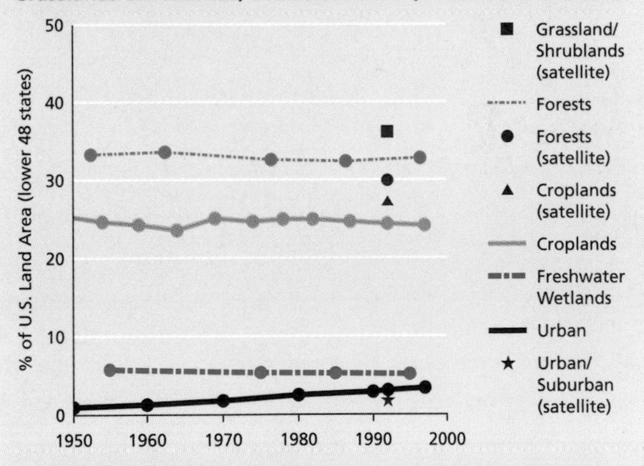

Data Source: USDA Forest Service (forests, current area, recent trends), USDA Economic Research Service (croplands trends, urban area trends), Multi-Resolution Land Characterization Consortium (MRLC; all satellite data, including current estimate of grass/shrub and urban/suburban area in top graph). Presettlement estimates are from Klopatek et al. (1979). Coverage: lower 48 states

Note: Because these estimates are from different sources, they do not sum to 100% of U.S. land area. Approximately 5% of lands are not accounted for by these data sources. They include some wetlands, some non-suburban developed areas, disturbed areas such as mines and quarries, and the like. In addition, freshwater wetlands currently occupy approximately 5% of the area of the lower 48 states, a reduction of about 50% since presettlement times. However, because they are found within forests, grasslands and shrublands, or croplands, they are not shown separately on the graph. See pp. 69 and 139. Finally, the "urban" trend line in this graph is based on a different definition from the one in this report (see p. 181) and is presented here to illustrate general trends. The definition used in this report was used to generate the "urban/suburban (satellite)" area estimate.

◐ Ecosystem Extent *(continued)*

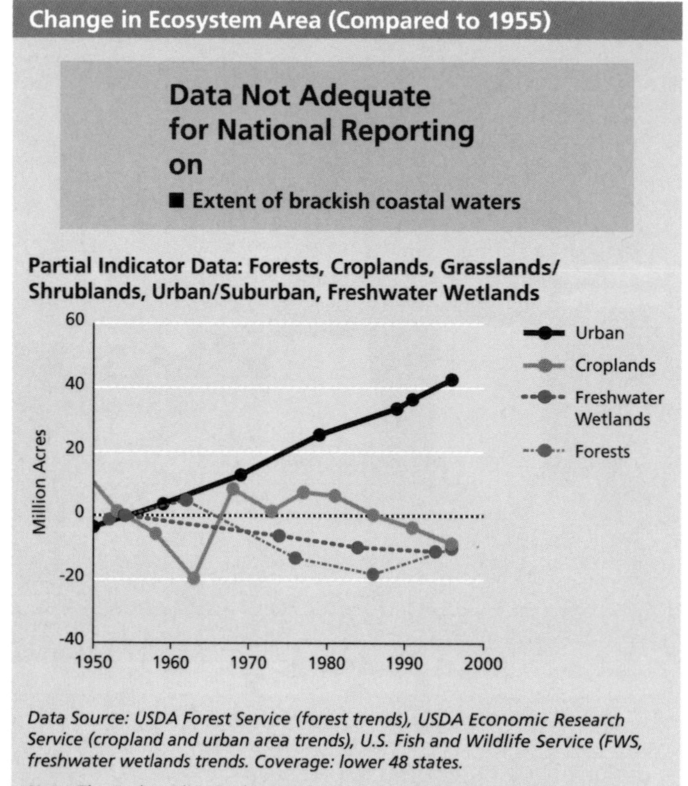

Data Source: USDA Forest Service (forest trends), USDA Economic Research Service (cropland and urban area trends), U.S. Fish and Wildlife Service (FWS, freshwater wetlands trends. Coverage: lower 48 states.

Note: The "urban" line in this graph is based on a different definition from that used in this report. It is presented here to illustrate general trends. The definition used in this report was used to generate the "urban/suburban (satellite)" area estimate in the preceding graph.

and between 900 million and 1 billion acres of grasslands and shrublands. Thus, each covered roughly half of the lower 48 states. While these estimates are necessarily imprecise, it is clear that croplands (including pastures) and urban and suburban areas—totaling together about 500 million acres—were created on lands that were either forests or grasslands and shrublands, causing the acreage of these ecosystems to drop. In addition, the area of freshwater wetlands has declined by about 50% since European settlement.

In reading these figures and the ones that follow, it is important to remember that the data presented here are from several sources; they do not add to 100% of the U.S. land area, and gains and losses cannot be tracked accurately from one system to another.

Coasts and Oceans include all waters in the U.S. Economic Exclusion Zone (EEZ), which extends 200 miles from the coastline. Because the area of the EEZ changes only when territory is acquired or international law changes, this indicator focuses on the dynamic area of mixed salt and fresh waters, or brackish waters, surrounding the U.S. coastline. Changes in the extent of brackish water reflect changes in the volume of freshwater runoff from the land, which can be altered by changes in climate and by modification of river flows by dams and other diversions. There are no current or historical data at a national scale on the area of brackish water. Another important aspect of the extent of coastal waters is the area covered by coastal wetlands, coral reefs, and shellfish and seagrass beds (see Coastal Living Habitats, p. 69).

Croplands, that portion of farmlands that is actively used for crop production (including pastures), occupy about 24% of the land area of the lower 48 states, or about 455 million acres. About 23 million fewer acres are in active farmland use than in 1949, but over this period, farmland area has fluctuated. American Indians had some lands under cultivation before European settlement, but there are no firm estimates of this amount. Satellite-based methods produce an estimate of just over 500 million acres of croplands in 1992. This report also identifies a "farmland landscape," which includes both croplands and intermingled and nearby forests, grasslands and shrublands, wetlands, and developed areas; see p. 92.

Forests cover about 33% of the land area of the lower 48 states, or just under 620 million acres. When Alaska is added in, the total is about three-quarters of a billion acres, down from just over 1 billion acres before European settlement. In the lower 48 states, forested area has declined by about 10 million acres since 1955. However, there is more forest now than in the middle of the 19th century (not shown in the illustrations), when many parts of the country were cleared for agriculture and settlement (see the forest area indicator, p. 117). Satellite-based methods produce an estimate of about 560 million acres of forest in the lower 48 states.

SYSTEM DIMENSIONS	CHEMICAL AND PHYSICAL	BIOLOGICAL COMPONENTS	HUMAN USES
Extent	Nutrients, Carbon, Oxygen	Plants and Animals	Food, Fiber, and Water
Pattern	Contaminants	Communities	Recreation and Other Services
	Physical	Ecological Productivity	

◐ Ecosystem Extent *(continued)*

Fresh waters include 94 million acres of wetlands in the lower 48 states, or about 5% of total land area. About half the freshwater wetlands that existed at the time of European settlement have been converted to other uses; about 10% of the wetlands existing in 1955 had been converted by the mid-1990s, although the rate of loss slowed after the 1980s. Comparable data do not exist for Alaska. Wetlands occur in many ecosystem types, so their area is often counted as part of the area of forests, grassland and shrublands, farmlands, and urban and suburban areas. Satellite-based methods estimate about 80 million acres of wetlands. While freshwater wetlands are a critical and highly visible aspect of the extent of freshwater systems, the area of lakes and ponds and the number of miles of streams are also important (see Freshwater Extent, p. 139).

Grasslands and shrublands, often called rangelands, occupy about 36% of the land area of the lower 48 states, or about 680 million acres. These figures do not include pastures and haylands. For this national estimate, these pastures and haylands—some of which resemble "natural" grasslands and shrublands and some of which are highly managed—are counted as croplands. In the chapter on grasslands and shrublands, however, these lands are included in the area estimates for this system (p. 161). If these less-managed (uncultivated) pastures were reported as grasslands and shrublands, the decline in grassland and shrubland area would be less than is indicated on the top graph on p. 41.

Urban and suburban areas take up about 32 million acres, or 1.7% of the area of the lower 48 states. This figure is based on a newly developed definition applied to satellite imagery; comparable satellite-based data from earlier periods are not available. To show trends, therefore, we also present a USDA estimate, based upon the Census Bureau definition of urban area. (This definition uses population, rather than the percentage of land area covered by buildings, roads, and the like, to define "urban.") Using the Census-based definition, urban areas cover 64 million acres—twice the area produced by the satellite-based method—and have grown by 40 million acres since 1955. Because it focuses on actual land cover, the satellite-based definition is more appropriate for this report and is used as the basis for the urban and suburban indicators (see Area of Urban and Suburban Lands, p. 181).

The technical note for this indicator is on page 207.

SYSTEM DIMENSIONS	CHEMICAL AND PHYSICAL	BIOLOGICAL COMPONENTS	HUMAN USES
Extent	Nutrients, Carbon, Oxygen	Plants and Animals	Food, Fiber, and Water
Pattern	Contaminants	Communities	Recreation and Other Services
	Physical	Ecological Productivity	

⑦ Fragmentation and Landscape Pattern

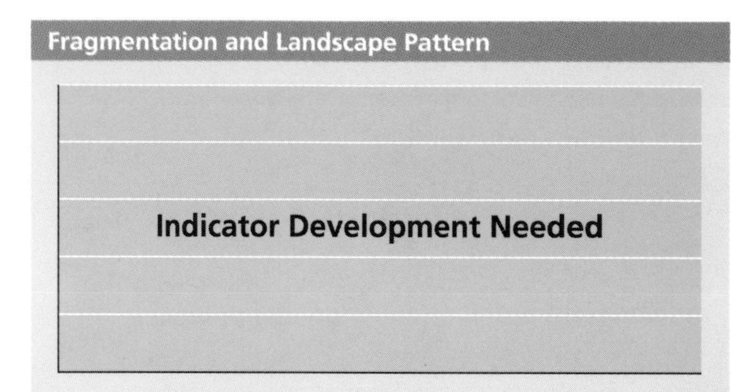

What Is This Indicator, and Why Is It Important? Fragmentation of ecosystems into small patches can reduce habitat for wildlife species that require larger, connected patches. It can hinder the movement of some species and introduce predators, parasites, and competitors associated with different land uses. Fragmentation can also alter the frequency and extent of fire and affect the dispersal and regeneration of plants. Suburban and urban development, farmlands, roads, railroads, powerline corridors, and other land uses cause various kinds and degrees of fragmentation.

Species that require large, unbroken expanses of habitat are often most sensitive to the effects of fragmentation. In some cases, the effects of fragmentation on sensitive species are a direct result of changes in the size and arrangement of suitable habitats across the landscape. In others, impacts are due mainly to more frequent interactions of species with humans, vehicles, or predators, or to other factors associated with an intruding land use.

People also react to changing landscapes. Areas that were primarily forest, grasslands, or shrublands but are now fragmented by other uses or bisected by roads provide a very different level of solitude and visual attraction. Likewise, the character of farm landscapes and communities changes radically when they are broken up by suburban development.

Human activity can also create landscapes that are less varied than the landscapes historically experienced by native species. Particularly in the West, natural fires create a patchy landscape, where forest and grasslands are intermingled in a mosaic that supports many different species. Fire suppression and the large fires that result after long periods of suppression can create broad expanses of very similar vegetation, with negative effects on species that thrive on the formerly varied landscape.

Landscape patterns affect people and other species in different ways and at different geographic scales. Some species are very sensitive to fragmentation, while others are more tolerant. Some effects, such as the changes that occur in farming communities undergoing suburbanization, operate at a county level, while other effects, such as those affecting forest birds, involve distances measured in feet or yards. The magnitude of fragmentation and its context are also important. A single incursion may not cause significant effects, but many such changes taken together may have a larger impact. Similarly, a modest amount of fragmentation in an abundant habitat may not be significant, but the same amount of fragmentation in a rare habitat may be cause for concern.

Why Can't This Indicator Be Reported at This Time? There are clear and obvious linkages between landscape patterns, the kinds of plants and animals that thrive in a region, and the ways in which people use the land. However, there are many different ways to characterize these patterns and the ways in which they are changing, and scientists do not agree on a single "best" measure. Additional work is necessary to select the specific features that should be measured, the geographic scale at which they should be monitored, and how they should be reported and interpreted.

SYSTEM DIMENSIONS	CHEMICAL AND PHYSICAL	BIOLOGICAL COMPONENTS	HUMAN USES
Extent	Nutrients, Carbon, Oxygen	Plants and Animals	Food, Fiber, and Water
Pattern	Contaminants	Communities	Recreation and Other Services
	Physical	Ecological Productivity	

⑦ Fragmentation and Landscape Pattern *(continued)*

This report includes measures of fragmentation or landscape pattern for grasslands and shrublands (p. 163), farmlands (pp. 93 and 94), forests (p. 120), and urban and suburban areas (pp. 182 and 183). Although some of these indicators require additional research, it is clear that there is more agreement among scientists on how to measure landscape pattern for specific ecosystem types than there is for an overall national measure.

What Steps Are Necessary To Achieve Reliable National Coverage? This is an area of active scientific investigation. Many possible indicators are being evaluated to determine which ones, or which combinations, provide the best view of the important changes that are occurring in the American landscape.

There is no technical note for this indicator.

The Movement of Nitrogen

Yield of Total Nitrogen from Major Watersheds (1996–1999)

Total Nitrogen (pounds of nitrogen per sq. mile per year)

- Data Not Available
- Less than 10
- 10-600
- 600-1,500
- 1,500-3,000
- 3,000-10,000

Data Source: U.S. Geological Survey National Stream Quality Network (NASQAN), National Water Quality Assessment (NAWQA), and Federal–State Cooperative Program. Coverage: selected areas of lower 48 states.

What Is This Indicator, and Why Is It Important? This indicator reports the yield of nitrogen from major watersheds: pounds of nitrogen per square mile of watershed area that enters rivers and streams through discharges, runoff, and other sources. It also reports the load of nitrate, a common form of nitrogen, from major rivers: tons of nitrate carried to the ocean each year by the four largest U.S. rivers.

Nitrogen is a component of protein and is essential to all life. Nitrate is an important plant nutrient and is often the most abundant form of nitrogen that is readily usable by aquatic plants, including algae. Nitrate and other forms of nitrogen occur both naturally and as a result of human activities.

In excess, however, nitrogen can cause significant water quality problems by stimulating the growth of algae. Overabundance of algae can reduce oxygen levels to near zero, especially in coastal waters (see Areas with Depleted Oxygen, p. 71). "Dead zones," or areas where oxygen levels are so low that fish and shellfish cannot live, are created when nutrients, particularly nitrate and other forms of nitrogen, are overabundant. The largest of these dead zones occurs every summer in the Gulf of Mexico, covering 5,000 or more square miles of one of the nation's most important commercial and recreational fisheries. Excess nitrogen in certain forms is also toxic to human beings and other animals.

Sources of nitrogen include wastewater treatment plants, runoff from fertilized lawns and cropland, failing septic systems, runoff from animal manure storage areas, and industrial discharges that contain corrosion inhibitors. Atmospheric deposition is also a significant source of added nitrogen in ecosystems. Burning of fossil fuels releases nitrogen into the atmosphere, where it can travel for long distances before being deposited in snow, rain, or dust.

Although this indicator reports on nitrogen in aquatic systems, excess nitrogen in soil, often derived from atmospheric deposition, can change the number and type of species in an ecosystem and otherwise alter the way the system functions.

What Do the Data Show? The map shows 1996–1999 average annual yield of total nitrogen from major watersheds for which data are available. Watersheds in the upper Midwest and the Northeast contribute the most nitrogen per square mile to rivers and streams ("yield").

The amount of nitrate carried by most major U.S. rivers ("load") has increased over the past several decades. The four largest rivers in the United States—the Mississippi, Columbia, St. Lawrence, and Susquehanna—together account for approximately 55% of all freshwater flow to the sea from the lower 48 states. The Mississippi has had the most striking increase in nitrate load. The Mississippi, which drains more than 40% of the area of the lower 48 states, carries roughly 15 times more nitrate than any other U.S. river, and this amount has approximately tripled since the 1950s. The increases in nitrate load for the Columbia and Susquehanna rivers are also significant, although some multiyear declines also occurred during the period.

The Movement of Nitrogen *(continued)*

The peaks and valleys within the overall upward trend generally reflect years with higher rainfall (peaks) and those with less rainfall (valleys). In wet years, increased runoff from land surface carries more nitrogen into streams, increasing nitrogen loads; the reverse is true in dry years.

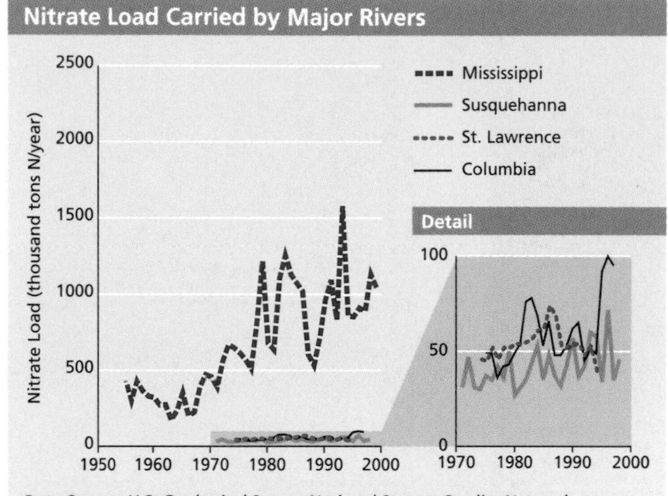

Data Source: U.S. Geological Survey National Stream Quality Network (NASQAN), National Water Quality Assessment (NAWQA), and Federal–State Cooperative Program. Coverage: selected major rivers.

Discussion Higher values for both loads and yields reflect greater "leakage" of nitrogen from a watershed, with potentially significant downstream effects, particularly on marine ecosystems.

Total nitrogen is the preferred form for reporting on the amount of nitrogen delivered from the U.S. landscape to our coastal waters, but because the historical record for it for the Mississippi River is short, we chose instead to present river nitrate loads. Nitrate is the largest component of total nitrogen and serves as a strong indicator of total nitrogen loads. The longer historical record for nitrate reveals the significant increases that have occurred over the past few decades. Future reports may present loads of total nitrogen.

Other indicators (see pp. 95, 122, 164, and 186) report on the amount of nitrate dissolved in streams or groundwater in farmlands, forests, grasslands and shrublands, and urban and suburban areas.

The technical note for this indicator is on page 210.

◐ Chemical Contamination

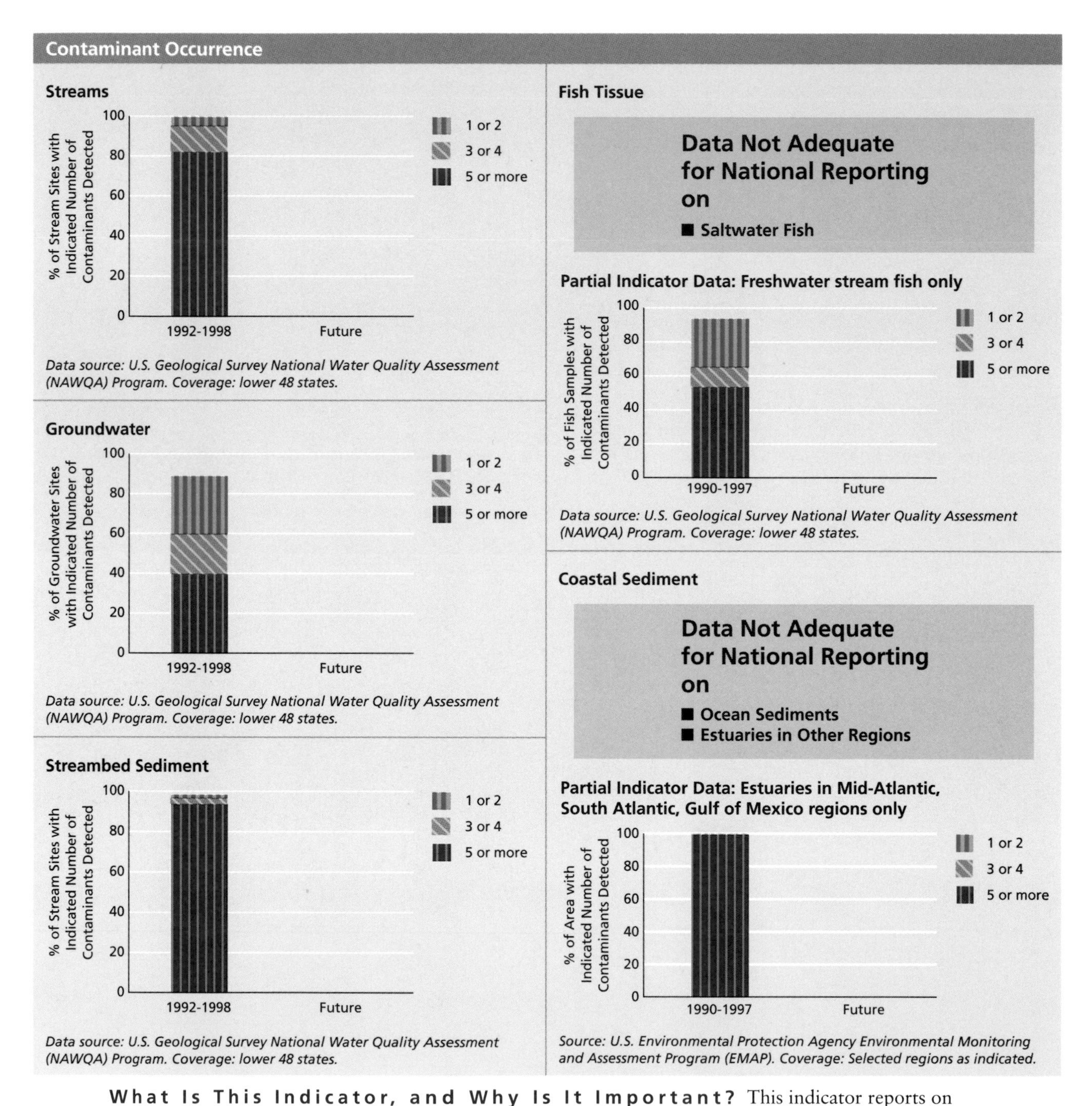

What Is This Indicator, and Why Is It Important? This indicator reports on contaminants found in streams, groundwater, sediment, and fish tissue. The graphs above report how often different numbers of contaminants are found, and those on p. 49 report how often these contaminants exceed standards and guidelines for the protection of human health and aquatic life. Contaminants reported here include many pesticides, selected degradation products, polychlorinated biphenyls (PCBs), polyaromatic hydrocarbons (PAHs), volatile organic compounds, other industrial contaminants, trace elements, nitrate, and ammonium. (Because nitrate, ammonium, and trace elements such as cadmium and chromium occur naturally, they are not included in the contaminant occurrence graphs.)

◐ Chemical Contamination *(continued)*

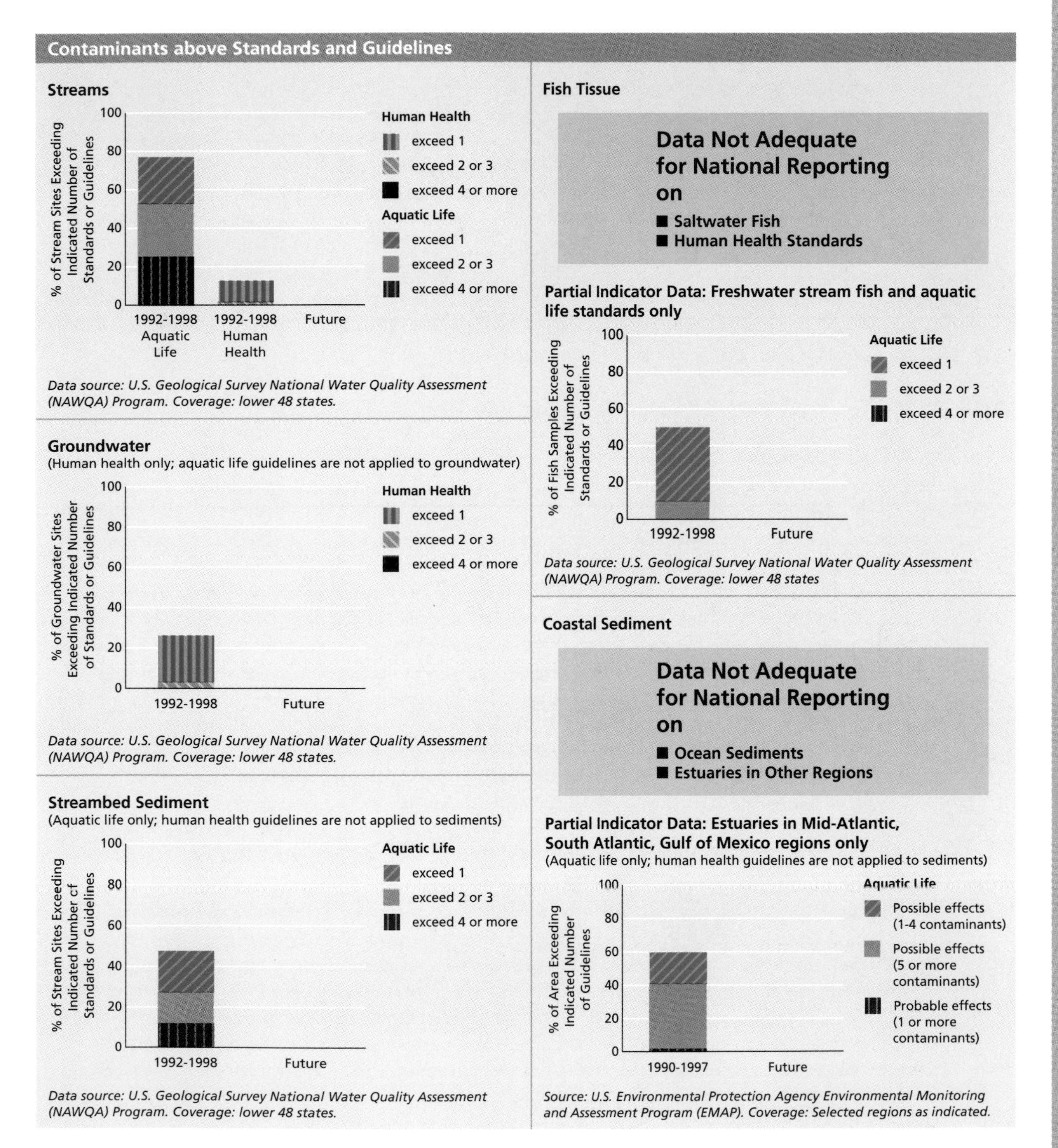

◐ Chemical Contamination *(continued)*

Synthetic chemicals, trace elements, and other contaminants can, in sufficient quantities, harm people as well as fish and other wildlife. Both the frequency of chemical contamination and the degree to which these contaminants exceed applicable standards and guidelines are important in understanding the extent and significance of chemical contamination. The number of contaminants found in streams, groundwater, and the like provides basic information on how widespread these compounds are in the environment. However, the presence of chemical contamination does not necessarily mean that the levels are high enough to cause problems; comparison to standards and guidelines provides a useful reference to help judge the significance of contamination.

There are no standards or guidelines for many contaminants. For example, drinking water standards and guidelines do not exist for 33 of the 76 pesticides analyzed in fresh waters, and there are no aquatic life guidelines for 48 of these 76 pesticides. Current standards and guidelines do not account for mixtures of chemicals and seasonal occurrences of very high concentrations. These gaps increase the importance of information on the occurrence of chemical contaminants. In addition, potential effects on reproductive, nervous, and immune systems, as well as on particularly sensitive people, are not yet well understood.

What Do the Data Show?

Streams. All tested streams averaged one or more contaminants at detectable levels throughout the year; about 80% averaged five or more contaminants at detectable levels. Three-fourths of streams tested had one or more contaminants at levels that exceeded guidelines for the protection of aquatic life; approximately one fourth had concentrations of four or more contaminants that exceeded these guidelines. Thirteen percent had at least one contaminant at levels that exceeded standards or guidelines for the protection of human health. Stream water was tested for pesticides, selected pesticide degradation products, and selected nutrients.

Groundwater. About 90% of groundwater wells tested had an average of one or more contaminants at detectable levels, and 40% had an average of five or more contaminants at detectable levels. About one fourth had contaminants at levels that exceeded human health standards or guidelines. Groundwater was tested for pesticides, selected pesticide degradation products, volatile organic contaminants, trace elements, and selected nutrients.

Stream sediments. Nearly all stream sediments tested had an average of five or more contaminants at detectable levels. About half had one or more contaminants at concentrations exceeding aquatic life guidelines. Stream sediments were tested for organochlorine pesticides, PCBs, PAHs, other industrial contaminants, and trace elements.

Freshwater fish. About half of fish tested had at least five contaminants at detectable levels, and approximately the same number had one or more contaminants at levels that exceeded standards for the protection of wildlife. Data are not available on exceedances of human health standards. Whole fish were tested for organochlorine pesticides, PCBs, and trace elements.

Although not shown on the graphs, all fish tested in the Great Lakes had five or more detected contaminants, and all Great Lakes fish had PCB concentrations that exceeded human health standards. (Great Lakes testing focuses on fish with a high likelihood of such contamination, such as coho salmon and lake trout.)

Coastal sediments. More than 99% of estuary sediments tested had five or more contaminants at detectable levels. About 60% of estuary sediments tested had contaminants above the levels designed to predict "possible effects" on aquatic life for one or more contaminants, and about 2% exceeded the level designed to predict "probable effects." Estuary sediments were tested for PCBs, PAHs, pesticides, and trace elements. Data on ocean sediments are not adequate for national reporting.

◐ Chemical Contamination *(continued)*

Discussion The data shown here do not represent assessments of the risks posed to people or ecosystems in any specific location, since they do not incorporate factors such as whether the water tested is actually used as a drinking water source or whether aquatic animals are biologically active at the time of year when the contaminants are found.

The standards and guidelines used in this indicator are useful reference points, but they must be interpreted carefully, since different standards reflect different levels of protection from harm. Furthermore, different standards and guidelines may apply to water, sediments, and fish tissue.

Guidelines for the protection of aquatic life are often numerically lower than standards and guidelines to protect human health. Aquatic animals spend much or all of their life in water, and may be more sensitive to specific contaminants.

People consume drinking water from both streams and groundwater, and they eat fish, so human health standards and guidelines apply to all three. Guidelines to protect aquatic life are not applied to groundwater, and standards and guidelines to protect human health are not applied to either stream or estuary sediments.

Different agencies and programs are responsible for the collection and analysis of data from freshwater systems (streams and groundwater) and estuaries. The objectives of these programs differ, leading to different site selection procedures, suites of contaminants measured, and collection and analysis procedures. Guidelines for freshwater fish are set to protect fish-eating wildlife, and aquatic life guidelines for coastal sediments differ from those for stream sediments. Thus, the results are not directly comparable.

The contaminants that were analyzed in different media (streams, groundwater, etc.) varied, depending on the chemical properties of the contaminants, known environmental occurrence, and potential for adverse effects on people or ecosystems. For example, volatile organic compounds were analyzed in groundwater but not in stream sediments because their chemical properties make it extremely unlikely that they would be found there.

Data are not available to compare either fresh or saltwater fish contaminant concentrations with human health/consumption guidelines.

See also the coastal, farmland, and urban contaminants indicators (pp. 72, 97, and 189).

The technical note for this indicator is on page 210.

◐ At-Risk Native Species

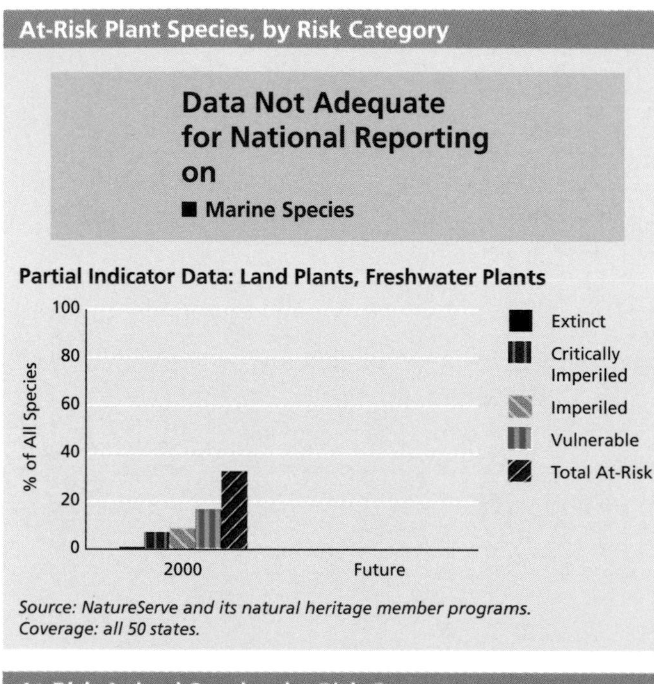

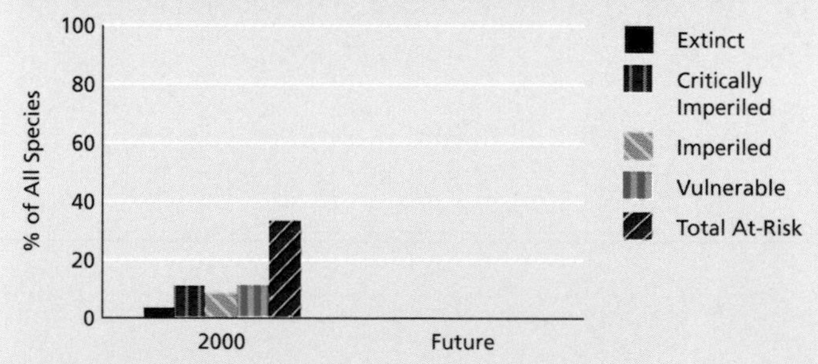

Source: NatureServe and its natural heritage member programs.
Coverage: all 50 states.

What Is This Indicator, and Why Is It Important? This indicator reports on the relative risk of extinction of native plant and animal species. The degree of risk for any particular plant or animal species varies considerably, from those species at little or no risk, to those that are in imminent danger of extinction. The data cover many of the best-known groups of native plants and animals, totaling about 22,000 native species.

Each species is ranked on such factors as the number and condition of individuals and populations, population trends, the area occupied by the species, and known threats. For example, "critically imperiled" species often are found in five or fewer places, may have experienced very steep declines, or show other evidence of very high risk. "Imperiled" species often are found in 20 or fewer places, may have experienced steep declines, or display other risk factors. "Vulnerable" species often are found in fewer than 80 places, may have recently experienced widespread decline, or show other signs of moderate risk. The remaining plant and animal species are regarded as "secure" or "apparently secure." In all cases, a wide variety of factors contribute to overall ratings.

Increased risk levels for a particular species may be due to historical or recent population declines, or they may reflect natural rarity; biologists often consider very rare species to be at risk even in the absence of recent declines or current threats.

Species are valued for a variety of reasons: they provide products, including food, fiber, and genetic materials; they serve as key elements of ecosystems, which provide valuable goods and services; and many people value them for their intrinsic worth or beauty.

What Do the Data Show? About 19% of native animal species and 15% of native plants species in the U.S. are ranked as "imperiled" or "critically imperiled," and another 1% of plants and 3% of animals may already be extinct—that is, they have not been located despite intensive searches. When "vulnerable" species are counted, about one-third of plant and animal species are considered to be "at risk."

Hawaii has a much higher percentage of at-risk plants and animals than any other region, followed by the Pacific Coast. In contrast, the Midwest and Northeast/Mid-Atlantic have the lowest percentages.

◐ At-Risk Native Species *(continued)*

Interpreting these figures is complicated because some species are naturally rare. Thus, the rankings are influenced by differences among regions and species groups in the number of naturally rare species, as well as by different types and levels of human activities that can cause species declines. Interpretation of these data will be greatly enhanced when information on population trends for these at-risk species becomes available.

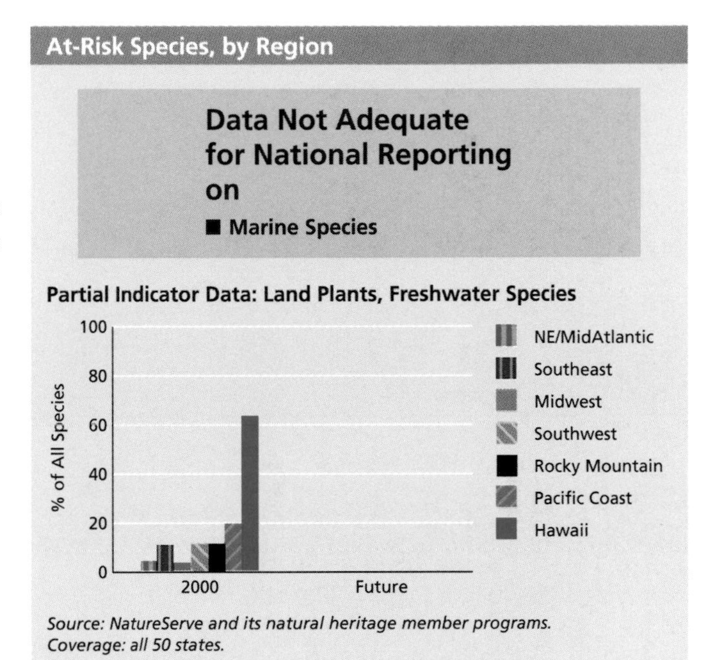

Source: NatureServe and its natural heritage member programs. Coverage: all 50 states.

Why Can't This Entire Indicator Be Reported? Data are not available on at-risk species in U.S. coastal waters.

Discussion At least 200,000 native plant, animal, and microbial species are thought to live in the United States, but little is known about the status and distribution of most of these. This indicator summarizes the status of 16,000 plant species and 6,000 animal species, which include all 22 species groups for which comprehensive status assessments are available. These species represent all higher plants, all terrestrial and freshwater vertebrates (mammals, birds, reptiles, amphibians, and freshwater fishes), selected invertebrate groups, including freshwater mussels and snails, crayfishes, butterflies and skippers, and about 2,000 species of grasshoppers, moths, beetles, and other invertebrates. This sample of species is believed to provide a powerful, yet practically manageable snapshot of the condition of U.S. species.

See http://www.natureserve.org/explorer/ranking for a description of these conservation status ranks and details of the assessment criteria.

See also the indicators for at-risk coastal (p. 75), forest (p. 124), freshwater (p. 144), and grassland and shrubland species (p. 168), as well as those for species in farmland (p. 103) and urban and suburban areas (p. 191).

The technical note for this indicator is on page 214.

⑦ Condition of Plant and Animal Communities

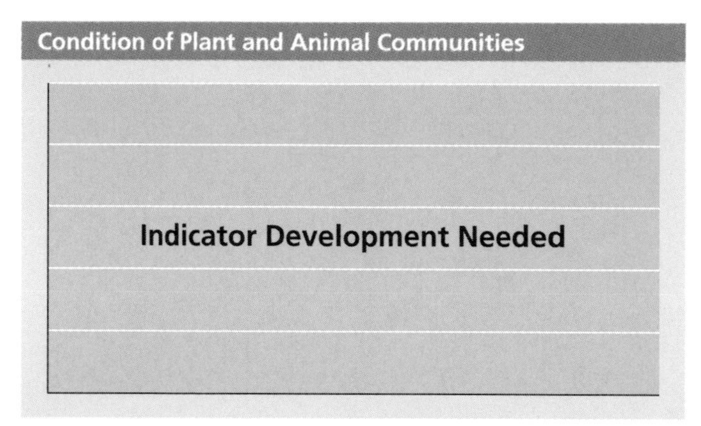

What Is This Indicator, and Why Is It Important? This indicator would report on the percentage of land area and stream and coastline length according to the level of disturbance, management, or physical alteration. Different levels of ecosystem alteration correspond to changes in both the type and number of species of plants and animals found in an area. Plants and animals in areas with high levels of alteration will be very different from those in similar areas that are relatively undisturbed.

The types of plants and animals found in areas that are highly managed or altered have in large part been determined by human activity. These areas are relatively easy to define, and more data about them are available:

- **Physically altered:** Areas in which a high percentage (for example, 30% or more) of the land surface is covered by asphalt, concrete, or buildings, is quarried or strip-mined, or, in the case of stream banks or shorelines, is "armored" with riprap or other materials.
- **Highly managed:** Areas in which human activity has directly and significantly altered the species, especially plants, found there; farms, plantation forests, golf courses, and intensively grazed grasslands and shrublands fall into this category.

In areas that are less substantially modified, the mix of plants and animals is less directly determined by people and more affected by ecological conditions. There are, however, no generally accepted methods for distinguishing between levels of alteration in these natural or semi-natural lands. This indicator presumes that such methods will be developed and that it will be possible to classify these areas into three broad categories:

- **Undisturbed:** Areas of relatively undisturbed biological communities where the types of plants and animals found are similar to what they would be without human influences. Examples might include wilderness areas and much of interior Alaska.
- **Disturbed:** Areas with a modified mix of plant and animal species. Examples might include areas with a high proportion of non-native species, or a different mix of native species as a result of the long-term exclusion of fire.
- **Less disturbed:** Areas with communities with changes intermediate between "disturbed" and "undisturbed."

The species that occur in a place strongly affect the goods and services an ecosystem provides. Areas that are highly managed or physically altered provide important and socially desired goods and services as a result of this management or alteration, but that set of goods and services is quite different from those provided by more natural communities.

Why Can't This Indicator Be Reported at This Time? Although there are data on the status of plants and animals across large regions of the country (see the At-Risk Native Species indicator, p. 52), there are few data on the mix of species found within smaller areas or stretches of stream and coastline. The best data are for land intensively used by people—highly managed and physically altered lands are distinct enough to be identified from satellite measurements.

⑦ Condition of Plant and Animal Communities *(continued)*

For semi-natural and natural areas, however, there are two problems. First, species composition is monitored for some systems (e.g., percent cover of non-native plants in forested areas), but not for all systems or for all types of species alterations. Just as important, ecologists have not agreed on how to classify a particular natural or semi-natural area as disturbed, less disturbed, or undisturbed. Again, measures exist for some ecosystem types (for example, see Status of Freshwater Animal Communities, pp. 147 and 193) but not for most.

Discussion The biological communities found on and in much of the nation's lands and waters today are very different from those of presettlement times. Much of this change is the result of deliberate human intervention: forests have been cleared for farms, streams dammed to form lakes or to generate power, and land covered with housing and roads. Clearly, the goods and services derived from these altered lands and waters differ considerably from those derived from wilderness areas and other lands in a natural or semi-natural condition.

Until better means are developed to determine what conditions exist on the two-thirds of U.S. lands that are natural to semi-natural, both reporting on and interpreting this indicator will remain difficult.

The technical note for this indicator is on page 215.

Plant Growth Index

Plant Growth Index: Lower 48 States, by Ecosystem

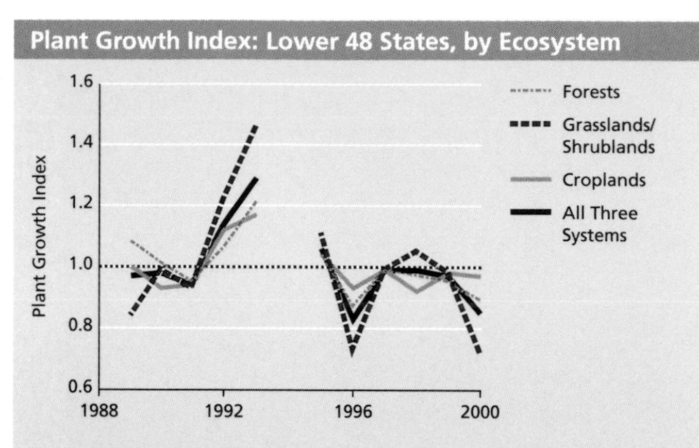

Data Source: U.S. Geological Survey; Multi-Resolution Land Characterization Consortium. Coverage: Lower 48 states.

Note: Because of satellite problems, no data are available for 1994.

Plant Growth Index: Southwest, Rocky Mountain, Pacific Regions

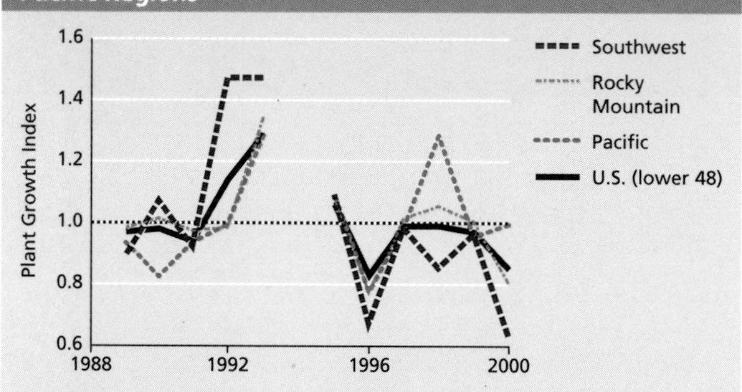

Data Source: U.S. Geological Survey; Multi-Resolution Land Characterization Consortium. Coverage: Lower 48 states.

Note: Because of satellite problems, no data are available for 1994.

Plant Growth Index: Northeast, Southeast, Midwest Regions

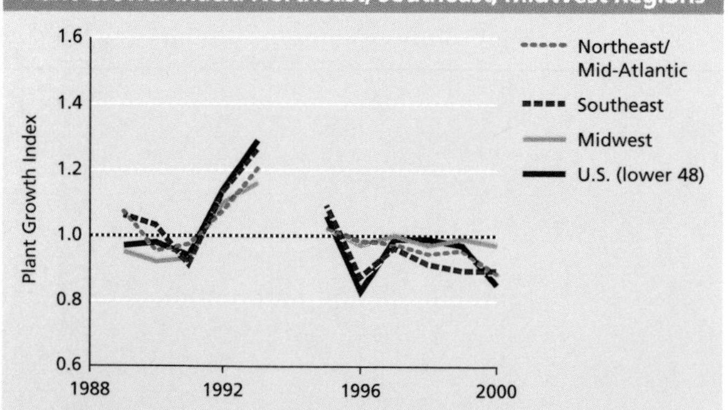

Data Source: U.S. Geological Survey; Multi-Resolution Land Characterization Consortium. Coverage: Lower 48 states.

Note: Because of satellite problems, no data are available for 1994.

What Is This Indicator, and Why Is It Important? This indicator reports a plant growth index, based on satellite measurements of the amount of solar energy absorbed by vegetation and potentially used for photosynthesis.

The index shows, for any given year, whether plant growth in a region or for an ecosystem type was above or below the 11-year average (1989 through 2000, with one missing year). An index value of 1.0 in any year means that the amount of solar energy captured by vegetation and used for photosynthesis in that region or system during that year was the same as the 11-year average.

Plants use energy from the sun to turn carbon dioxide from the air, plus water and nutrients, into plant matter. This process, photosynthesis, drives and sustains virtually all life on earth. The amount of sunlight absorbed by plants is a key factor in determining the amount of photosynthesis and thus the amount of plant growth that occurs in a year. Changes in the amount of energy captured by plants over very large regions, as reported in this measure, may signal significant changes in ecosystem functioning. These changes could lead to increases or decreases in yield of products such as crops or wood and possibly changes in the number and types of species that live in a region. Changes in climate (including temperature and timing and amount of precipitation), as well as factors such as ground-level ozone, increased atmospheric deposition of nitrogen, and increased levels of carbon dioxide, might cause or contribute to changes in plant growth.

What Do the Data Show? No overall trend in plant growth can be seen for this 11-year period, either nationally or within any of the regions or ecosystem types. The similarity in year-to-year variation among regions and systems, however, is striking. For example, in 1993 all regions and systems had higher than average growth index values; in 1996, the opposite was true. The reason for this is not clear.

Year-to-year variability of the plant growth index is high nationally, within all six regions, and within all three ecosystem types. Year-to-year variability was greater in grasslands and shrublands than in either forests or farmlands. Variation was also greater in the West,

Plant Growth Index *(continued)*

particularly in the Pacific and Rocky Mountain regions, than in the East or Midwest.

During 2000, the plant growth index nationwide was lower than the 11-year average. The index was about average in the Pacific states and the Midwest and lower than the 11-year average in the other four regions. The index was farthest below the 11-year average in the Southwest.

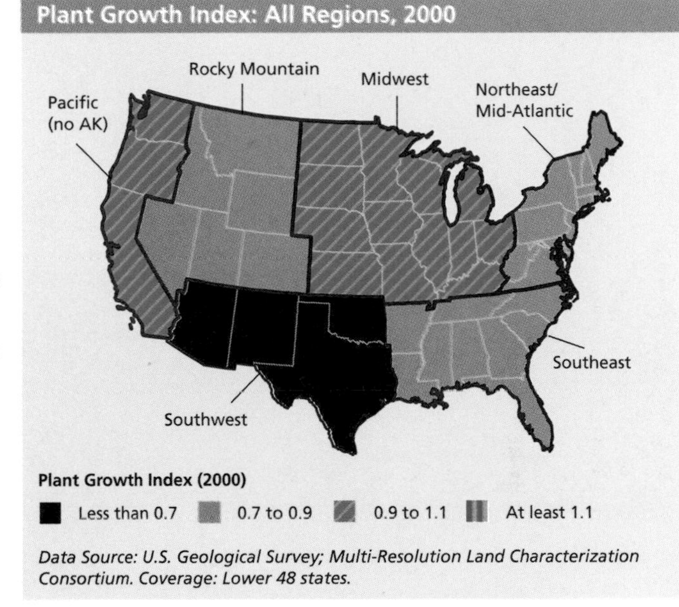

Data Source: U.S. Geological Survey; Multi-Resolution Land Characterization Consortium. Coverage: Lower 48 states.

Discussion The energy brought into an ecosystem is an overall measure of its performance. How much energy a system absorbs can be affected by factors such as climate and weather, pollution, and how farms, forests, and other areas are managed, to name a few. Long-term changes in the amount of energy absorbed can have significant implications for the way an ecosystem functions.

Some ecosystem types naturally capture more energy than others; that is, they are more productive. Rather than comparing the absolute amount of energy captured, the plant growth index compares each year's growth at a particular location with the long-term average at that location.

Given natural year-to-year variability, the 11 years for which data are available are not enough to determine whether there are any regional or system-specific trends (data for 1994 are not available because of satellite failure). The particular satellite measurement used for this analysis, Normalized Difference Vegetation Index (NDVI), correlates well, but by no means perfectly, with ground measurements of plant productivity. Measurements are taken every two weeks and summed over the entire growing season.

Data for this measure are available only for the land area of the lower 48 states. The Coasts and Oceans section of this report includes a measure related to productivity of algae in coastal waters (p. 80), but that indicator focuses on seasonal peaks rather than annualized measurements, as reported here. In addition, it is possible to measure the plant growth or productivity of freshwater lakes, but these data are not available on a consistent basis nationwide.

The technical note for this indicator is on page 216.

SYSTEM DIMENSIONS	CHEMICAL AND PHYSICAL	BIOLOGICAL COMPONENTS	HUMAN USES
Extent Pattern	Nutrients, Carbon, Oxygen Contaminants Physical	Plants and Animals Communities Ecological Productivity	**Food, Fiber, and Water** Recreation and Other Services

Production of Food and Fiber and Water Withdrawals

Production of Food and Fiber and Water Withdrawals: Entire United States

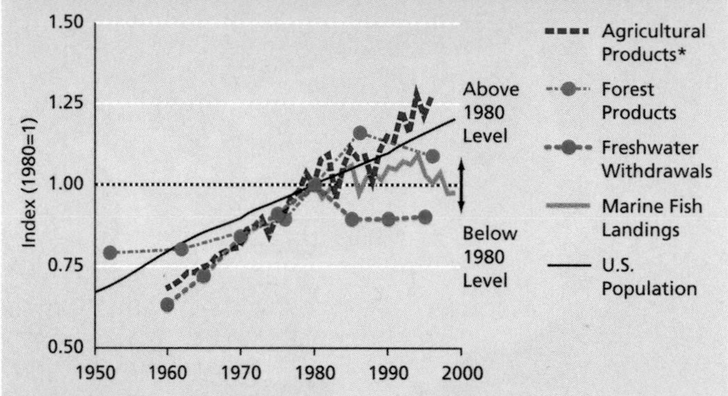

** Including livestock raised on grasslands and shrublands.*

Note: Fish landings prior to 1978 do not include foreign landings, thus are not displayed. See p. 81 for more detail.

What Is This Indicator, and Why Is It Important? This indicator reports the production of food and fiber and the withdrawals of water, using an index with 1980 as the base year. Values above 1.0 indicate that production or withdrawals were greater than in 1980; values below 1.0 indicate that production or withdrawals were lower than in 1980.

Products from U.S. ecosystems meet much of the nation's food, fiber, and water needs. Changes in the quantities of these goods signal fundamental changes in the direct benefits we receive from ecosystems.

What Do the Data Show? Over the past half-century or so, agricultural and forest production and freshwater withdrawals have all increased. But the rates of increase—and in some cases, periods of decline—vary from system to system. Agricultural production has grown the fastest. Except for a few periodic downturns, growth in agricultural production has generally been faster than the growth in U.S. population. Forest production has generally grown more slowly than population growth, except for a decade of more rapid growth during the late 1970s to 1980s. Forest production has declined since the mid-1980s. Freshwater use increased faster than population through 1980, declined by about 10% by the mid-1980s, and has grown slowly since then. Marine fish landings grew slowly from the late 1970s, when reliable statistics became available, through the mid-1990s, but have declined recently.

Most of the regional patterns of food and fiber production and water withdrawals match the national patterns above, with a few notable exceptions:

- Regional agricultural production generally follows the national growth trends (regardless of regional population growth).
- The recent modest decline in forest production nationally is the result of large declines in the Pacific Coast and Rocky Mountain regions being partially offset by increases in forest production in the Southeast.
- While freshwater withdrawals declined relative to population growth in most regions, withdrawals increased at about the same rate as population in the Southwest.
- Since the late 1970s, increased marine fish landings in the Pacific Coast region have offset declines in the Northeast/Mid-Atlantic and Southeast regions.

Discussion This indicator allows comparison between the amounts of a single good produced in two time periods. So, for example, an index value for agricultural products that is greater in 1994 than in any other year means that the nationwide harvest in 1994 was greater than at any other time in this 50-year series. The index value for 1994 is approximately 1.25, which means that the harvest in 1994 was about 25% greater than the 1980 harvest.

The index also allows comparison of the rate of growth or decline in production of two different goods. This can be seen, for example, by comparing agricultural production to marine fish landings since 1980. A steadily increasing line, such as in agricultural products, indicates that the amount of products we obtain from that ecosystem continues to grow. In contrast, marine fish landings grew until the mid-1990s, but have since declined to about 1980 levels.

The technical note for this indicator is on page 217.

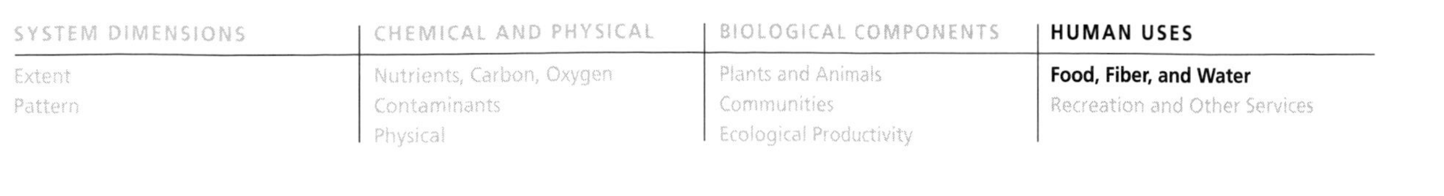

Production of Food and Fiber and Water Withdrawals *(continued)*

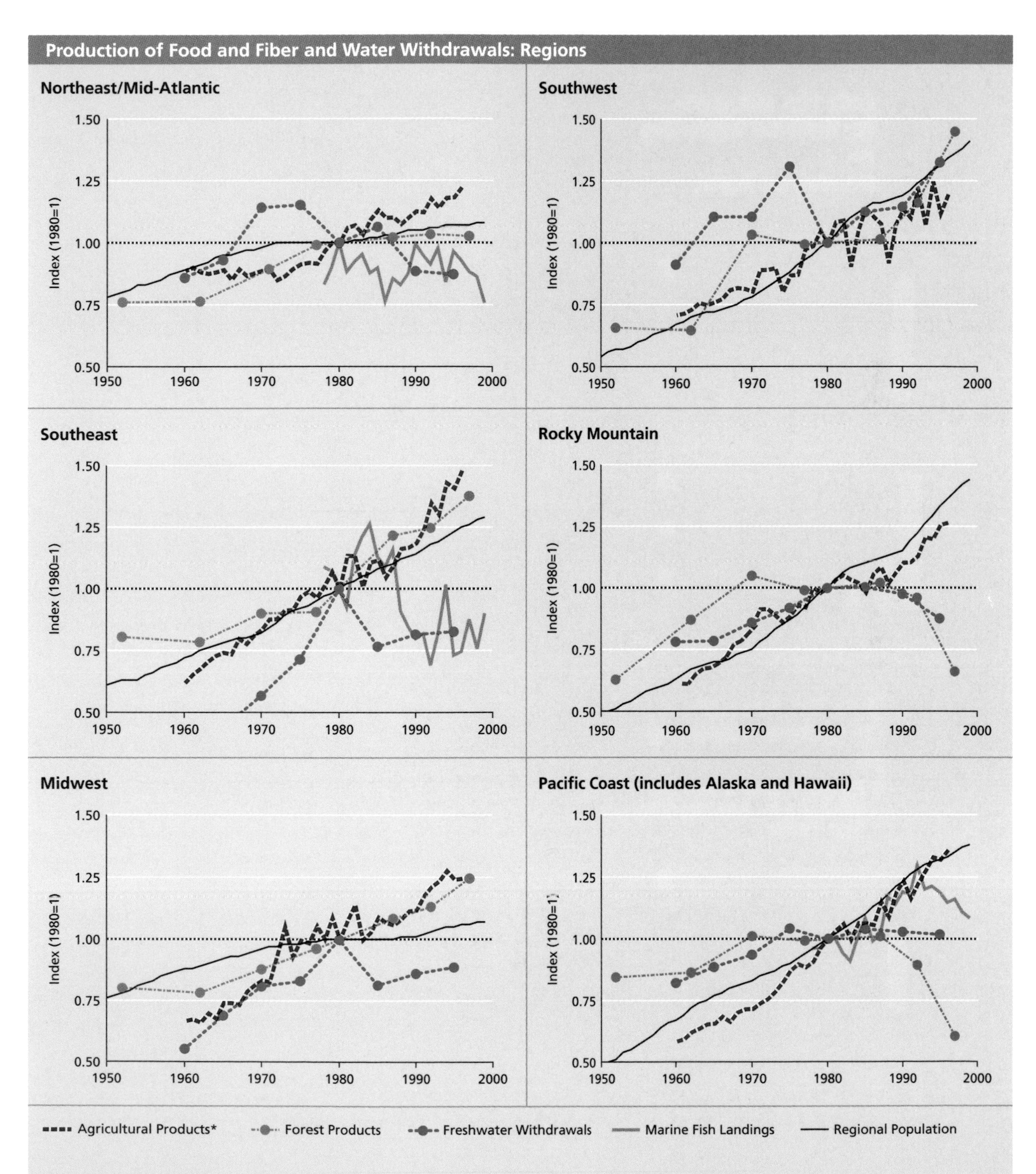

** Including livestock raised on grasslands and shrublands*

Data Source: USDA Economic Research Service, USDA Forest Service, US Geological Survey, National Marine Fisheries Service, U.S. Census Bureau. Coverage: all 50 states. .

◐ Outdoor Recreation

Participation in Outdoor Recreation Activites

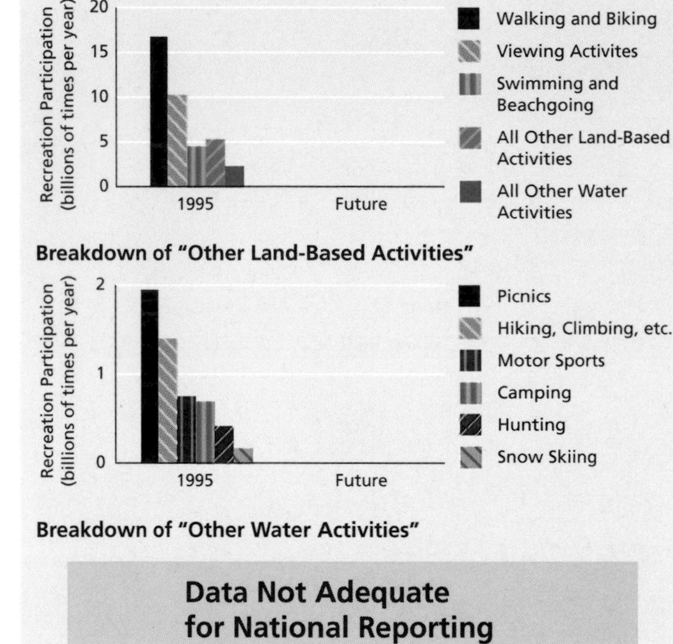

Breakdown of "Other Water Activities"

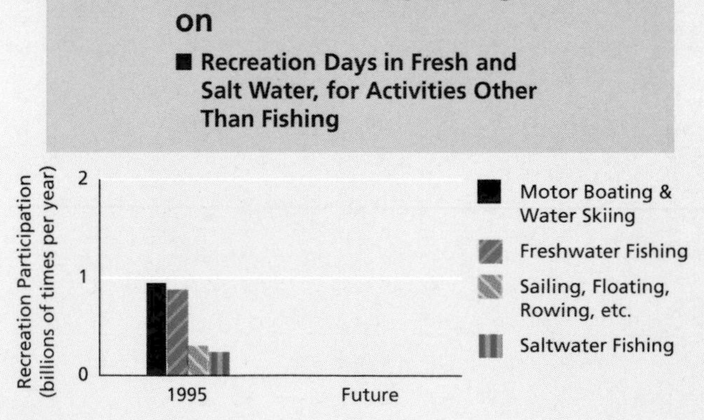

Data Source: USDA Forest Service. Coverage: all 50 states.

Note: The scale for the top graph is considerably different from that used on the second two graphs.

What Is This Indicator, and Why Is It Important? This indicator reports the number of times Americans over the age of 15 took part in a variety of outdoor recreational activities. (Each time someone took part in an activity is counted: if the activity took place over multiple days, each day counts as a separate event, and if a person took part in several activities on a single day, each activity is counted as a separate event.)

Outdoor recreation is highly popular, with many people taking part in at least one of the listed activities over the course of the year. Recreation is a benefit that is derived from ecosystems, in much the same way as we derive products such as food, fiber, and water (p. 58) from these systems.

What Do the Data Show? Walking and biking are by far the most common outdoor recreation activity for which information is available. Americans over the age of 15 walk outdoors or bike about 17 billion times per year. Nature viewing and swimming and beachgoing are next in terms of overall popularity, totaling another 15 billion times per year. The lower graphs break out annual participation in "all other land-based activities," and " all other water activities," showing participation in outdoor social activities like picnics and family gatherings, and in hiking, boating, and fishing.

Why Can't This Entire Indicator Be Reported at This Time? The data presented here are from an extensive national survey. However, the list of recreational activities about which data are collected is not exhaustive, and the survey currently does not distinguish between fresh and salt water for many aquatic activities.

Discussion This indicator reports the number of times people participated in various activities, not how long they spent, so an hourlong walk and a day at the beach count the same (as noted above, each day in a multiday trip is counted as a separate event). Therefore, the fact that people participated more frequently in some activities (such as walking) does not necessarily mean that they spent more time on this than another activity.

The technical note for this indicator is on page 217.

⑦ Natural Ecosystem Services

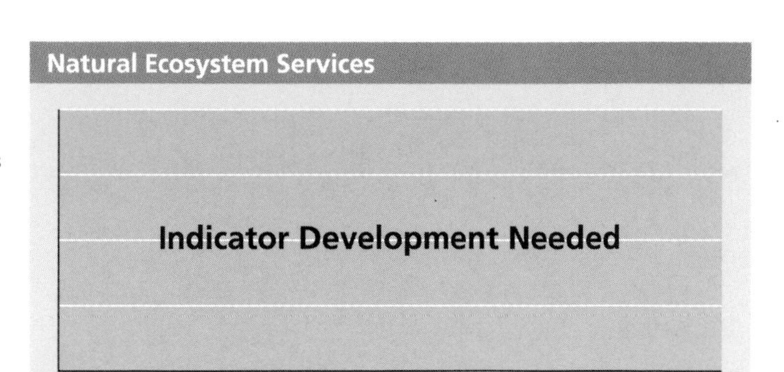

What Is This Indicator, and Why Is It Important? This indicator would report on the levels of key services provided by "natural" ecosystems—forests, grasslands and shrublands, fresh waters, and coasts and oceans. The goods, or products, these ecosystems provide—such as fish, wood products, and food—can be counted, and a monetary value often placed upon them (key ecosystem products are described on p. 58).

Some services, such as recreation, are also fairly easily quantified (pp. 60, 109, 132, 153, and 174). But many of the services provided by natural ecosystems are less tangible and more difficult to quantify, including such vital processes as purification of air and water, detoxification and recycling of wastes, regulation of climate through storage of carbon dioxide, regeneration of soil fertility, and maintenance of the earth's startling variety of plants and animals, which we use to sustain ourselves, but which we also enjoy for their own sake. Natural ecosystem processes reduce the severity of floods, promote pollination of crops and natural vegetation, ensure dispersal of seeds, control agricultural pests, and protect coasts and hillsides from erosion.

These services are often unrecognized, or at best taken for granted—until conversion or loss of the ecosystem results in loss of the services. For example, wetlands and floodplains can play a vital role in minimizing flood peaks, but this was often not recognized until downstream flooding increased following upstream conversion and filling. Or a steep hillside, formerly stabilized by trees and shrubs, slides downward, taking with it the houses that replaced the trees. Indeed, one of the greatest environmental, social, and economic disasters in the nation's history—the Dust Bowl—occurred when the intangible services provided by the natural grassland ecosystem were lost as a result of widespread agricultural conversion.

Land can also change from agricultural use into a more natural condition (this occurs less often for urban lands). For example, demographic and economic changes in New England have replaced farmland production with forest ecosystem services, and the Conservation Reserve Program (which removes environmentally sensitive farmlands from production) implicitly acknowledges that the ecosystem services provided by these lands can outweigh the value of their agricultural production.

Why Can't This Indicator Be Reported at This Time? We report indirectly on some ecosystem services by reporting on changes in the extent of major ecosystem types. Since many ecosystem services are lost or exchanged for other, different, services when natural ecosystems are converted to farmland or urban/suburban use, or when wetlands are filled, tracking changes in ecosystem extent is the best way we currently have of quantifying changes in ecosystem services.

Although it is the best we have, it is not good enough, because changes in the condition of an ecosystem—short of outright conversion to another land use—can alter the amount and type of services the system provides. An alternative, but also unsatisfactory, approach involves very detailed studies of individual systems and services. Neither the broad-brush surrogate method nor the tightly focused individual service approach allows measurement of broad categories of ecosystem services, such as would be necessary for national reporting.

What Steps Are Necessary To Achieve Reliable National Coverage? There is substantial scientific uncertainty about ecosystem services—not about whether they exist or whether they are important to society—but about how to measure them, which ones to track, and the like. This is an area of active research among ecologists and ecological economists.

There is no technical note for this indicator.

Coasts and Oceans

What Indicators Are Used To Describe Coasts and Oceans?			Can we report trends? Are there other useful reference points?
SYSTEM DIMENSIONS			
◐	Coastal Living Habitats	What is the area of coastal wetlands, coral reefs, seagrasses, and shellfish beds?	Trends (where data are available)
◐	Shoreline Types	How much of the nation's shoreline is composed of beach, sand or mudflats, steep cliffs, wetlands and mangroves, and how much has bulkheads or riprap?	Current data only, regional comparison
CHEMICAL AND PHYSICAL CONDITIONS			
⊖	Areas with Depleted Oxygen	How extensive are areas with low dissolved oxygen levels?	No data reported.
◐	Contamination in Bottom Sediments	How contaminated are bottom sediments in estuaries and coastal ocean waters?	Current data only, federal guidelines
⊖	Coastal Erosion	How much of the nation's coastline is eroding?	No data reported
●	Sea Surface Temperature	What is the temperature of the ocean's surface?	Trends, regional comparison
BIOLOGICAL COMPONENTS			
⊖	At-Risk Native Marine Species	How many native marine species are at different levels of risk of extinction?	No data reported
?	Non-native Species	What is the extent of invasion by non-native species?	No data reported
◐	Unusual Marine Mortalities	How many marine mammals, turtles, and other animals die in unusual mortality events?	Trends
?	Harmful Algal Blooms	How frequent and extensive are harmful algal blooms?	No data reported
◐	Condition of Bottom-Dwelling Animals	What is the condition of small bottom-dwelling animals (worms, clams, snails, and shrimplike animals)?	Current data only, comparison to "undisturbed" conditions, regional comparison
◐	Chlorophyll Concentrations	What is the concentration of chlorophyll in coastal waters?	Current data only, regional comparison
HUMAN USES			
●	Commercial Fish and Shellfish Landings	What is the size of the commercial fish catch from U.S. waters?	Trends, regional comparison
◐	Status of Commercially Important Fish Stocks	What is the condition of commercially important fish stocks in U.S. waters?	Trends, regional comparison
⊖	Selected Contaminants in Fish and Shellfish	What is the concentration of DDT, PCBs, and mercury in fish caught in U.S. waters?	No data reported
⊖	Recreational Water Quality	How often are bacteria associated with human and animal waste found in bathing water?	No data reported

● All Necessary Data Available ◐ Partial Data Available ⊖ Data Not Adequate for National Reporting ? Indicator Development Needed

Chapter 5:

Indicators of the Condition and Use of Coasts and Oceans

The coasts and oceans of the United States extend from the narrow ribbon of shoreline that defines the water's edge out some 200 miles into the open ocean. The cold and rocky coast of Maine and the mangrove swamps of Florida, the glacial fjords of Alaska and the black lava cliffs of Hawaii, the seagrass beds of the Chesapeake Bay and the pebble beaches of California—all these and more are found along the thousands upon thousands of miles of U.S. coastline. Offshore, kelp beds, coral reefs, and the open ocean provide habitat for fish, shellfish, birds, and mammals, as well as recreational and economic opportunities for many Americans (more than half of all Americans live within 50 miles of the coast). This vast expanse and the myriad plant and animal species that inhabit it are defined by the interaction between land and sea, between fresh water and salt water, an interaction that produces a rich mix of species and also of human activities.

What can we say about the condition and use of U.S. coasts and oceans?

Sixteen indicators describe the condition and use of America's coasts and oceans. Partial or complete data are available for nine of the indicators. Of these, five have a data record that is long enough to judge trends, and three have a federally adopted reference point or other type of benchmark for comparison. For seven indicators, we report no data. In five of these cases, some data exist, but they are of uncertain coverage or consistency and have not been aggregated for national reporting. Two indicators require additional refinement or other development before reporting is possible. Eight of the indicators are, or should be, reported on a regional basis.

After the following brief summaries of the findings and data availability for each indicator, the remainder of this chapter consists of the indicators themselves. Each indicator page offers a graphic representation of the available data, defines the indicator and explains why it is important, and describes either the available data or the gaps in those data.

Each of the indicators in this section focuses on some part of the overall "coasts and oceans" system: estuaries, bays, and the like; shorelines; waters within 25 miles of the coast; waters out to 200 miles; and combinations of these four components. See Table 5.1 (p. 68) for the reporting area for each indicator.

System Dimensions

Tracking changes in selected types of coastal land and water habitat is important for understanding the goods and services that this system can provide. There are two basic indicators of coastal system dimensions. The first tracks the changes in area of such key habitat types as coastal wetlands, coral reefs, and seagrasses; the second focuses on the nature of the shoreline itself—beach, wetlands, cliff, bulkhead, and so on.

- **What is the area of coastal wetlands, coral reefs, seagrasses, and shellfish beds?** These features are key habitat for many species of crabs, fish, and seabirds, as well as for the smaller creatures that serve as food for these larger animals. These habitats are unique in that they are created by living (or once-living) organisms, such as mangrove trees and coral. From the mid-1950s to the mid-1990s, wetland acreage on the Atlantic and Gulf coasts declined by about 400,000 acres, or about 8%, with the rate of loss slowing in the 1990s. Data are not adequate for national reporting on wetlands in other regions or on seagrasses, shellfish beds, or coral reefs.

- **How much of the nation's shoreline is composed of beach, sand or mudflats, steep cliffs, wetlands and mangroves, and how much has bulkheads or riprap?** More than two-thirds of the 37,000 miles of shoreline mapped to date is coastal wetlands, most of which are in the South Atlantic region. Sixteen percent, or 6,000 miles, is beach. Steep shorelines and mud and sand flats each make up about 8% of the total, and armored shorelines account for about 11%. (Some areas, such as sandy beaches backed by steep cliffs, may be counted twice.) These data are for the Pacific and South Atlantic coasts; data for other regions have not yet been analyzed.

Chemical and Physical Condition

Four quite varied indicators describe the chemical and physical condition of the nation's coasts and oceans. Oxygen and temperature are two key determinants of the kinds of marine plants and animals that can inhabit a region. Thus, we track the area of coastal waters with abnormally low oxygen levels and changes in regional sea surface temperature. Chemical contamination is also of concern, so we track changes in harmful man-made chemicals that can accumulate in bottom sediments. The fourth indicator will track the percentage of the nation's shoreline that is eroding.

- **How extensive are areas with low dissolved oxygen levels?** Low-oxygen (hypoxic) and no-oxygen (anoxic) conditions can cause mass mortalities among aquatic animals and disrupt migration patterns. Data are not adequate to report on the extent of these areas.
- **How contaminated are bottom sediments in estuaries and coastal ocean waters?** About 60% of the area of estuaries on the Mid-Atlantic, South Atlantic, and Gulf Coasts has levels of contaminants that might harm fish or wildlife, and 2% has levels that probably will harm these organisms. Neither trend data nor data on other regions are available.
- **How much of the nation's coastline is eroding?** Erosion can damage coastal properties and decrease the recreational value of beaches. Data are not adequate for national reporting on erosion and the opposite process, accretion, for the U.S. coastline.
- **What is the temperature of the ocean's surface?** Plants and animals are accustomed to certain water temperature ranges, and changes in temperature may cause species to disappear (or appear) in certain areas. Data for a 14-year period show neither warming nor cooling trends for waters within 25 miles of the U.S. coast.

Biological Components

Six indicators describe biological conditions within coastal waters. As in other ecosystems, one indicator tracks species that are at risk of extinction. Another records unusual "mortality events" among such marine animals as whales, sea turtles, seabirds, and fish, and a third considers the condition of worms, snails, and other bottom-dwelling animals. Three indicators, two still under development, focus on undesirable species or conditions. One of the indicators that needs further development would track the "blooms" of several toxic algae harmful to people or marine animals; the other would report on invasions of non-native species that can supplant more desirable natives. The last of the indicators measures the concentration of chlorophyll in coastal waters—chlorophyll is a measure of the presence of algae, which in excess can be harmful to fish and other animals and plants and interfere with swimming and other recreation.

- **How many native marine species are at different levels of risk of extinction?** The nation's coastal waters are home to a staggering diversity of plants and animals, from microscopic organisms to the world's largest animals. However, we know the status of only a very few of these species; data are not adequate for national reporting on marine species at risk of extinction.
- **What is the extent of invasion by non-native species?** More work is needed to develop this indicator, which will consider both the number of non-native species and what fraction of available habitat they occupy.

- **How many marine mammals, turtles, and other animals die in unusual "mortality events"?** For both major groups of marine mammals (whales, dolphins, and porpoises; and seals, sea lions, sea otters, and manatees), there were at least three years out of the last ten in which no unusual mortalities occurred. Years with high mortalities included 1992 (more than 2500 sea lions) and 1999 (215 harbor porpoises and 270 gray whales). Data are not adequate for national reporting on sea turtles, seabirds, fish, and shellfish.
- **How frequent and extensive are harmful algal blooms?** Harmful algae produce toxins that pose a danger to people as well as to marine animals. Data are not adequate for national reporting on this indicator.
- **What is the condition of small bottom-dwelling animals (worms, clams, snails, and shrimplike animals)?** About half the estuary area in along the Mid-Atlantic, South Atlantic, and Gulf coasts has bottom-dwelling communities that are considered to be in "undegraded" condition when compared to a relatively undisturbed site. About 20% are in "degraded" condition. Data are not available for reporting on other regions or for oceans.
- **What is the concentration of chlorophyll in coastal waters?** Chlorophyll is a measure of the abundance of algae, which are the source of food, directly or indirectly, for most marine animals, but too many algae can be harmful to other marine life. Only very short-term data are available for ocean waters (3 years); additional data are needed to establish trends. Data are not adequate for national reporting on estuaries.

Human Use

Four indicators describe the way people use the coasts and oceans. Three indicators focus on commercially important fish and shellfish: trends in commercial fish landings; trends in commercially important fish populations; and trends in chemical contaminants found in fish and shellfish, which might affect human health. The fourth indicator tracks contamination of beaches by bacteria from human or animal waste—a constraint to beach use that complements the core national outdoor recreation indicator (p. 60).

- **What is the size of the commercial fish catch from U.S. waters?** Since the late 1970s, landings of fish and shellfish from U.S. waters have totaled around 5 million tons per year. Over this time, and for most regions, landings have remained more or less constant (the catch in Alaska from U.S. boats has increased). Estimates of catches before the late 1970s are uncertain because of the presence then of large foreign fleets, which are no longer permitted to fish in U.S. waters.
- **What is the condition of commercially important fish stocks in U.S. waters?** Overall, about 40% of stocks with known populations were declining in size and 20% were increasing in size. However, data are not available on the population trends of about three-fourths of all U.S. fish stocks.
- **What is the concentration of DDT, PCBs, and mercury in fish and shellfish caught in U.S. waters?** Seafood containing high levels of these contaminants can be harmful to human health, but data are not adequate for national reporting on this indicator.
- **How often are bacteria associated with human and animal waste found in bathing water at the nation's beaches?** Swimming in sewage-contaminated waters can cause disease. Data are not adequate for national reporting on this indicator.

What do we mean by "coasts and oceans"?

"Coasts and oceans" consists of three components: estuaries, ocean waters under U.S. jurisdiction, and the shoreline along both estuaries and oceanfront areas.

Estuaries are partially enclosed bodies of water (often referred to as bays, sounds, lagoons, fjords, and the like), where fresh water from the land is mixed with salt water from the ocean. They are generally considered to begin at the upper end of tidal or saltwater influence and end where they meet the ocean, although major rivers often have plumes of brackish water (mixed fresh and salt) that extend

Table 5.1. Reporting Areas for Coasts and Oceans Indicators

Shorelines	Estuaries	Estuaries and Ocean Waters within 25 Miles of Shore[a]	Estuaries and Ocean Waters to 200 Miles	Ocean Waters within 25 Miles of Shore[a]
■ Shoreline Types ■ Coastal Erosion ■ Recreational Water Quality	■ Non-native Species ■ Condition of Bottom-Dwelling Animals	■ Areas with Depleted Oxygen ■ Contamination in Bottom Sediments ■ Chlorophyll Concentrations	■ Coastal Living Habitats ■ At-Risk Marine Species ■ Unusual Marine Mortalities ■ Harmful Algal Blooms ■ Commercial Fish and Shellfish Landings ■ Status of Commercially Important Fish Stocks ■ Selected Contaminants in Fish and Shellfish	■ Sea Surface Temperature

[a] While it would be preferable in many cases to adjust the width of the reporting zone to conform to the extent of brackish water, the lack of consistent national monitoring of the extent of brackish water makes this impractical at this time. Because of this, these indicators focus on the area within 25 miles of the coast, a relatively conservative value for the width of this zone.

Map 5.1. Regions Used for Reporting Selected Coasts and Oceans Indicators

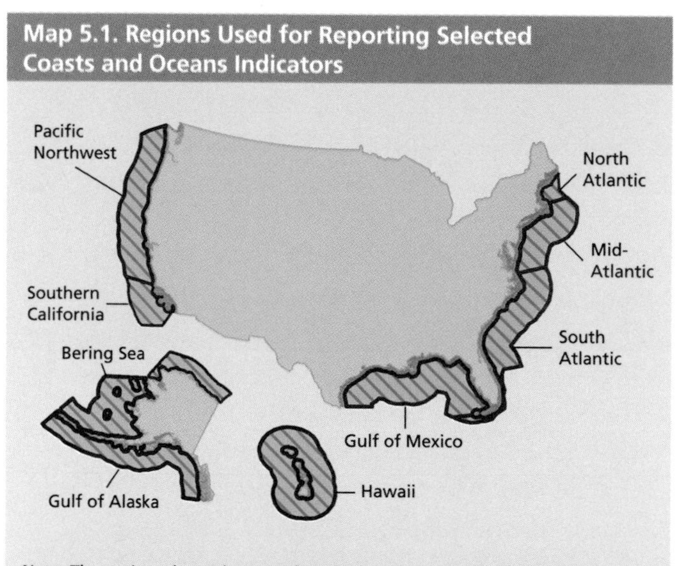

Note: The regions shown here conform to those used by the National Oceanic and Atmospheric Administration in its National Estuarine Eutrophication Assessment and the Environmental Protection Agency's Environmental Monitoring and Assessment Program; they also match the regional structure established for regional marine research under Public Law 101-593. For some indicators, regions are combined for reporting purposes.

for great distances. Many estuaries are highly productive, highly variable environments, and many have been greatly affected by human activities.

In general, ocean waters along the coast are largely influenced by the deep ocean, while terrestrial ecosystems are the main influence on estuaries. Both estuaries and ocean encompass a wide variety of habitats, including salt and brackish water, subtidal habitats (e.g., soft and hard bottom communities, coral and oyster reefs, and beds of seagrasses and kelp) and intertidal habitats (rocky shores, mud flats, marshes, and mangrove forests).

By definition, U.S. waters extend to the boundaries of the 200-mile Exclusive Economic Zone (EEZ),[1] but not all indicators report on this entire zone. In reporting the extent of coastal waters (see the core national extent indicator, p. 40), we have selected the area of "brackish water"—the area in which the influence of fresh water from rivers and groundwater reduces salinity below that of the open ocean. The width of this area varies; along the Pacific Coast it is relatively narrow, while along parts of the Atlantic Coast it may be as wide as 200 miles. Table 5.1 shows the reporting area for each of the 16 indicators.

A Note about Regions

Eight of the sixteen Coasts and Oceans indicators are reported on a regional basis, and they all make use of the same regional definitions (see Map 5.1). These indicators are shoreline types (p. 70); areas with depleted oxygen (p. 71), sea surface temperature (p. 74), at-risk species (p. 75), condition of bottom-dwelling animals (p. 79), chlorophyll concentration (p. 80), commercial fish and shellfish landings (p. 81), and the status of commercially important fish stocks (p. 82).

---

[1] *The Exclusive Economic Zone of the United States was established in 1983 by presidential proclamation (#5030). See text at* http://www.nara.gov/fedreg/codific/procs/p05030.html. *According to the United Nations Convention on the Law of the Sea, nations have sovereign rights in a 200-nautical-mile exclusive economic zone (EEZ) with respect to natural resources and certain economic activities, and they exercise jurisdiction over marine science research and environmental protection in the EEZ.* http://www.un.org/Depts/los/convention_agreements/convention_overview_convention.htm *(accessed November 21, 2001).*

SYSTEM DIMENSIONS	CHEMICAL AND PHYSICAL	BIOLOGICAL COMPONENTS	HUMAN USES
Extent Pattern	Nutrients, Carbon, Oxygen Contaminants Physical	Plants and Animals Communities Ecological Productivity	Food, Fiber, and Water Recreation and Other Services

◐ Coastal Living Habitats (Coral Reefs, Wetlands, Seagrasses, and Shellfish Beds)

What Is This Indicator, and Why Is It Important? This indicator reports the acreage of coastal habitats whose defining feature is that they are composed of living organisms (such as seagrasses, mangrove forests, and coastal wetlands) or are built by them (such as coral reefs or shellfish beds). These areas provide habitat for many other organisms, and in some cases (such as shellfish beds) they continue to do so even after the animals that built them are no longer living.

Loss of habitat is a major cause of the decline of coastal species. The habitats described here are critical for many species of crabs, fish, and seabirds, as well as for smaller animals that provide food for these larger creatures. When these habitats decline in area, organisms that depend on them are lost or displaced.

What Do the Data Show? From the mid-1950s to the mid-1990s, wetland acreage on the Atlantic and Gulf coasts declined by about 8%. Four hundred thousand acres of coastal wetlands, out of a total of 5 million acres, were lost, although the rate of loss slowed in the 1990s.

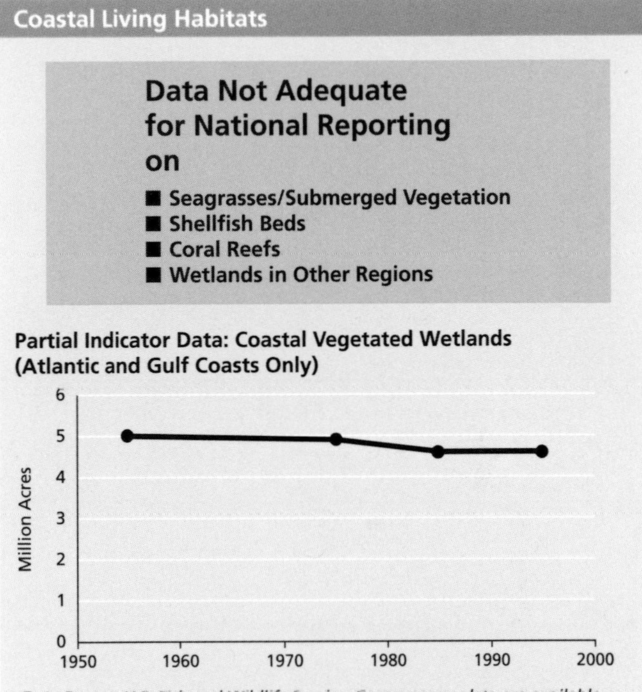

Data Source: U.S. Fish and Wildlife Service. Coverage: no data are available for the Pacific Coast, Alaska, Hawaii, U.S. Virgin Islands, or Puerto Rico.

Why Can't This Entire Indicator Be Reported at This Time? Data for coral reefs and seagrasses and other "submerged aquatic vegetation" are available for many areas, but these data have not been synthesized to produce national estimates. (A federal task force has developed a 5–7-year plan for mapping all coral reefs in U.S. waters.) Data on the area of shellfish beds are available, but changes in the area covered by monitoring programs may obscure changes in the area of shellfish beds. Data on vegetated wetlands are available only for the East (Maine to Florida) and Gulf coasts.

The technical note for this indicator is on page 218.

SYSTEM DIMENSIONS	CHEMICAL AND PHYSICAL	BIOLOGICAL COMPONENTS	HUMAN USES
Extent Pattern	Nutrients, Carbon, Oxygen Contaminants Physical	Plants and Animals Communities Ecological Productivity	Food, Fiber, and Water Recreation and Other Services

◐ Shoreline Types

Shoreline Types

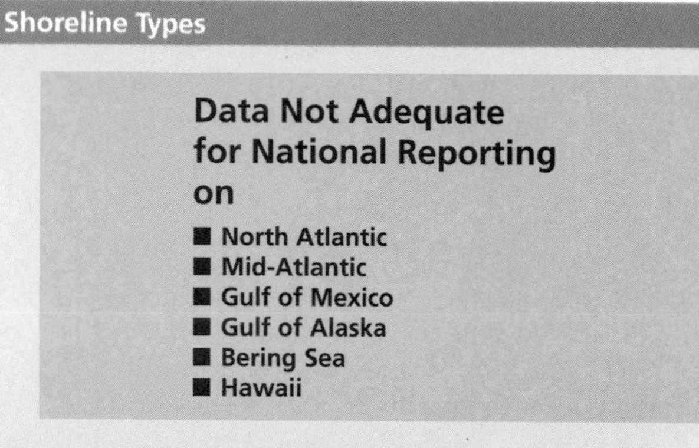

Partial Indicator Data: Shoreline Types (Pacific Northwest, Southern California, and South Atlantic Regions Only)

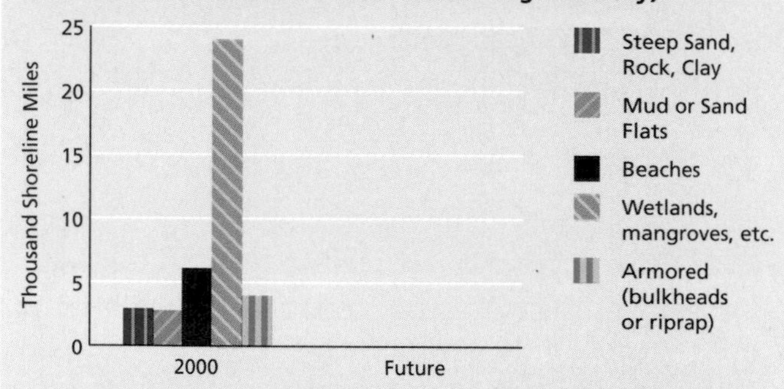

Partial Indicator Data: Shoreline Types, by Region (2000) (Pacific Northwest, Southern California, and South Atlantic Regions Only)

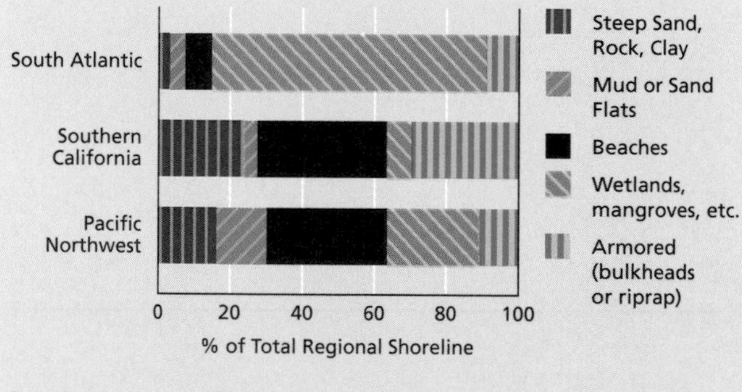

Data Source: National Oceanic and Atmospheric Administration.

What Is This Indicator, and Why Is It Important? This indicator reports the miles of coastline in several categories, including beach; mud or sand flats; steep sand, rock, or clay cliffs; wetlands; and coastline "armored" with bulkhead or riprap. The coastline includes ocean-front areas and the shoreline of estuaries and bays.

Whether a shoreline is, for example, beach, mudflat, or bulkhead determines how people and wildlife will use that shoreline. Armoring is usually intended to stabilize a beach or shoreline in an attempt to reduce erosion and property loss from storms, coastal flooding, and other processes (see Coastal Erosion, p. 73).

What Do the Data Show? Over two-thirds of the mapped shoreline (37,000 miles) in these three regions is coastal wetlands (24,000 miles), most of which are in the South Atlantic region. Sixteen percent, or 6,000 miles, of the mapped shoreline is beach. Steep shorelines and mud and sand flats each make up about 8% of the total (2,800 miles), while armored shorelines make up about 11% of the total (about 4,000 miles). (These numbers exceed the total shoreline miles because some locations contain multiple shoreline types, e.g., sandy beach backed by a steep cliff.)

Beaches account for about a third of the shoreline of both Southern California and the Pacific Northwest, but these regions differ greatly in other respects. Southern California has a much lower percentage of wetlands and mud or sand flats and a much higher proportion of both steep shorelines and armored shorelines. Three-quarters of the South Atlantic region's shoreline is wetlands, and nearly 10% is armored.

The National Oceanic and Atmospheric Administration is analyzing data for other regions, but the analysis is not yet complete.

Discussion Besides the benefits mentioned above, bulkheads and other "armoring" can have negative effects on natural coastlines, by isolating coastal wetlands from tidal influence, for example, which can dramatically alter the wetlands. In addition, these structures may provide only temporary erosion control and can ultimately result in complete loss of the beach.

The technical note for this indicator is on page 219.

⊖ Areas with Depleted Oxygen

What Is This Indicator, and Why Is It Important? This indicator will report the percentage of area of estuaries and coastal waters out to 25 miles whose lowest oxygen levels fall within one of several concentration ranges for at least 1 month. These ranges are: anoxic (no oxygen present), hypoxic (up to 2 parts per million, or ppm), low (between 2 and 4 ppm), and adequate (more than 4 ppm). In addition, for each region the percentage of coastal and estuarine waters that are hypoxic for at least 1 month will be reported.

Most animals that live in the water need oxygen, and, except for air-breathing animals like turtles and whales, most use oxygen dissolved in the water. Natural processes and human pollution can cause serious reductions in dissolved oxygen. Both anoxia (no oxygen) and hypoxia (very low oxygen) are harmful to fish, shellfish and other marine animals. These conditions can result in mass mortalities (see p. 77) and increases in predation, reduce the area of suitable habitat, and form barriers through which migratory species such as striped bass and salmon cannot pass, keeping them from their spawning grounds.

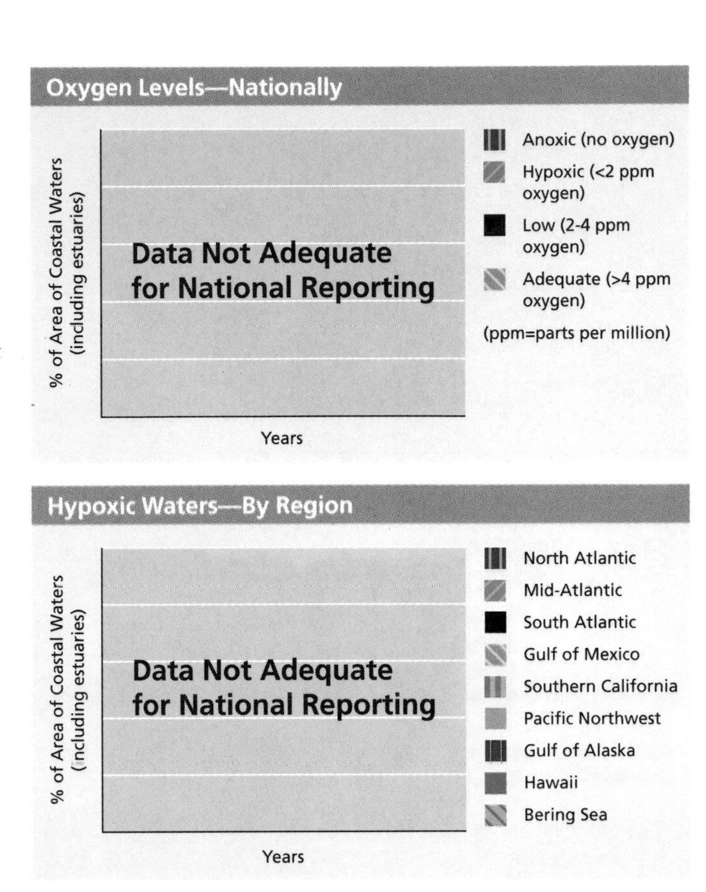

Why Can't This Indicator Be Reported at This Time? Too few estuaries and waters of the U.S. coastal ocean are sampled frequently or thoroughly enough to report on this indicator at a regional or national scale.

Discussion High algae growth, often fueled by nutrients from runoff, sewage treatment plants, or deposition of airborne pollutants, can lead to increased bacterial activity (as bacteria decompose the algae); this increased activity can deplete available oxygen. Low oxygen levels generally affect bottom waters first and most severely. See the chlorophyll indicator, p. 80, and the national nitrogen indicator, p. 46.

The technical note for this indicator is on page 220.

◐ Contamination in Bottom Sediments

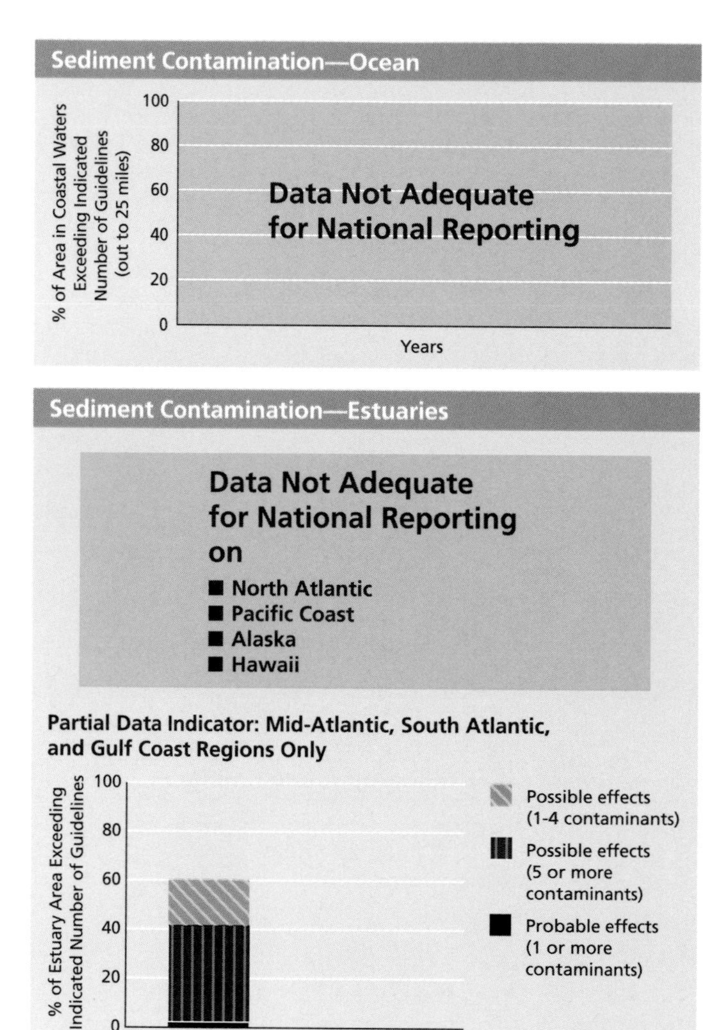

Data Source: U.S. Environmental Protection Agency, Environmental Monitoring and Assessment Program (EMAP). Guidelines are from the National Oceanic and Atmospheric Administration (NOAA).

What Is This Indicator, and Why Is It Important? This indicator reports the percentage of sediments that exceed federal guidelines for concentrations of four major classes of contaminants—pesticides, polychlorinated biphenyls (PCBs), polycyclic aromatic hydrocarbons (PAHs), and heavy metals. The indicator reports on estuaries and ocean waters within 25 miles of the coast that have bottom sediments with varying degrees of contamination, the lowest indicating *possible* effects on fish and other aquatic organisms from 1 to 4 contaminants and the highest indicating *probable* effects from at least one contaminant.

Polluted sediments are a starting point for contamination throughout the food chain, potentially damaging marine life and affecting human health (see Selected Contaminants in Fish and Shellfish, p. 83). Pollutants from industrial discharges, burning of fossil fuels, and runoff from farms and urban and suburban areas are carried to coastal waters by rivers, rainfall, and wind, where they accumulate on the bottom. Small organisms incorporate these contaminants into their bodies, and when they are eaten by other organisms, the contaminants may move up the food chain (bioaccumulation). Areas with contaminated sediments may also be unsafe for swimming and other recreation.

Why Can't This Entire Indicator Be Reported at This Time? No program exists to provide nationally consistent data on sediment contamination in ocean waters along the coast. Data for estuaries in the North Atlantic, Southern California, and Pacific Northwest will be available in the future.

What Do the Data Show? Sediment contaminant levels in about 60% of the area of U.S. estuaries monitored are high enough to potentially harm fish and other aquatic organisms. In 19% of sediments, the concentration of 1 to 4 contaminants exceeds the guideline for *possible* harmful effects; in 39%, 5 or more contaminants exceed this level; and in 2%, contaminant levels exceed the guideline for *probable* harmful effects. (Note that all sites with contaminants exceeding the *probable* effects guidelines also had 5 or more compounds exceeding the *possible* effects level.)

Discussion The NOAA guidelines used here were developed as informal interpretive tools and are intended as the basis for regulatory decisions. The *possible* effects guidelines identify concentrations below which negative effects rarely occur, and thus levels above which such effects may occur. The *probable* effects guidelines indicate levels above which negative effects are likely.

The technical note for this indicator is on page 220.

SYSTEM DIMENSIONS	CHEMICAL AND PHYSICAL	BIOLOGICAL COMPONENTS	HUMAN USES
Extent Pattern	Nutrients, Carbon, Oxygen Contaminants **Physical**	Plants and Animals Communities Ecological Productivity	Food, Fiber, and Water Recreation and Other Services

⊖ Coastal Erosion

What Is This Indicator, and Why Is It Important? The indicator will report how much of the U.S. coast is managed in an attempt to control erosion and how much remains in a "natural" state, with no erosion control. For unmanaged areas, the indicator reports what fraction is eroding, accreting (gaining land area), or stable.

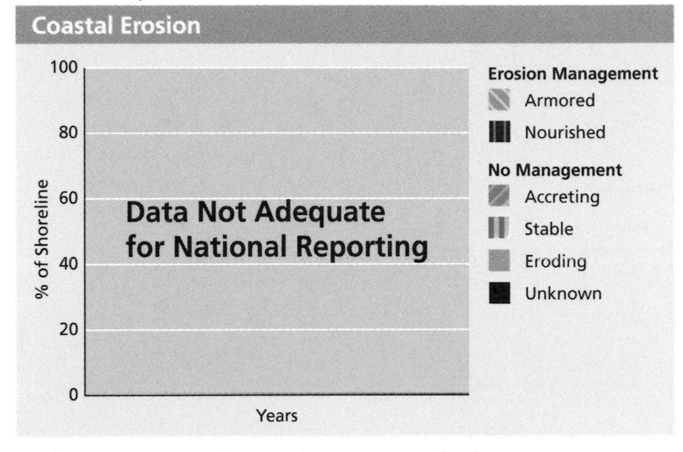

Management methods include replacement of sand (often called "beach nourishment") and construction of bulkheads or other "armoring." Neither approach necessarily eliminates future erosion, but the effects of armoring generally last longer.

Coastal erosion costs hundreds of millions of dollars a year, including damage caused by storms and flooding, costs of erosion prevention, and expenses to dredge channels and harbors. Poorly designed or sited development can lead to erosion, while measures to control erosion in one place may exacerbate it in others and may have significant environmental impacts of their own. Accretion may also create problems, as when inlets fill in, interfering with navigation. Also, many experts predict that continued global warming will be accompanied by rising sea levels, resulting in increased coastal erosion worldwide.

Why Can't This Indicator Be Reported at This Time? Assessments of shoreline stability are now conducted as short-term or single-purpose projects that are neither regional nor national in scope. Local assessments often use different methods, which makes it difficult to combine results into an accurate national picture.

Discussion Scientists and coastal managers will need to agree on numerical definitions of "eroding" or "accreting" (this is likely to be in the range of from one-half to several feet horizontally per year). Further, how long a beach that has been nourished should be reported as "managed" needs to be determined.

Priority should be given to using the large amount of existing local data, which will require assessment of coverage, quality, and comparability. Also, standard methods and definitions should be developed for nationwide use, ensuring the compatibility of data collected in the future.

The technical note for this indicator is on page 221.

● Sea Surface Temperature

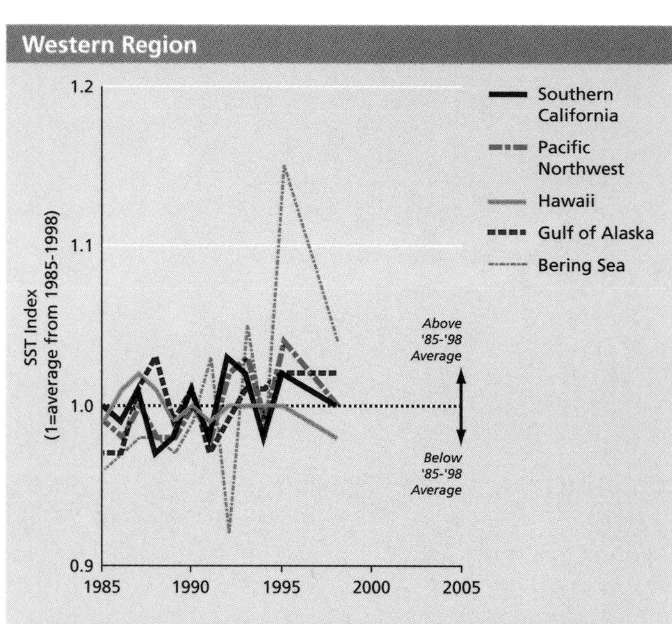

Data Source: National Oceanic and Atmospheric Administration, National Aeronautics and Space Administration. Coverage: ocean waters within 25 miles of the coast, not including estuaries.

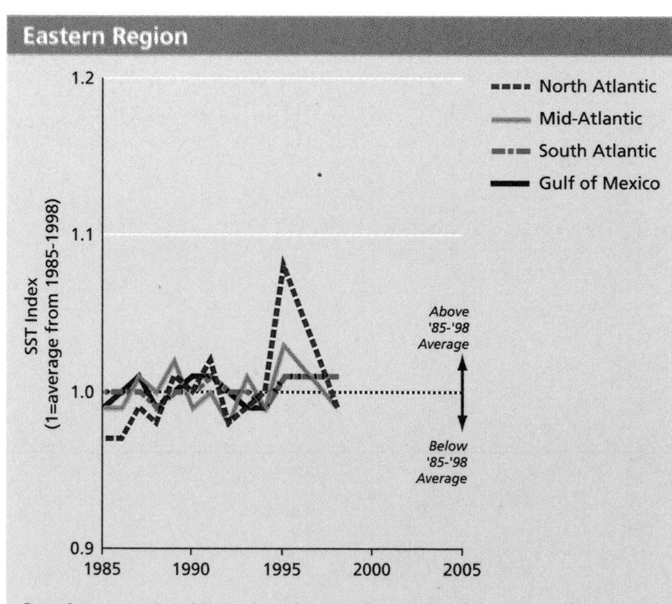

Data Source: National Oceanic and Atmospheric Administration, National Aeronautics and Space Administration. Coverage: ocean waters within 25 miles of the coast, not including estuaries.

Note: The South Atlantic and Gulf of Mexico lines on the graph coincide for several years.

What Is This Indicator, and Why Is It Important? This indicator describes whether sea surface temperature (SST) is above or below average. Using an index, the indicator tracks how much regional average temperatures in any given year deviate from the average for the 14-year period, for waters within 25 miles of the coast. Thus, a "1.1" on the graph means that the SST for that region in that year was 10% warmer than the 14-year average for that region. The indicator defines "average SST" for a region as the average temperature for the warmest season in that region.

Water temperature directly affects the species of plants (such as algae, seagrasses, marsh plants, and mangroves) and animals (microscopic animals, larger invertebrates, fish, and mammals) that live in a particular region. In addition, increases in temperature are thought to be associated with the degradation of coral reefs (bleaching) and may increase the frequency or extent of blooms of harmful algae (see Harmful Algal Blooms, p. 78). There is widespread concern that global climate change may lead to increases in SST. Such changes could, in turn, lead to increases in the strength and frequency of storms and changes in ocean currents, such as the Gulf Stream, that would in turn lead to shifts in regional climate.

What Do the Data Show? While SST varies noticeably from year to year, and there are individual reports of gradually increasing temperatures in several of these ocean regions (see the technical note for citations), the data presented here do not show any trends.

The technical note for this indicator is on page 222.

⊖ At-Risk Native Marine Species

What Is This Indicator, and Why Is It Important? This indicator will report on the relative risk of extinction of native marine species, both plants and animals. The risk categories are based on such factors as the number and condition of individuals and populations, the area occupied by the species, population trends, and known threats. Degrees of risk to be reported here range from very high ("critically imperiled" species are often found in five or fewer places or have experienced very steep declines) to moderate ("vulnerable" species are often found in fewer than 80 places or have recently experienced widespread declines). Species ranked as "secure" or "apparently secure" would not be reported. The data would also be presented on a regional basis for estuaries and coastal waters out to 200 miles.

Species are valued for a variety of reasons: they provide products, including food, fiber, and genetic materials; they are key elements of ecosystems, which themselves provide valuable goods and services; and many people value them for their intrinsic worth or beauty.

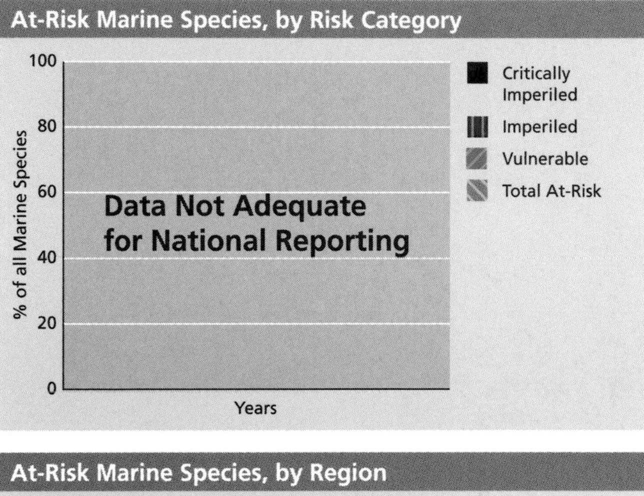

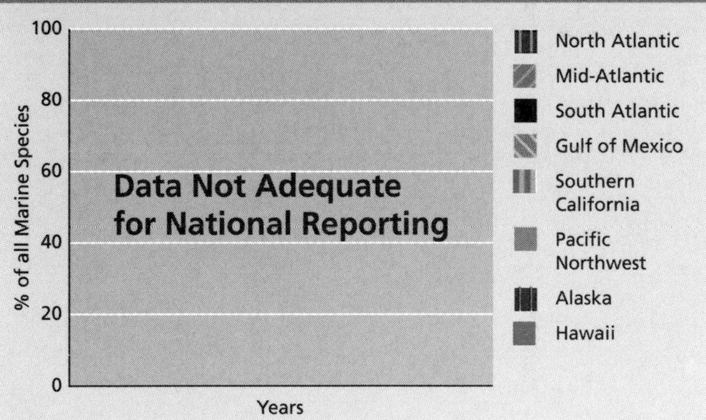

Why Can't This Indicator Be Reported at This Time? Data are available on the status of only a relatively small number of marine species, including those of commercial interest (see p. 82) and those that are listed for protection under the Endangered Species Act and Marine Mammal Protection Act. However, these programs do not address the status of a broad cross-section of marine species, as is needed for this indicator.

NatureServe and its member natural heritage programs (see www.natureserve.org) report on the status of about 22,000 U.S. species (see the forest (p. 124), freshwater (p. 144), grasslands and shrublands (p. 168), and core national (p. 52) at-risk species indicators). These programs provide a useful framework for reporting on marine species, but so far their datasets contain information on only a relatively small number of marine species.

There is no technical note for this indicator. The technical note for the core national indicator for at-risk species (p. 214) describes NatureServe's natural heritage programs.

SYSTEM DIMENSIONS	CHEMICAL AND PHYSICAL	**BIOLOGICAL COMPONENTS**	HUMAN USES
Extent	Nutrients, Carbon, Oxygen	**Plants and Animals**	Food, Fiber, and Water
Pattern	Contaminants	Communities	Recreation and Other Services
	Physical	Ecological Productivity	

⑦ Non-native Species

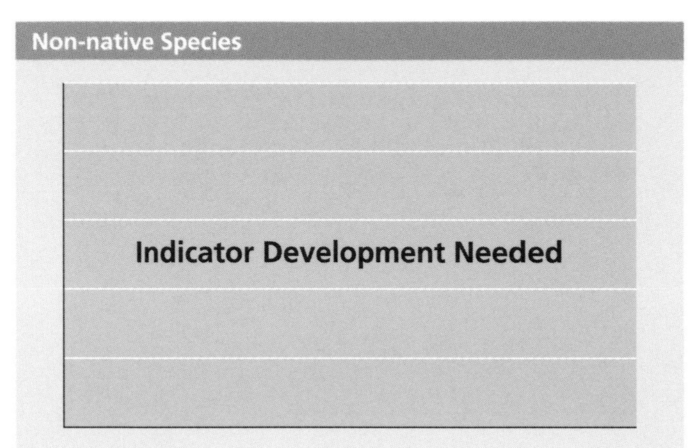

What Is This Indicator, and Why Is It Important? This indicator will report the percentage of major estuaries with high, medium, or low influence by non-native species. Ratings of the degree of influence should incorporate both the number of different species present and the degree to which they occupy available habitat.

Non-native species often spread aggressively and crowd out species native to a region; they may act as predators or parasites of native species, cause diseases, compete for food or habitat, and alter habitat. These species—whose spread has been promoted by increased travel and trade—may also pose threats to human health (e.g., exotic diseases and harmful algae) and economic well-being (e.g., loss of shellfish production). Non-native species are also called nonindigenous, exotic, alien, or introduced species; particularly aggressive species are termed "invasive."

U.S. estuaries are now home to many non-native species. These include the Asian clam and the veined, or Asian, rapa whelk, which cause economic and ecological damage as they displace native clams and mussels, and the European green crab, which is blamed for the collapse of the soft-shelled clam industry in Maine. The problem is both worldwide and apparently growing: an introduced North American jellyfish has devastated the anchovy fishery in the Black Sea, and in San Francisco Bay three or four new non-native species are established each year.

Why Can't This Indicator Be Reported at This Time? There are neither nationwide monitoring programs for coastal non-native species nor agreed-upon methods for combining information on the number of species and the area they occupy into a single index. Individual studies have documented the occurrence of non-native species in major estuaries, but this information has not been gathered regularly or on a broad scale.

Discussion Several more decisions about the scope of this indicator are required: whether to focus on all non-natives or only on invasive species; whether North American species that are found outside their normal range should be treated as non-natives; and whether there is a time (e.g,. 50 or 100 years) after which an introduced species is considered to be native.

The technical note for this indicator is on page 222.

◐ Unusual Marine Mortalities

What Is This Indicator, and Why Is It Important? This indicator reports the occurrence of "unusual" mortalities of marine animals. Unusual mortality events (UME) are characterized by an abnormal number of dead animals or by the appearance of dead animals in locations or at times of the year that are not typical for that species. For larger animals like whales, dolphins, porpoises, seals, sea lions, sea otters, manatees, and sea turtles, where a small number of deaths is significant, the indicator will report the actual number of dead individuals. For smaller, more abundant, animals (seabirds, fish, and shellfish), the indicator will report the number of mortality events, rather than number of individual deaths.

Factors that may contribute to unusual mortalities include infectious diseases, toxic algae (see Harmful Algal Blooms, p. 78), and uncommon weather patterns. Trends in unusual mortalities are generally believed to reflect the integrity of an ecosystem.

Unusual Mortalities: Marine Mammals and Sea Turtles

Data Not Adequate for National Reporting on
■ Sea Turtles

Partial Indicator Data: Marine Mammals

Number of Mortalities: 0, 200, 400, 600, 2500
Years: 1990, 1995, 2000, 2005
Whales, Dolphins & Porpoises
Seals, Sea Lions, Sea Otters & Manatees

Data Source: National Marine Fisheries Service and Dierauf and Gulland (2001). Coverage: all U.S. waters.

Why Can't This Entire Indicator Be Reported at This Time? National data on turtle, seabird, fish, and shellfish mortality events are not available. Further work is required to define the criteria for UMEs for seabirds, fish, and shellfish.

Unusual Mortalities: Seabirds, Fish and Shellfish

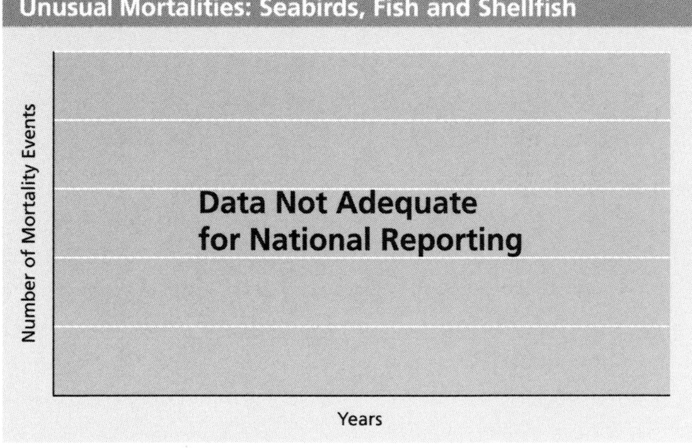

What Do the Data Show? Over 2,500 California sea lions were involved in a UME in 1992—more than 10 times the total number of seals, sea lions, sea otters, and manatees lost in UMEs for any year since. The deaths of 150 manatees off the Florida coast during 1996 and 185 California sea lions in 1997 were the next largest events for this group. For whales, dolphins, and porpoises, perhaps the most striking finding is the peak in 1999; the 576 deaths in that year reflect the deaths of 215 harbor porpoises and 270 gray whales off the West Coast (unusual gray whale deaths continued over the next two years, during which some 400 more animals died).

Discussion Instead of reporting all observed mortalities, this indicator reports unusual events. By restricting reporting in this way, the indicator focuses on events that raise more serious concern about the state of the marine environment than would more typical mortalities, which may be caused by old age or "normal" interactions with people, such as recreational boat strikes or entanglement in fishing nets.

The technical note for this indicator is on page 223.

⑦ Harmful Algal Blooms

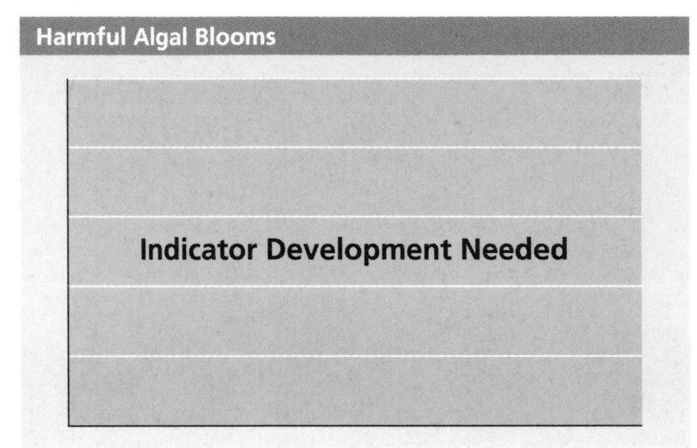

What Is This Indicator, and Why Is It Important? This indicator will report the number of harmful algal blooms (HABs) of low, medium, and high intensity for estuaries and ocean waters within 200 miles of shore. Harmful algal blooms are defined as an increased abundance of algae species that cause illness in people or marine animals or the actual occurrence of algae-caused illnesses.

HABs can cause mass mortalities of marine organisms (p. 77), are a public health risk, and can cause economic damage through declines in tourism, shellfish bed closures, and reductions in the market value of seafood. There are indications that HABs may be occurring more frequently, both in the United States and worldwide. The causes of HABS are not fully known, but changes in sea surface temperature (see p. 74) and nutrient inputs (see the national nitrogen indicator, p. 46) are believed to increase the likelihood of such events.

Why Can't This Indicator Be Reported at This Time? There are no nationwide monitoring or reporting programs for harmful algal blooms, nor are there generally accepted definitions of low, medium, and high intensity. High-intensity events might be defined as those that last for more than a month or affect an area of 40 square miles or more, low-intensity events as those that last for less than a week or affect less than 4 square miles and medium-intensity events as those that are intermediate in either size or duration. Because these definitions apply to a classic "bloom" event, they would have to be refined to include events that are characterized by illness in people or marine animals.

Discussion Algae, also called phytoplankton, are directly or indirectly the source of food for virtually all marine animals, including commercial and sport fish. Most species are not toxic, and most algal blooms do not involve species that produce toxins harmful to people or animals; however, they may reduce oxygen in coastal waters, which can harm fish and other animals (see the hypoxia indicator, p. 71). This indicator targets the most common species known to produce toxins; these species are listed in the technical note.

The technical note for this indicator is on page 224.

◐ Condition of Bottom-Dwelling Animals

What Is This Indicator, and Why Is It Important? This indicator describes the condition of worms, clams, snails, and shrimplike animals in bottom sediments ("benthic communities") by reporting the percentage of area in which these communities are in "undegraded," "moderate," and "degraded" condition. The indicator is calculated by comparing the number and kinds of animals found in a sampling site with those that would be expected in an undisturbed area of similar character (a reference site). The indicator would be reported for estuaries and for ocean areas within 25 miles of the coast.

Benthic communities reflect the influence of contaminants, oxygen levels, physical changes in habitat (such as from trawling), and shifts in temperature or salinity. They are a good indicator because contaminants accumulate in bottom sediments and hypoxia (lack of oxygen) is most severe there. Also, these animals live several years, so their response reflects exposure to these stresses over a long period, and they are fairly immobile, so their condition strongly reflects conditions at the site where they were collected (see the depleted oxygen and sediment contamination indicators, pp. 71 and 72).

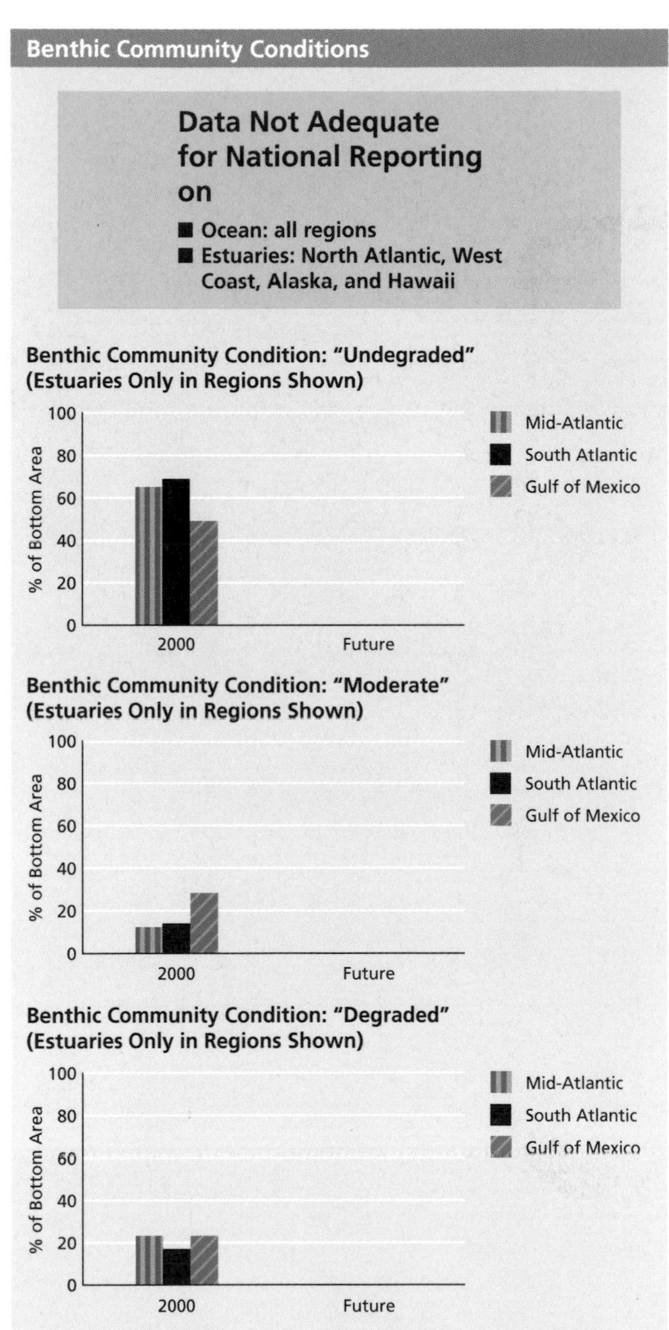

Data Source: U.S. EPA Environmental Monitoring and Assessment Program.

Why Can't This Entire Indicator Be Reported at This Time? Data are available from estuaries in most regions of the country, but the tools needed to compare benthic communities with those in undisturbed sites have been developed for only three regions. Additional work is also needed to ensure that the indiators developed for different regions are comparable. Only limited data are available for ocean waters out to 25 miles.

What Do the Data Show? At least half the estuary area in the Mid-Atlantic, South Atlantic, and Gulf of Mexico regions has "undegraded" bottom-dwelling animal communities. About 20% of estuary area in these regions has "degraded" bottom-dwelling animal communities.

Discussion "Undegraded" means that the benthic animals found at a site are similar in number and type of species to those expected in an undisturbed site in that region. "Degraded" means that the animals found are quite different from those at a reference site, reflecting one or more negative influences.

The technical note for this indicator is on page 225.

◐ Chlorophyll Concentrations

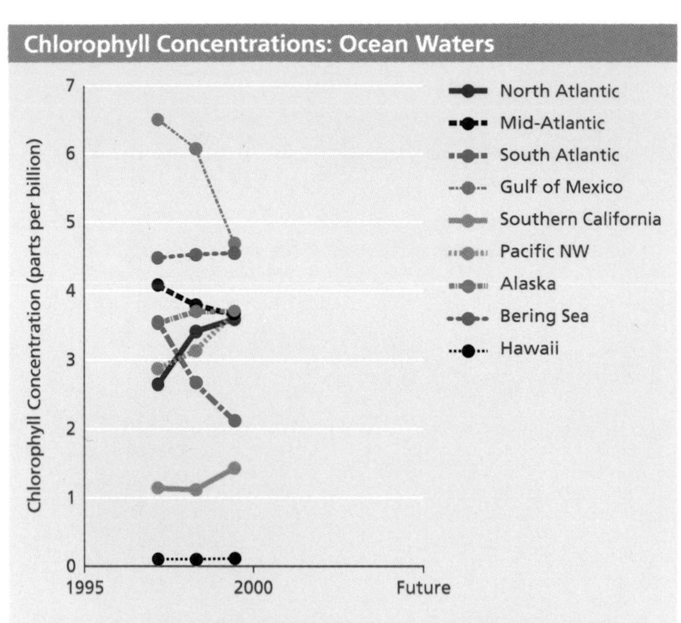

Data Source: National Ocean Service, National Oceanic and Atmospheric Administration; National Aeronautics and Space Administration. Coverage: all U.S. waters, including Alaska and Hawaii, within 25 miles of the coast.

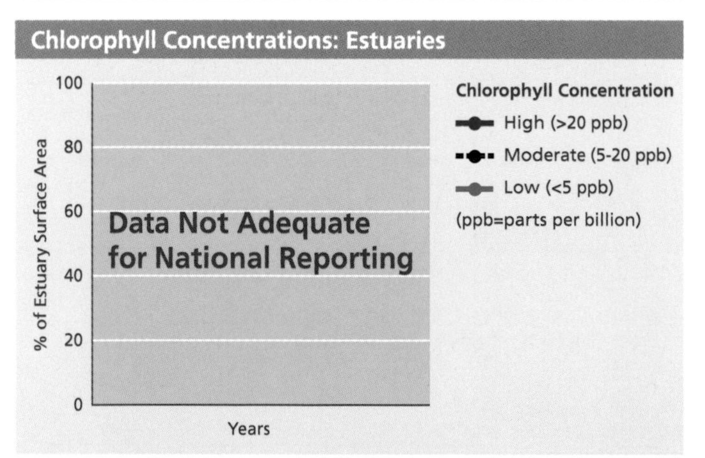

What Is This Indicator, and Why Is It Important? This indicator reports the chlorophyll concentration in estuaries and ocean waters within 25 miles of shore. For ocean waters, the indicator reports the average value for the season with the highest concentration, for each region. For estuaries, the indicator will report the percentage of area in three ranges: below 5 parts per billion (ppb), between 5 and 20 ppb, and above 20 ppb, using data for the season with the highest average concentration.

Chlorophyll concentration is a measure of the abundance of algae, also called phytoplankton, which account for most of the plant production in the ocean. Phytoplankton are difficult to measure directly, yet they are the direct or indirect source of food for most marine animals.

Although increasing algae growth (as measured by chlorophyll) tends to support larger fish populations, excessive growth often leads to degraded water quality—for example, decreases in water clarity, noxious odors, oxygen depletion (see p. 71), and fish kills (see p. 77)—and may be linked to harmful algal blooms (see p. 78). Excessive algae growth appears to occur as a consequence of increases in nutrient inputs (especially nitrogen—see the national nitrogen indicator, p. 46) and in response to declines in the abundance of filter-feeding organisms like oysters, clams, and mussels.

Why Can't This Entire Indicator Be Reported at This Time? Most estuaries are not sampled frequently enough or thoroughly enough to produce comparable data on seasonal chlorophyll levels.

What Do the Data Show? Ocean data from 1998–2000 suggest that chlorophyll levels are higher in the Gulf of Mexico than in the waters off Hawaii and Southern California; differences between other regions may not be meaningful. The time series is too short to establish trends.

The technical note for this indicator is on page 226.

● Commercial Fish and Shellfish Landings

What Is This Indicator, and Why Is It Important? This indicator reports the weight of fish, shellfish, and other products taken from U.S. waters. Landings, plus certain aquaculture harvests, are shown for five regions that cover all waters out to the 200-mile territorial limit.

The amount of fish and shellfish caught for food, meal, and oil is a measure of society's reliance on the seas for these products. Of total landings, about 70% is for human consumption, about 20% is for meal, oil, and other industrial purposes, and the remainder is used for bait and animal feed.

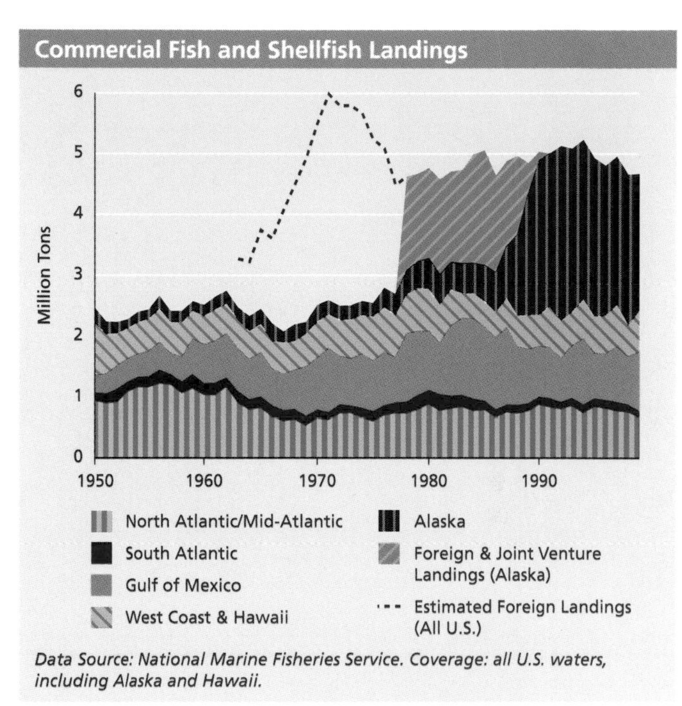

Data Source: National Marine Fisheries Service. Coverage: all U.S. waters, including Alaska and Hawaii.

What Do the Data Show? Since the late 1970s, yearly landings of fish and shellfish from U.S. waters have totaled nearly 5 million tons. In the late 1970s, the United States established a Fishery Conservation Zone (FCZ), covering hundreds of thousands of square miles of formerly international waters. Foreign fishing in these waters was eliminated, except in Alaska, where it was phased out, ending completely in 1991. The total foreign catch in the FCZ is uncertain, as indicated by the dotted line on the graph, and pre-1963 estimates are not available. In most regions, landings by U.S. vessels have remained more or less constant over the past 30 years. In Alaska, an expanding fleet has substantially increased U.S. landings.

Discussion This indicator does not provide information on the condition of fish stocks (see Status of Commercially Important Fish Stocks, p. 82). Furthermore, these aggregate landing figures do not reveal that, over the years, fishing efforts have repeatedly shifted from species that have been depleted or overfished to others that have been relatively unexploited.

In 1999, about 84% of landings were fish, about 14% shellfish, and about 2% other products, including sea urchins and worms. These data include some aquaculture production.

The technical note for this indicator is on page 226.

◐ Status of Commercially Important Fish Stocks

What Percent of Stocks Were Increasing?

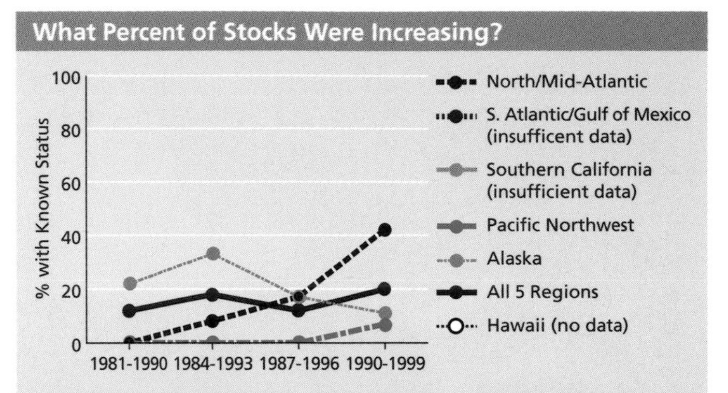

Data Source: NOAA National Marine Fisheries Service, analyzed by Natural Resources Consultants, Inc. Coverage: does not include Hawaii. Nearshore stocks (generally in state waters within 3 miles of shore) are excluded.

Note: "All 5 regions" includes some data from South Atlantic/Gulf of Mexico and Southern California, but these data are insufficient to provide regional trends.

What Percent of Stocks Were Decreasing?

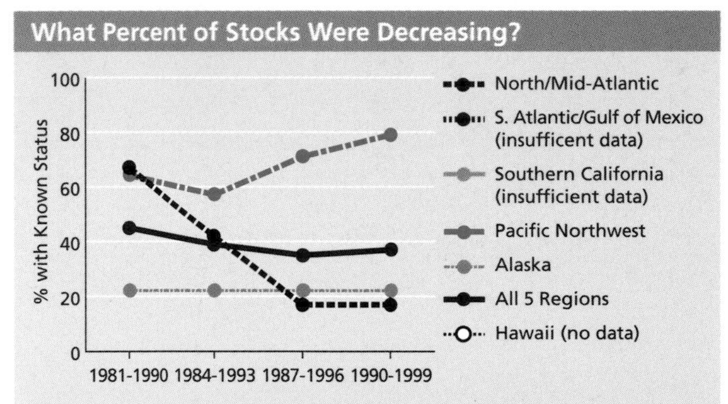

Data Source: NOAA National Marine Fisheries Service, analyzed by Natural Resources Consultants, Inc. Coverage: does not include Hawaii. Nearshore stocks (generally in state waters within 3 miles of shore) are excluded.

Note: "All 5 regions" includes some data from South Atlantic/Gulf of Mexico and Southern California, but these data are insufficient to provide regional trends.

What Percent of Stocks Had Known Population Trends? (1991–1999)

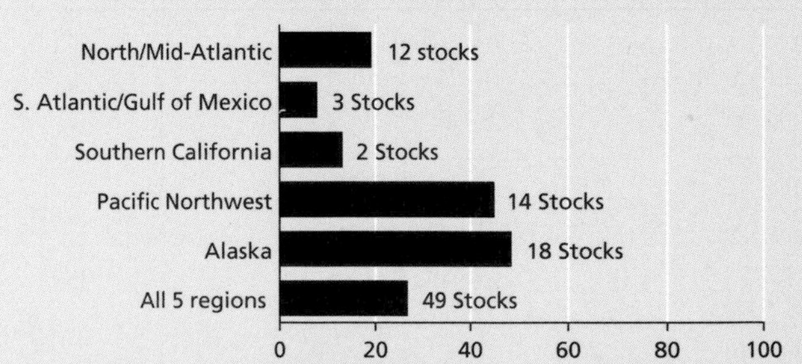

Data Source: NOAA National Marine Fisheries Service, analyzed by Natural Resources Consultants, Inc. Coverage: does not include Hawaii. Nearshore stocks (generally in state waters within 3 miles of shore) are excluded.

Note: "All 5 regions" includes some data from South Atlantic/Gulf of Mexico and Southern California, but these data are insufficient to provide regional trends.

What Is This Indicator, and Why Is It Important? This indicator tracks the percentage of commercially important fish species, or "stocks," that are increasing or decreasing in size. Only stocks whose population increased or decreased by at least 25% are reported. Trends are based on the estimated weight, or "biomass," of the entire stock.

Americans take large amounts of fish from U.S. waters (see Commercial Fish and Shellfish Landings, p. 81). Landings of a given stock cannot be maintained indefinitely if that stock's population declines. If declines persist, stocks can become too small to fish, with attendant economic and social consequences; declines may also lead to significant changes in the marine ecosystem.

What Do the Data Show? The North/Mid-Atlantic region stands out as having, over time, more fish stocks with increasing populations and fewer stocks with declining populations. However, we know trends for only 20% of the stocks in this region. In contrast, the number of declining stocks went up in the Pacific Northwest, where we know trends for more than 40% of the stocks; by the 1990s, about 80% of Pacific Northwest stocks with known trends were declining. There are no clear trends in the other regions. However, when all five regions are considered together, about 40% of stocks had decreasing trends over the time period, while about 20% of stocks had increasing population trends.

Discussion An increasing population trend may signal an increased ability of a stock to support commercial fishing, or it may reflect the recovery of an overfished stock. This latter case is likely in the Northeast, where strict catch restrictions have been imposed in response to severe stock declines.

While the data presented here represent only about 25% of all commercial fish stocks, the stocks for which population trends do exist, and which are reported here, account for about 75% of the weight of fish caught each year in the United States.

The technical note for this indicator is on page 227.

⊖ Selected Contaminants in Fish and Shellfish

What Is This Indicator, and Why Is It Important? This indicator measures the concentration of PCBs, mercury, and DDT in the edible tissue of seafood from U.S. coastal waters. For comparison, the graphs would also include information on the levels at which the Environmental Protection Agency (EPA) and Food and Drug Administration (FDA) recommend that action (such as consumption advisories) be taken.

These compounds can reach concentrations that are harmful to humans, especially in larger fish. Many coastal environments are contaminated with synthetic toxic substances like DDT and PCBs, and mercury is ubiquitous in the marine environment. Bottom-dwelling organisms that ingest these contaminants are eaten by fish that are in turn eaten by larger fish—a process called bioaccumulation. Elevated concentrations of both PCBs and DDT are a concern both in bottom-feeding fish and shellfish and in predators such as tunas, swordfish, and some sharks, while mercury is concentrated primarily in predators.

While the manufacture and distribution of PCBs and DDT has been banned in the United States since the 1970s, historical deposits in coastal watersheds and sediments continue to provide an active source of contamination. Mercury can come from industrial releases, abandoned mines, the burning of fossil fuels for electric power generation, and the weathering of rock. Human health risk assessments have shown that consumption of certain species of fish in certain locations produces a measurable risk of cancer from one or more of these contaminants. These risk assessments are the basis of consumption advisories that suggest limiting the intake of particular species, especially for groups at higher risk, such as children, pregnant women, and nursing mothers.

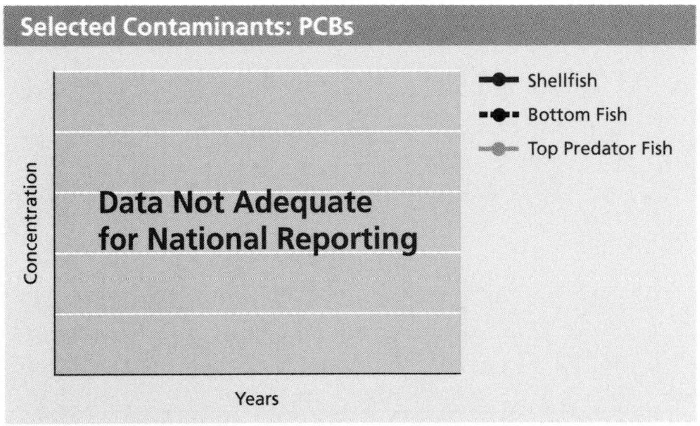

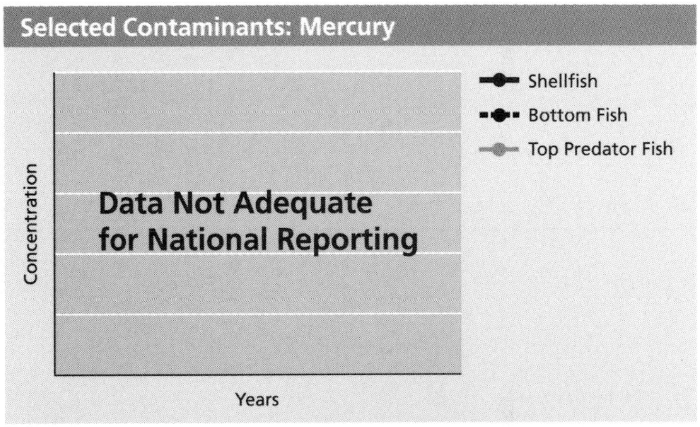

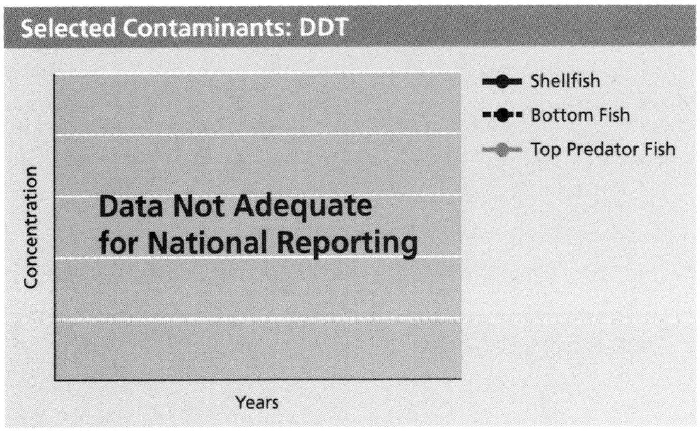

Selected Contaminants: PCBs

Concentration

Data Not Adequate for National Reporting

Years

Shellfish
Bottom Fish
Top Predator Fish

Selected Contaminants: Mercury

Concentration

Data Not Adequate for National Reporting

Years

Shellfish
Bottom Fish
Top Predator Fish

Selected Contaminants: DDT

Concentration

Data Not Adequate for National Reporting

Years

Shellfish
Bottom Fish
Top Predator Fish

Why Can't This Indicator Be Reported at This Time? While the FDA, EPA, and state governments have a variety of monitoring and reporting programs in place, these programs do not provide the basis for national reporting on contaminant concentrations.

The technical note for this indicator is on page 228.

SYSTEM DIMENSIONS	CHEMICAL AND PHYSICAL	BIOLOGICAL COMPONENTS	**HUMAN USES**
Extent Pattern	Nutrients, Carbon, Oxygen Contaminants Physical	Plants and Animals Communities Ecological Productivity	Food, Fiber, and Water **Recreation and Other Services**

⊖ Recreational Water Quality

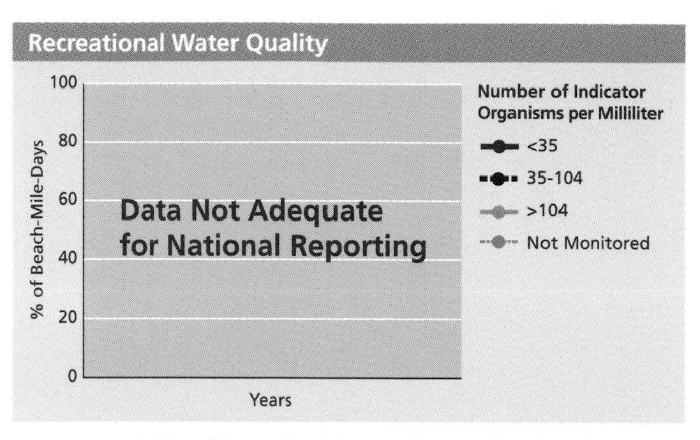

What Is This Indicator, and Why Is It Important? This indicator will report the percentage of "beach-mile-days" affected by various levels of *Enterococcus,* a bacterium that indicates contamination with human or animal waste. A "beach-mile-day" is one mile of beach affected for one day—100 miles of beach affected for one day would count the same as 1 mile affected for 100 days.

Swimming in sewage-contaminated waters can cause minor ailments, like sore throats and diarrhea, as well as more serious, even fatal, illnesses like severe gastroenteritis, meningitis, and encephalitis. Beach-based activities, like sunbathing, surfing, and swimming, are popular (see the national recreation indicator, p. 60), add billions of dollars to the economy, and contribute to the value of coastal properties. Poor water quality threatens these benefits.

Why Can't This Indicator Be Reported at This Time? A great deal of information is collected on coastal recreational water quality, but the data are scattered, incomplete, and inconsistent. Beach monitoring is typically conducted by city or county health departments, which frequently use different methods, while many areas choose not to monitor at all. Recent federal legislation provides increased incentives to monitor using nationally consistent methods, so data for this indicator should be available in the future.

Discussion There is no national standard for closing beaches because of sewage contamination; such decisions are made locally, using many different standards. This indicator reports the most commonly used indicator organism (*Enterococcus*), which is also recommended by EPA, but some monitoring relies upon other organisms. There are other aspects of water quality, such as the presence of contaminated sediments (see p. 72), that are not addressed by this indicator.

The contamination reported by this indicator may be caused by sewage treatment plant malfunctions, overflow of combined sewer systems during rain storms, discharges from boats, leaking septic systems, and runoff after heavy rains that may contain animal waste from farms, urban lawns, and streets.

The technical note for this indicator is on page 228.

Farmlands

	What Indicators Are Used To Describe Farmlands?		Can we report trends? Are there other useful reference points?
SYSTEM DIMENSIONS			
●	Total Cropland	How much land is used directly for production of crops and livestock?	Trends
●	The Farmland Landscape	How much of the farmland landscape is forest, grasslands and shrublands, wetlands, or urban and suburban?	Current data only, regional comparison
⊖	Fragmentation of Farmlands Landscapes by Development	How intermingled are croplands and urban and suburban development?	No data reported
⊖	Shape of "Natural" Patches in the Farmland Landscape	How much of the "natural" area in farmlands is in patches of different shapes?	No data reported
CHEMICAL AND PHYSICAL CONDITIONS			
●	Nitrate in Farmland Streams and Groundwater	How much nitrate is there in farmland streams and groundwater?	Current data only, federal standard, cross-ecosystem comparison
●	Phosphorus in Farmland Streams	How much phosphorus is there in farmland streams?	Current data only, federal guideline, cross-ecosystem comparison
●	Pesticides in Farmland Streams and Groundwater	How many pesticides are found in farmland streams and groundwater, and how often do they exceed federal standards and guidelines?	Current data only, federal standards and guidelines
⊖	Soil Organic Matter	How much organic matter is there in cropland soils?	No data reported
●	Soil Erosion	How much cropland is subject to erosion by wind or water?	Trends, national map
⊖	Soil Salinity	How much cropland soil has high salt levels?	No data reported
BIOLOGICAL COMPONENTS			
⊖	Soil Biological Condition	What is the condition of the microscopic animal communities in cropland soils?	No data reported
?	Status of Animal Species in Farmlands Areas	What is the condition of wildlife in areas that are heavily dominated by farmlands?	No data reported
?	Native Vegetation in Areas Dominated by Croplands	In areas that are heavily dominated by croplands, is most of the remaining non-cropland vegetation native or non-native?	No data reported
?	Stream Habitat Quality	What is the quality of the habitat in farmland streams?	No data reported
HUMAN USES			
●	Major Crop Yields	How has the per-acre yield of major crops changed over time?	Trends
●	Agricultural Inputs and Outputs	How have farm output and the inputs (pesticides, fertilizers, labor, land, etc.) needed to produce that output, changed over time?	Trends
●	Monetary Value of Agricultural Production	What is the value of the nation's production of crops and livestock?	Trends, national map
⊖	Recreation on Farmlands	How much recreation takes place on farmlands?	No data reported

● All Necessary Data Available ◐ Partial Data Available ⊖ Data Not Adequate for National Reporting ? Indicator Development Needed

Chapter 6:
Indicators of the Condition and Use of Farmlands

America's farmlands are part of a larger "farmland landscape," a landscape they both define and are defined by. The farmland landscape includes fields and orchards, pastures and vineyards, which we refer to as "croplands." It also includes the hedgerows, streams, ponds, wetlands, prairies, and woodlots that enliven an agricultural setting, as well as lands set aside under government programs. All over the nation, from the endless wheatfields of the Midwest to the picturesque dairy farms of Pennsylvania Dutch country to the avocado groves of California, the farmland landscape provides Americans, and the world, with an abundance of food and fiber, along with an image of beauty and order that figures large in the American imagination.

What can we say about the condition and use of U.S. farmlands?

Eighteen indicators describe the condition and use of farmlands in the United States. Full data are available for nine of these indicators, a larger percentage than for any other ecosystem type. Five of these nine have a long enough data record from which to judge trends, and three can be compared to a regulatory standard or similar benchmark. For six indicators, we report no data, and three measures require additional refinement or other development before data availability can be assessed.

After the following brief summaries of the findings and data availability for each indicator, the remainder of this chapter consists of the indicators themselves. Each indicator page offers a graphic representation of the available data, defines the indicator and explains why it is important, and describes either the available data or the gaps in those data.

System Dimensions

The goods and services that we obtain from farmlands depend on both the acreage of land producing crops and other farm products and the acreage and pattern of the forests, grasslands, and urban areas mixed within the farmland landscape. Four key indicators describe the dimensions of the farmland system.

- **How much land is used directly for production of crops and livestock?** Croplands, including pasture and haylands, cover between 430 and 500 million acres (estimates from different agencies vary), or about a quarter of the land area of the United States (excluding Alaska) in 1997. Cropland acreage has declined since the 1950s, but because official estimates vary, it is difficult to determine exactly how much farmland has been converted to other uses.
- **How much of the farmland landscape is forest, grassland or shrubland, wetlands, or developed land?** Some noncropland areas provide wildlife habitat or serve as streamside buffers or windbreaks, and all these areas add to the visual character of the farmland landscape. In all regions but the Midwest, croplands make up 50–60% of the farmland landscape; the remainder is forest, wetlands, or grasslands and shrublands. In the Midwest, croplands make up about 75% of the farmland landscape.
- **How intermingled are croplands and urban and suburban development?** Increased development in farming areas can interfere with traditional farming practices and may make farming economically unviable. For example, new residents are often opposed to long-standing farming practices like field application of manure, and rising property values, and property taxes, may drive farmers out of business. Data are not adequate for national reporting on this indicator.

- **How much of the "natural" area in farmlands is in patches of different shapes?** The size and shape of these "natural" patches help determine the ecological services they provide, including erosion control and wildlife habitat. Data are not adequate for national reporting on this indicator.

Chemical and Physical Condition

Six indicators describe the chemical and physical condition of farmlands, three that characterize farmland streams and groundwater and three that tell us the state of the soil. (We complement these measures with two biological indicators related to water and soil—see Biological Components, below.)

To describe the condition of farmland streams and groundwater, we include measures of pesticides, and of nitrate and phosphorus—two important nutrients that, if present in excess, can cause problems. An indicator measuring the quality of stream habitat complements these measures (see Biological Components, below). To characterize the suitability of soils for growing crops, our indicators track changes in soil organic matter, in the potential of the soil to erode by wind and water, and in soil salinity. A complementary indicator describes the microscopic animals in cropland soils (see Biological Components, below).

- **How much nitrate is there in farmland streams and groundwater?** High levels of nitrate in drinking water—especially untreated well water—are a human health concern, and nitrate from the nation's rivers contributes to algal blooms in coastal waters. About 20% of the groundwater wells and 10% of the stream sites tested had nitrate concentrations that exceeded federal drinking water standards. Monitored streams and groundwater in farmland regions have higher concentrations of nitrate than those in urban and suburban or forested areas.
- **How much phosphorus is there in farmland streams?** About three-fourths of farmland stream sites had phosphorus concentrations that exceeded the level recommended by the Environmental Protection Agency to protect against excess algae growth. Concentrations of phosphorus in monitored farmland streams were similar to those in urban/suburban streams, and much higher concentrations than streams in forested areas.
- **How many pesticides are found in farmland streams and groundwater, and how often do they exceed federal standards and guidelines?** Eighty-three percent of monitored streams in farmland areas had at least one pesticide whose concentration exceeded aquatic life guidelines; 4% had at least one compound that exceeded human health standards or guidelines. All streams had at least one pesticide at detectable levels throughout the year, and 75% had an average of five or more. Fewer than 1% of groundwater sites in farmland areas had pesticides in concentrations that exceeded human health standards, and 40% of groundwater sites had no detectable pesticides.
- **How much organic matter is there in cropland soils?** Organic matter improves the ability of soils to hold water, provides nutrients for crops, reduces erosion, and can help to support soil microorganisms. Data are not adequate for national reporting on this indicator.
- **How much cropland is subject to erosion by wind or water?** From 1982 to 1997, the acreage of U.S. farmland with the greatest potential for wind erosion decreased by nearly a third, to about 63 million acres, or about 15% of U.S. croplands. The area with the greatest potential for water erosion also decreased by nearly a third, to 89 million acres, or about 22% of U.S. croplands.
- **How much cropland soil has high salt levels?** High-salinity soils, which typically result from irrigation in arid climates, can reduce the ability of soils to support plant growth. Data are not adequate for national reporting on this indicator.

Biological Components

Four indicators describe the biological condition of farmlands. Continuing from the three soil indicators noted above, the first biological indicator measures the condition of microscopic animals in cropland soils. The second indicator focuses on the wildlife that live in the farmland landscape, a third reports on native and non-native plants in those parts of the farmland landscape that are not used for production, and a fourth measures stream habitat quality—the sediments in the streambed, the stability of stream banks, and similar physical attributes. The latter three indicators require additional development.

- **What is the condition of the microscopic animal communities in cropland soils?** The condition of nematodes (roundworms) in the soil is a good indicator of overall soil condition. Data are not adequate for national reporting on this indicator.
- **What is the condition of wildlife in areas that are heavily dominated by farmlands?** Additional work is necessary to develop an indicator that describes the condition of species that prosper in the farmland landscape and of those that depend on the kind of habitat that existed before conversion to agriculture.
- **In areas that are heavily dominated by croplands, is most of the remaining noncropland vegetation native or non-native?** Non-native vegetation often provides less suitable wildlife habitat. This indicator requires further development.
- **What is the quality of the habitat in streams in farmland regions?** Stream habitat quality often reflects the effects of activities, including farming practices, in the watershed. This indicator requires further development.

Human Use

Four indicators measure the human use of farmlands. Three focus on aspects of production: the first tracks changes in the yield per acre for five major crops; the second tallies total agricultural output and changes in the inputs, such as fertilizer and labor, used to produce farm goods; and the third focuses on the dollar value of farm sales, which depends on both the amount of goods produced and the prices farmers receive. The fourth indicator focuses on another human use of farmlands, recreation.

- **How has the per-acre yield of major crops changed over time?** Since 1950, per-acre yields of corn, wheat, and cotton have more than doubled, with corn yield increasing almost fourfold. Of major crops, soybean yields went up the least, but still nearly doubled.
- **How have farm output and the inputs (pesticides, fertilizers, labor, land, etc.) needed to produce that output changed over time?** U.S. agricultural output has been increasing steadily since 1950, while the major inputs required to produce a unit of output—with the exception of pesticides—have decreased. Pesticide use has leveled off since 1980.
- **What is the value of the nation's production of crops and livestock?** The gross value of agricultural output (adjusted for inflation) was about $180 billion in 1999, or about 10% more than in 1950. Over the past half-century, however, there have been major fluctuations, from a low of $140 billion in 1959 to a high of about $260 billion in 1973. Livestock sales have consistently accounted for about half of all agricultural value.
- **How much recreation takes place on farmlands?** A considerable amount of recreation takes place on farmlands—hunting and fishing, for example—and some farmers depend on income from such activities. Data are not adequate to report nationally on this indicator.

What do we mean by "farmlands" or the "farmland landscape"?

Lands used for production of annual and perennial crops and livestock—croplands—are the heart of the farmland ecosystem. However, croplands are found within a larger landscape that includes field borders and windbreaks, small woodlots, grassland or shrubland areas, wetlands, farmsteads, small villages and other built-up areas, set-aside lands, and similar areas not used for production. This overall landscape is referred to as the *farmland landscape* in this report. Some indicators describe the condition of cropped lands, while some describe the more broadly defined farmland landscape.

In general, we have excluded lands enrolled under the Conservation Reserve Program from the estimates presented here for croplands; these lands are included, however, in indicators dealing with the larger farmland landscape. In addition, lands used for intensive animal raising or feeding, often called feedlots or confined animal feeding operations, are clearly a part of the overall agricultural landscape, and some of these areas are included in the estimates of cropland that we present. However, it is also likely that some, located in otherwise nonagricultural settings and not owned by farmers/producers, are not included. The acreage involved is believed to be negligible compared to other types of cropland.

The farmland landscape inevitably overlaps with other ecosystems. Most notably, pastures are considered "croplands," since they are clearly part of farming operations. They are also considered part of the grasslands/shrublands ecosystem, since they are grass-covered (perhaps with scattered trees or shrubs) and thus provide some of the services and values and share many characteristics of that ecosystem. (The fact that some farmers harvest hay from native prairies further blurs the distinction between these two ecosystems.) In addition, by defining the farmland landscape to include noncropland areas surrounding and intermingled with croplands, we obviously incorporate lands covered with forest, grassland or shrubland, wetlands, and suburban development.

The production of livestock is clearly an agricultural activity, but not all land used for livestock production is considered as part of the farmland landscape. For example, while pastures are included as croplands, many cattle spend significant portions of their lives grazing on grasslands or shrublands that are not subject to significant management and that we report as grasslands / shrublands, not croplands. Clearly, distinguishing between these lands is at times difficult.

Map 6.1. Natural Resources Conservation Service Regions

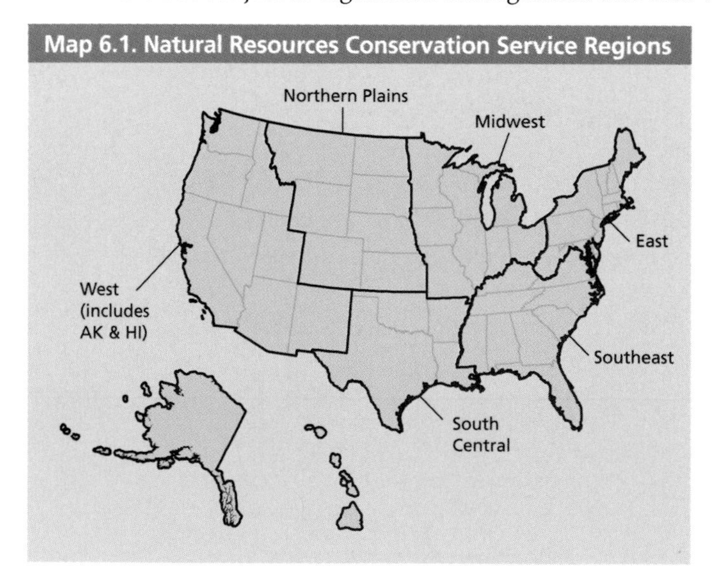

A Note about Regions

One indicator (Farmland Landscape, p. 92) in this section presents data using the USDA Natural Resources Conservation Service (NRCS) regions (see Map 6.1). The data presented in the indicator do not include Alaska and Hawaii, but, when such data become available, these two states will be included. Two indicators (Soil Erosion and Monetary Value of Agricultural Production, pp. 100 and 108) present their data using maps at a finer scale of resolution.

If data were available, several indicators—Fragmentation of Farmlands Landscapes by Development; Size and Shape of "Natural" Patches in the Farmland Landscape; Soil Organic Matter; Soil Salinity; and Soil Biological Condition—would also be presented using the NRCS regions.

SYSTEM DIMENSIONS	CHEMICAL AND PHYSICAL	BIOLOGICAL COMPONENTS	HUMAN USES
Extent	Nutrients, Carbon, Oxygen	Plants and Animals	Food, Fiber, and Water
Pattern	Contaminants	Communities	Recreation and Other Services
	Physical	Ecological Productivity	

● Total Cropland (including pasture and hayland)

What Is This Indicator, and Why Is It Important? This indicator reports the amount of land used for crops, including pasture and hay. Acreage that is enrolled in long-term set-aside programs, such as the Conservation Reserve Program (CRP) is not considered to be part of this indicator.

Agriculture is a major component of the U.S. economy, and land is the most basic resource in farming. In addition, the size of and fluctuations in the agricultural land base provide important baseline information for other indicators, such as Farmland Landscape (p. 92).

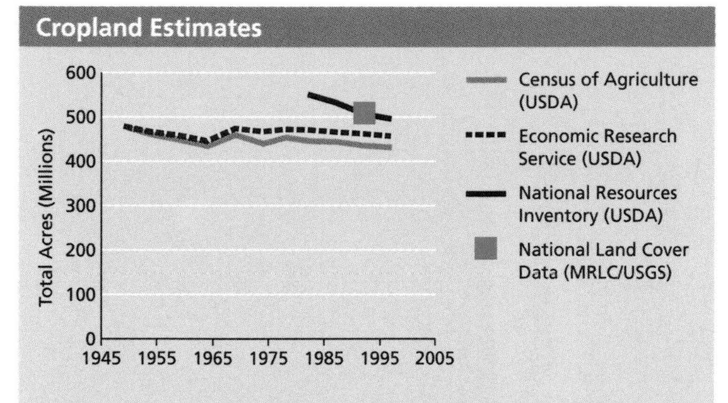

Data Source: USDA National Resources Conservation Service, National Resources Inventory (NRI) program; USDA National Agricultural Statistical Service, Census of Agriculture; USDA Economic Research Service (ERS); Multi-Resolution Land Characterization Consortium (MRLC) and the U.S. Geological Survey. Coverage: lower 48 states. Conservation Reserve Program acreage has been removed from all but the ERS data set; also, some CRP lands may be included in the National Land Cover Data.

What Do the Data Show? Cropland, including pasture and haylands, covered between 430 and 500 million acres, or about a quarter of total U.S. land area (excluding Alaska) in 1997. Cropland acreage has declined over the past half century. Unfortunately, the multiple sources of cropland acreage information provide estimates that are not always consistent. Two sources—the Census of Agriculture and the Economic Research Service (ERS)—show a decline in acreage from a peak in 1949 (about 5% for ERS and 10% for the Census). In contrast, the National Resources Inventory (NRI) reports greater cropland acreage and a 10% drop over the much shorter time period from 1982 to 1997. The National Land Cover Data agree with the NRI for the single time point available.

Discussion Note that even if overall acreage changes little, different parts of the country may experience sharp increases or decreases in cropland acreage. In addition, even apparently small changes in total acreage may involve millions of acres of land (see the national extent indicator, p. 40).

Estimates of the amount of land devoted to farming differ because different programs use different methods to acquire, define, and analyze their data. We are aware of no overall reconciliation among these estimates.

Cropland is a flexible resource: it may be used for crops one year, be left idle for one or many years, and then returned to production. Changes in government programs or crop prices may cause land to be idled for short periods or to be used for different crops. In contrast, long-term changes in cropland acreage may result from conversion of land to other uses, including CRP set-asides and development.

The technical note for this indicator is on page 229.

SYSTEM DIMENSIONS	CHEMICAL AND PHYSICAL	BIOLOGICAL COMPONENTS	HUMAN USES
Extent	Nutrients, Carbon, Oxygen	Plants and Animals	Food, Fiber, and Water
Pattern	Contaminants	Communities	Recreation and Other Services
	Physical	Ecological Productivity	

● The Farmland Landscape

Croplands in the Farmland Landscape

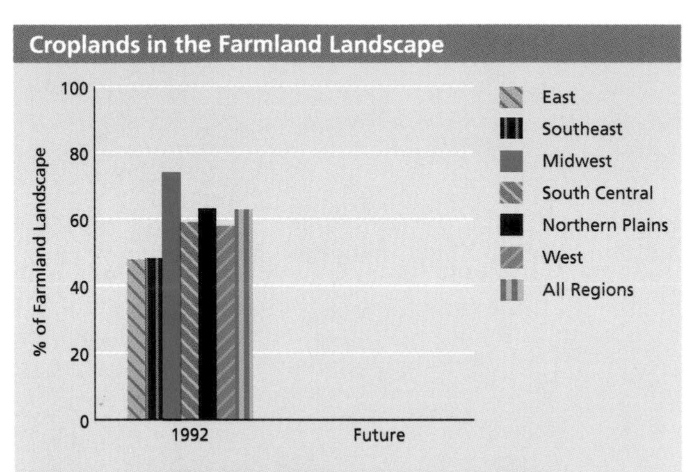

Data Source: Multi-Resolution Land Characterization Consortium; U.S. Geological Survey EROS Data Center. Coverage: lower 48 states.

Noncropland Components of the Farmland Landscape, 1992

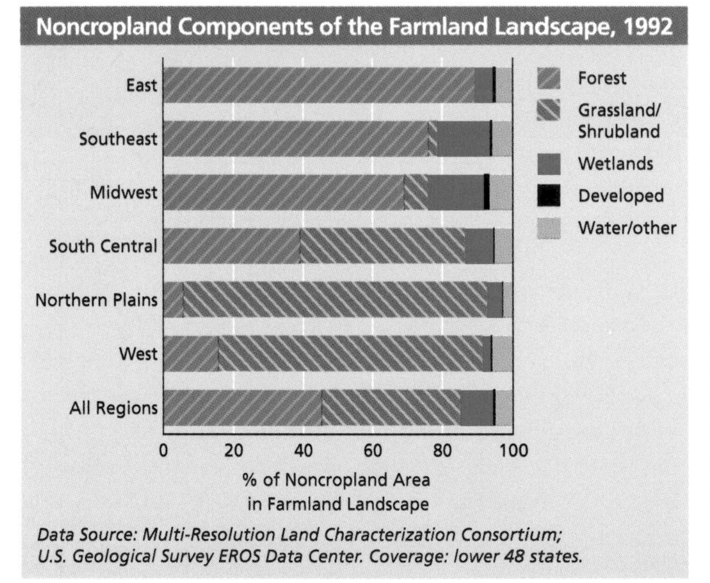

Data Source: Multi-Resolution Land Characterization Consortium; U.S. Geological Survey EROS Data Center. Coverage: lower 48 states.

What Is this Indicator, and Why Is It Important? This indicator reports the percentage of the farmland landscape that is actively used for crop production, pasture, or haylands (i.e., croplands, see p. 91). The "farmland landscape" includes croplands and the forests or woodlots, wetlands, grasslands and shrublands, and the like that surround or are intermingled with them. This indicator describes the degree to which croplands dominate the landscape, or, conversely, the degree to which these other lands are intermingled.

This indicator also describes the composition of the noncropland portion of the farmland landscape by reporting the percentage of these lands that are forests, grasslands and shrublands, wetlands, developed areas, and other lands and waters.

The noncropland elements of the farmland landscape (other than developed) provide wildlife habitat, serve as streamside buffers and windbreaks, and lend a distinctive visual character to the landscape. (Pasture and haylands are intermediate in character between "natural" grasslands and cultivated croplands; for this indicator, they are counted as croplands.)

What Do the Data Show? In the East and Southeast, croplands make up about half of the overall farmland landscape; most of the remainder is forest and, in the Southeast, wetlands. In the Midwest, only about a quarter of the farmland landscape is something other than croplands; forests and wetlands dominate the noncropland areas in this region as well.

About 60% of the farmland landscape is croplands in the South Central, Northern Plains, and Western regions. Grasslands and shrublands dominate the noncropland portion of the Western and Northern Plains regions; in the South Central region, forests and grasslands and shrublands are about equal in area.

Discussion This indicator should, over time, be sensitive to the expansion of urban and suburban land use into farmland areas as well as to the conversion of forest, grassland, or other land cover to cropland. However, the data reported here do not measure very low density "exurban" development (more than scattered rural settlements, but less dense than "suburban").

The farmland landscape reported here is defined using satellite land cover data. Areas dominated by cropland are included, along with their immediate surroundings (see the technical note for details). Note also that identifying wetlands on croplands is difficult; wetlands data should be interpreted cautiously.

The technical note for this indicator is on page 231.

SYSTEM DIMENSIONS	CHEMICAL AND PHYSICAL	BIOLOGICAL COMPONENTS	HUMAN USES
Extent	Nutrients, Carbon, Oxygen	Plants and Animals	Food, Fiber, and Water
Pattern	Contaminants	Communities	Recreation and Other Services
	Physical	Ecological Productivity	

⊖ Fragmentation of Farmland Landscapes by Development

What Is This Indicator, and Why Is It Important? This indicator would report the degree to which suburban development and other built-up areas break up (fragment) the farmland landscape (croplands plus intermingled "natural" areas such as forests, wetlands, and grasslands and shrublands). Areas with a mosaic of cropland and intermingled natural areas—but little or no development—would be rated as "low" on the "fragmentation index" used for this indicator, while those in which small patches of cropland are mixed into a backdrop of suburban development would be rated as "high." These data would be presented nationally, as above, and by region for the most current year.

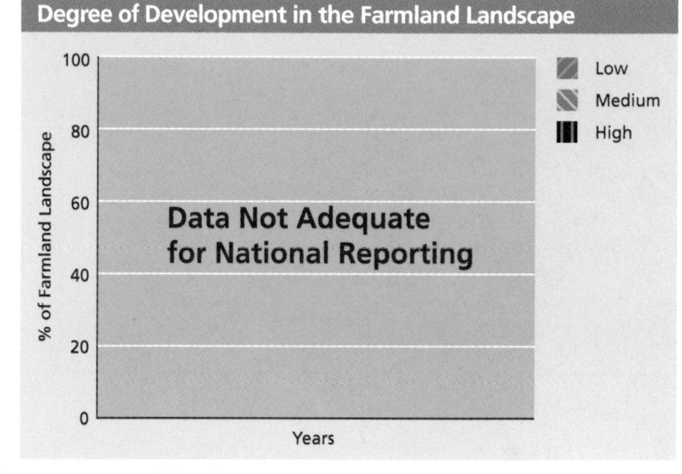

Housing and other development in farmland areas may compromise the economic viability of farming. Low-density, scattered development requires a great deal of surface area for roads and infrastructure, spreading over a relatively large proportion of the farmland landscape. Commuter traffic on rural roads produces dangerous conflicts with slow-moving farm machinery, and new residents may object to long-standing farm practices such as manure spreading. Development also typically increases nearby land values and, in some states, the property taxes on farmland, thereby increasing incentives for farmers to sell their land for further development. Finally, some development can diminish the aesthetic quality and recreation potential of formerly pastoral landscapes.

This indicator was selected to address the ability of farmland landscapes to produce goods for human benefit. It does not address how such fragmentation affects wildlife or other environmental values; for this, see p. 94.

Why Can't This Indicator Be Reported at This Time? The satellite land cover data necessary to report this index are available, but the data have not yet been analyzed.

The technical note for this indicator is on page 231.

SYSTEM DIMENSIONS	CHEMICAL AND PHYSICAL	BIOLOGICAL COMPONENTS	HUMAN USES
Extent Pattern	Nutrients, Carbon, Oxygen Contaminants Physical	Plants and Animals Communities Ecological Productivity	Food, Fiber, and Water Recreation and Other Services

⊖ Shape of "Natural" Patches in the Farmland Landscape

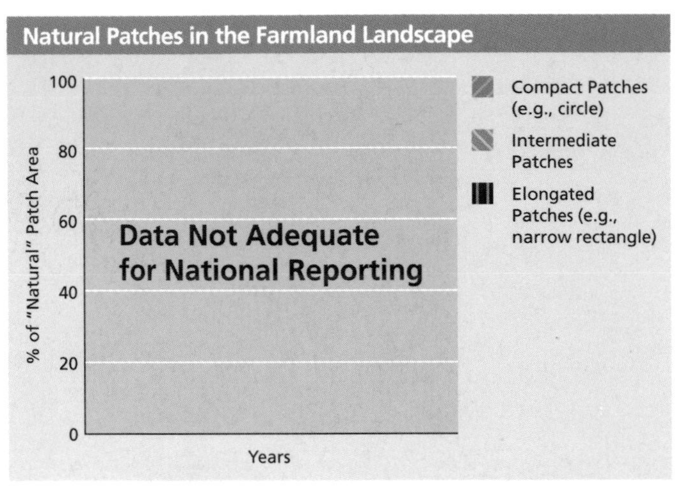

What Is This Indicator, and Why Is It Important? This indicator describes the shape of patches of "natural" lands in the farmland landscape, by reporting on the percentage of patch area that is found in "compact" patches (e.g., like a circle), "elongated" patches (e.g., like a long narrow rectangle), and an intermediate class of patch shape. These classes are defined based on the ratio of the perimeter, or edge, of each patch to its area; these perimeter-to-area ratios will be divided by patch area for the sake of comparison. "Natural" areas include forest, grasslands and shrublands, wetlands, and lands enrolled in the Conservation Reserve Program (CRP). These data would be presented nationally and by region for the most current year.

Natural lands within the farmland landscape control erosion, facilitate groundwater recharge, provide critical habitat for wildlife, and serve other important ecological functions. The size and shape of these often small and isolated remnants, along with restored conservation areas (e.g., CRP land), directly influence the amount and type of ecosystem services provided. Habitat fragmentation may create new kinds of habitats that are colonized by generalist native species or exotic species. For example, small patches and long narrow ones may have little or no "interior" habitat. Since some species thrive only in interior habitat—where there is a relatively large and contiguous area of forest, grassland, or other natural cover (see the forest fragmentation indicator, p. 120), small narrow areas may not provide habitat for these species. On the other hand, narrow strips may function quite well for erosion and sediment control.

Why Can't This Indicator Be Reported at This Time? As is the case for the development indicator (p. 93), the land cover data necessary to report this index are available, but have not been analyzed.

The technical note for this indicator is on page 232.

● Nitrate in Farmland Streams and Groundwater

What Is This Indicator, and Why Is It Important? This indicator reports on the concentration of nitrate in representative farmland streams and groundwater sites. Specifically, the indicator reports the percentage of streams and groundwater wells with average nitrate concentrations in one of four ranges, in areas that are primarily farmland.

Nitrate is a naturally occurring form of nitrogen and an important plant nutrient; it is often the most abundant of the forms of nitrogen that are readily usable by plants, including algae. Increased nitrate in streams that ultimately empty into coastal waters can lead to algal blooms in those waters; these blooms decrease recreational and aesthetic values and help deplete oxygen needed by fish and other animals (see the national nitrogen indicator and the hypoxia indicator, pp. 46 and 71). Elevated nitrate in drinking water is a health threat to young children and is of particular concern for people using household groundwater wells; municipal water supply systems typically take steps to remove nitrate.

Sources of nitrate in farmlands streams and groundwater include chemical fertilizers and runoff from manure associated with animal raising operations. If more fertilizer is applied than can be used by plants or stored in the soil, nitrates will seep into groundwater or drain into streams.

What Do the Data Show? Just over half the stream sites and 45% of groundwater wells sampled in areas where agriculture is the primary land use have concentrations of nitrate below 2 parts per million (ppm). About 20% of the groundwater wells and about 10% of stream sites had concentrations that exceed the federal drinking water standard (10 ppm).

Groundwater samples from areas dominated by agricultural use have higher concentrations of nitrate than either urban or forested areas, with forested lands having the lowest of the three. Only for farmland areas (and 3% of urban groundwater sites) did nitrate exceed the 10 ppm federal drinking water standard.

There is also a core national indicator for nitrogen (p. 46).

The technical note for this indicator is on page 232.

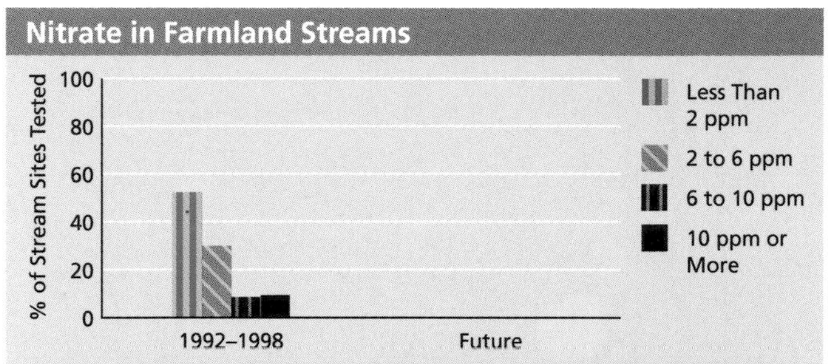

Data Source: USGS National Water Quality Assessment. Coverage: lower 48 states. Each sampling area was sampled intensively for approximately 2 years during 1992–1998.

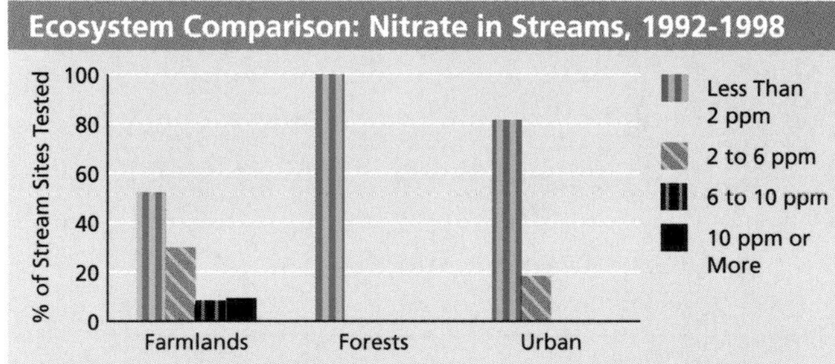

Data Source: USGS National Water Quality Assessment. Coverage: lower 48 states. Each sampling area was sampled intensively for approximately 2 years during 1992–1998.

Nitrate in Farmland Groundwater

% of Groundwater Sites Tested: 100, 80, 60, 40, 20, 0

1992–1998 Future

Less Than 2 ppm; 2 to 6 ppm; 6 to 10 ppm; 10 ppm or More

Data Source: USGS National Water Quality Assessment. Coverage: lower 48 states. Each sampling area was sampled intensively for approximately 2 years during 1992–1998.

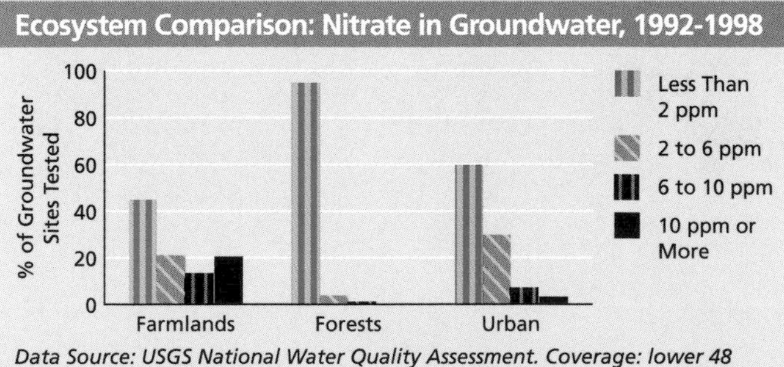

Data Source: USGS National Water Quality Assessment. Coverage: lower 48 states. Each sampling area was sampled intensively for approximately 2 years during 1992–1998.

Phosphorus in Farmland Streams

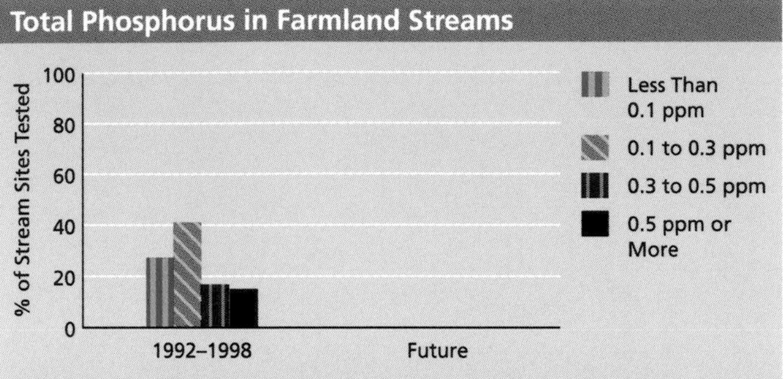

Data Source: USGS National Water Quality Assessment. Coverage: lower 48 states. Each sampling area was sampled intensively for approximately 2 years during the period from 1992 to 1998.

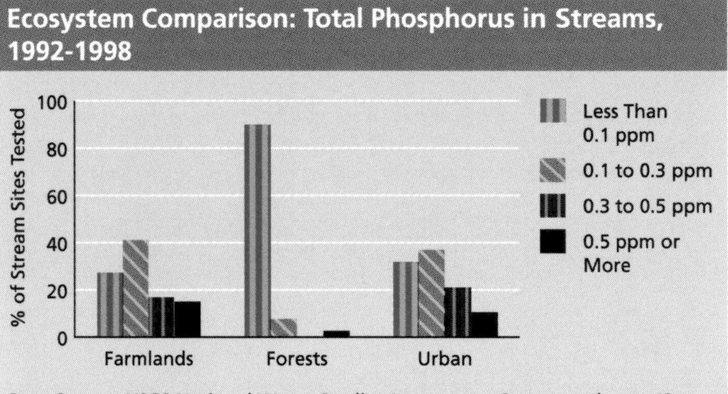

Data Source: USGS National Water Quality Assessment. Coverage: lower 48 states. Each sampling area was sampled intensively for approximately 2 years during the period from 1992 to 1998.

What Is This Indicator, and Why Is It Important? This indicator reports on the concentration of phosphorus in representative farmland streams. Specifically, the indicator reports the percentage of streams with average annual concentrations in one of four ranges, for streams draining watersheds that are primarily farmland.

Phosphorus is an essential nutrient for all life forms and occurs naturally in soils and aquatic systems; phosphate is the most biologically active form of phosphorus. At high concentrations in freshwater systems, however, phosphorus can lead to algal blooms, which can decrease recreational and aesthetic values and help deplete oxygen needed by fish and other animals.

Sources of phosphorus in farmlands streams include chemical fertilizers and runoff from manure associated with animal-raising operations. If more fertilizer is applied than can be used by plants or stored in the soil, phosphorus will drain into adjacent streams.

What Do the Data Show? About three-fourths of farmland stream sites had concentrations of phosphorus that were at least 0.1 part per million (ppm), and about 15% of farmland stream sites had phosphorus concentrations of at least 0.5 ppm.

Average phosphorus concentrations in farmland streams are similar to concentrations in streams draining urban watersheds (p. 187) and much lower than streams draining forested watersheds.

Discussion The U.S. Environmental Protection Agency (EPA) has recommended 0.1 ppm as a goal for preventing excess algae growth in streams. In 2000, EPA took steps to facilitate development of regional criteria, but these regional criteria have not yet been adopted. There is no federal drinking water standard for phosphorus.

The technical note for this indicator is on page 232.

● Pesticides in Farmland Streams and Groundwater

What Is This Indicator, and Why Is It Important? This indicator reports on pesticides found in farmland streams and groundwater. The graphs on the top show the average number of pesticides detected throughout the year in streams and shallow groundwater wells. The graphs on the bottom show the percentage of streams and shallow groundwater wells with pesticide concentrations that exceeded standards and guidelines (benchmarks) set for the protection of human health or aquatic life. These graphs report currently used agricultural pesticides and selected breakdown products of these pesticides, as well as selected organochlorine insecticides that were widely used in the past but whose use is no longer permitted in the United States.

The number of pesticides detected is important, but the presence of pesticides does not necessarily mean that the levels are high enough to cause problems. Comparison with benchmarks provides a useful reference to help judge the significance of contamination.

However, drinking water standards or guidelines do not exist for 33 of the 76 pesticides analyzed, and aquatic life guidelines do not exist for 48 of the 76 compounds. Current benchmarks do not account for mixtures of chemicals and seasonal events involving high concentrations. In addition, potential effects on the reproductive, nervous, and immune systems, as well as on particularly sensitive people, are not yet well understood.

What Do the Data Show? All monitored streams in farmland areas had at least one pesticide at detectable levels throughout the year, and about 75% had an average of five or more. Eighty-three percent of streams had at least one pesticide whose concentration exceeded aquatic life guidelines; about 4% had one or more compounds that exceeded human health standards or guidelines.

About 60% of groundwater wells in farmland areas had at least one pesticide at detectable levels, and less than 1% had any pesticides that exceeded human health standards or guidelines.

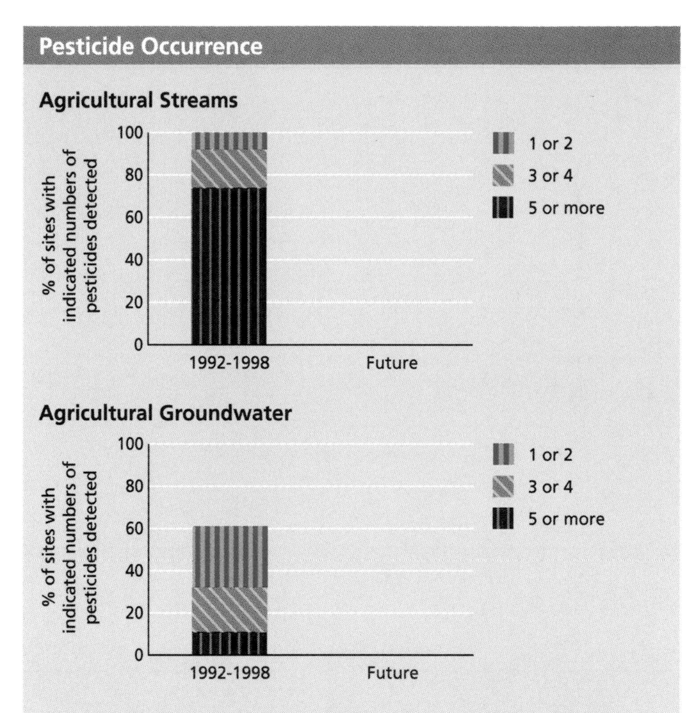

Data Source: U.S. Geological Survey National Water Quality Assessment Program (NAWQA). Coverage: lower 48 states.

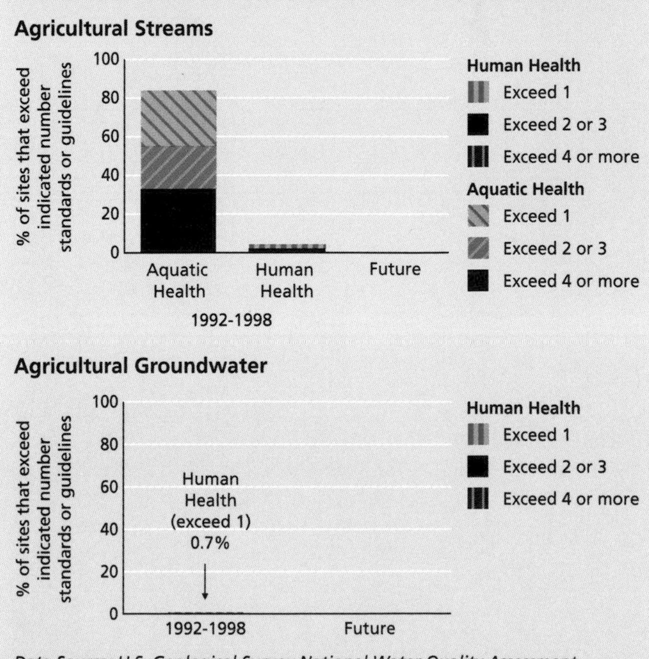

Data Source: U.S. Geological Survey National Water Quality Assessment Program (NAWQA). Coverage: lower 48 states.

Note: aquatic life standards are not currently applied to groundwater.

● Pesticides in Farmland Streams and Groundwater *(continued)*

Discussion The data shown here do not represent assessments of the risks posed to people or ecosystems in any specific location, since they do not incorporate factors such as whether the water tested is actually used as a drinking water source and the time of year when the pesticides are found, relative to when animals are most active.

Guidelines for the protection of aquatic life are often numerically lower than standards and guidelines to protect human health. Aquatic animals spend much or all of their life in water, and may be more sensitive than people to specific contaminants. People consume drinking water from both streams and groundwater, thus human health standards and guidelines apply to both. Guidelines to protect aquatic life are not applied to groundwater.

The pesticides reported here are generally associated with agriculture, but some may have other uses (currently or in the past). Thus, not all contamination is necessarily attributable to agricultural use.

See also the national, coastal, and urban contaminants indicators (pp. 48, 72, and 189).

The technical note for this indicator is on page 234.

⊖ Soil Organic Matter

What Is This Indicator, and Why Is It Important? This indicator reports how much organic matter—partially decayed plant and animal matter—there is in the top 4–6 inches of cropland soil. This will be reported nationally over time, and by region for the most recent year of data.

Organic matter helps the soil hold water and supplies nutrients, which are crucial for crop production; it also protects against erosion and helps support a healthy and diverse set of microscopic plants and animals. Organic matter content, erosion (p. 100), soil salinity (p. 101), and soil biological condition (p. 102) are key indicators of soil quality, reflecting the effect of agriculture on soils and the influence of changing crop and soil management practices.

Soil organic matter is usually measured as the percentage of organic matter (by dry weight) in the top 4–6 inches of the soil, where human activities have most influence on soil condition. While there are large regional differences in soil organic matter content because of climate and other factors, changes in this indicator nationally and within regions will provide important information on the effect of cropland management.

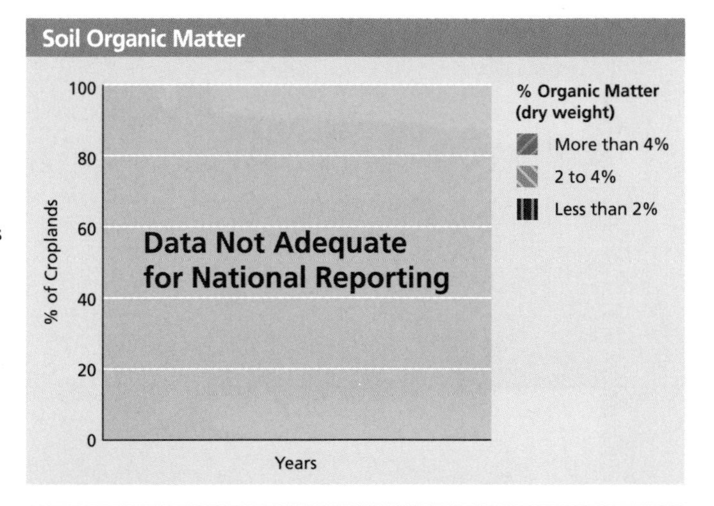

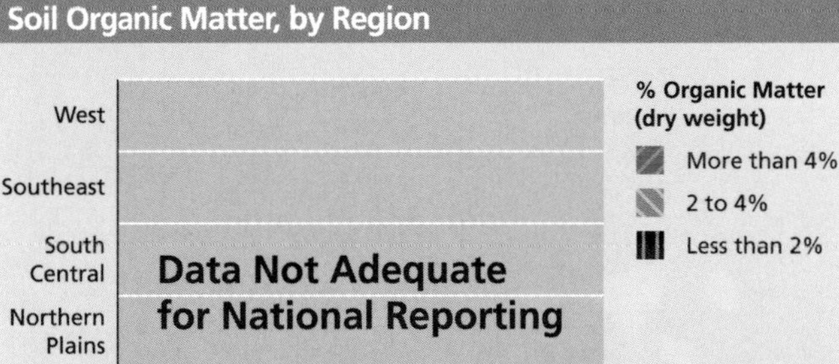

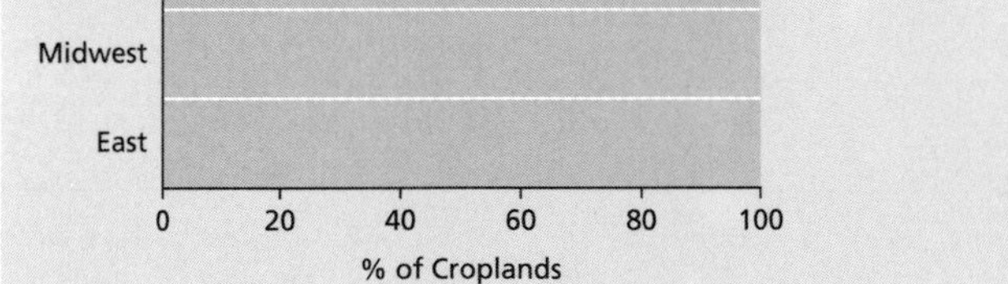

Why Can't This Indicator Be Reported at This Time? There are baseline estimates of the amount of organic matter in soils across the United States through Soil Survey reports produced by the USDA Natural Resources Conservation Service, but there is no mechanism for systematic monitoring of changes in these amounts. Long-term observations of changes in organic matter resulting from different management practices are under way in a number of research plots and other locations, but these do not provide an adequate basis for nationwide monitoring. In addition, efforts are under way to develop techniques to use satellite data to estimate organic matter in surface soils.

The technical note for this indicator is on page 234.

SYSTEM DIMENSIONS	CHEMICAL AND PHYSICAL	BIOLOGICAL COMPONENTS	HUMAN USES
Extent	Nutrients, Carbon, Oxygen	Plants and Animals	Food, Fiber, and Water
Pattern	Contaminants	Communities	Recreation and Other Services
	Physical	Ecological Productivity	

● Soil Erosion

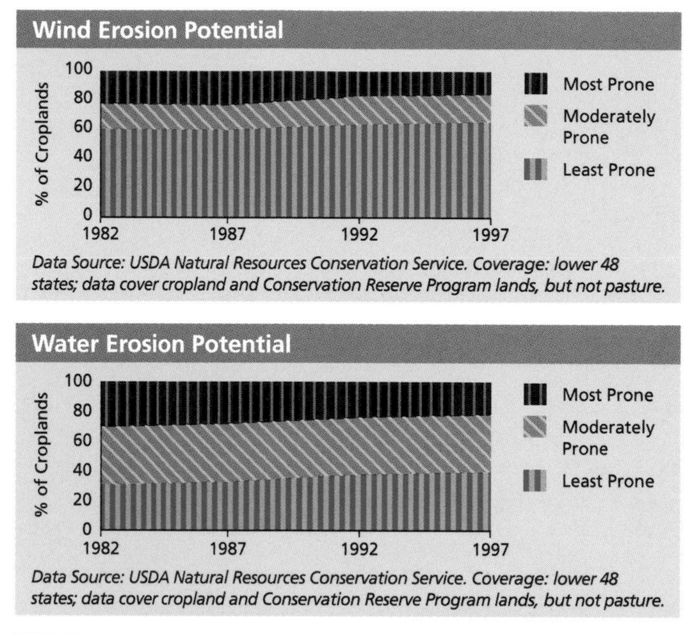

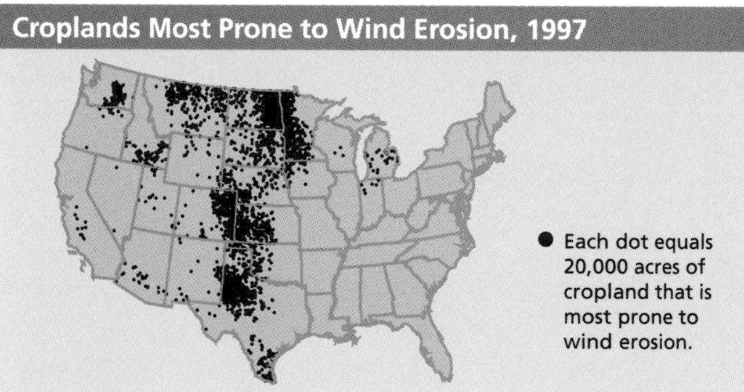

Data Source: USDA Natural Resources Conservation Service. Coverage: lower 48 states; data cover cropland and Conservation Reserve Program lands, but not pasture.

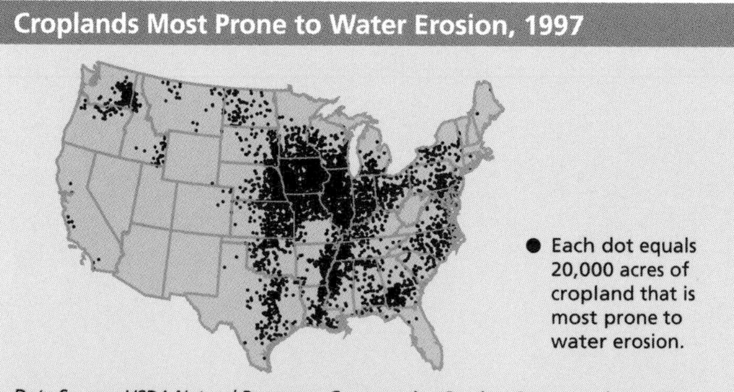

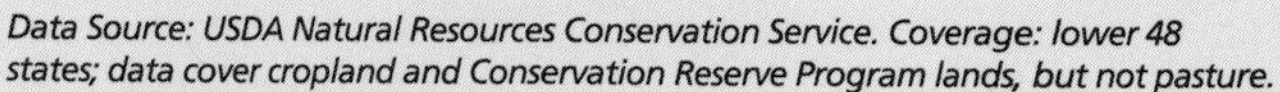

Data Source: USDA Natural Resources Conservation Service. Coverage: lower 48 states; data cover cropland and Conservation Reserve Program lands, but not pasture.

What Is This Indicator, and Why Is It Important? This indicator reports the percentage of U.S. farmlands according to their potential for erosion by wind or water. These data are based on an index that combines information on soil characteristics, topography, and management activities such as tillage practices and whether crop residue is left on the field or not. This indicator covers croplands (excluding pastures) and Conservation Reserve Program (CRP) lands. In addition, those croplands most prone to wind and water erosion are mapped for 1997.

Agricultural soil erosion reduces soil quality and degrades water quality. Even relatively small movements—for example, from the top of a slope to the bottom—cause changes in soil structure that can reduce fertility and make normal cropping practices difficult. When soil moves further, eventually ending up in streams and lakes, it causes water quality problems, in part because eroded sediments often carry both fertilizers and pesticides. Even without such pollution, sedimentation alone imposes significant costs on reservoirs and water treatment facilities, navigation, and other water and waterway users. Erosion, organic matter content (p. 99), soil salinity (p. 101), and soil biological condition (p. 102) are key indicators of soil quality; changes to crop and soil management practices affect soil quality.

What Do the Data Show? From 1982 to 1997, the acreage of U.S. farmland with the greatest potential for wind erosion decreased by nearly one-third, to about 63 million acres, or about 15% of U.S. croplands. The area with the greatest potential for water erosion also decreased by nearly one-third, to 89 million acres, or about 22% of U.S. croplands.

Although both water and wind erosion occur throughout the United States, high levels of water erosion are more common in the eastern half of the nation, and wind erosion is more likely in the West.

Discussion Reductions in erosion can result from changes in management practices; common practices used to reduce soil erosion are no-till or minimum tillage, installation of terraces and field wind breaks, and contour farming. In addition, removal of highly erosion-prone lands from cultivation, (for example, enrollment in the Conservation Reserve Program) typically lowers its erosion potential.

The technical note for this indicator is on page 235.

SYSTEM DIMENSIONS	CHEMICAL AND PHYSICAL	BIOLOGICAL COMPONENTS	HUMAN USES
Extent	Nutrients, Carbon, Oxygen	Plants and Animals	Food, Fiber, and Water
Pattern	Contaminants	Communities	Recreation and Other Services
	Physical	Ecological Productivity	

⊖ Soil Salinity

What Is This Indicator, and Why Is It Important? This indicator would report the percentage of cropland with different levels of salt content, measured in decisiemens per meter (dS/m). A map showing the percentage of land in each major cropland region with elevated salt levels (i.e., over 4 dS/m), would accompany the nationwide data.

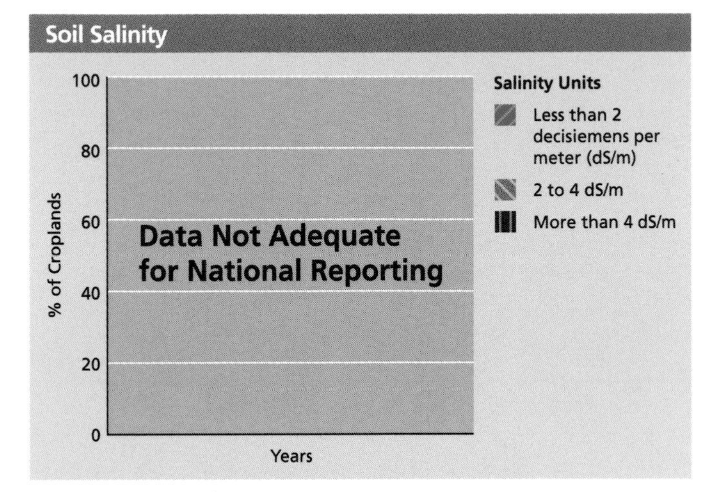

Excess salt has the same effect on plants as drought: too much salt in soil reduces the ability of plants to take up water, which interferes with their growth and reduces their vitality. Excess salt in soils can also enter groundwater and surface water. Highly saline water is hazardous to freshwater fish, and waterfowl accustomed to freshwater avoid it. Some salts, like those containing sodium, can change the physical condition of the soil, reducing infiltration, increasing runoff and erosion, and impairing biological activity. Soil salinity, along with organic matter content (p. 99), erosion (p. 100), and soil biological condition (p. 102), is a key indicator of soil quality.

Soil salinization often results from irrigated agriculture, and it is generally a problem in arid areas. Water used for irrigation contains small amounts of salt, and when water evaporates from the soil surface or from the leaves of plants, it leaves salt behind in the soil. In arid areas, these salts can accumulate and cause problems. In areas with greater rainfall, salts are drained from the soil by the larger volumes of water flowing through the soil, and tend not to accumulate to high levels.

Although much less widespread, salinization can occur in the absence of irrigation. Some areas have naturally high salt content in their soil, while saline seeps can occur when water moves through the soil, picking up salts, and then emerges at a seep or spring.

Why Can't This Indicator Be Reported at This Time? Salinity measurements are often included in routine soil tests conducted by farmers, government agencies, and researchers. However, there is no unified effort to collect these data and incorporate them into a national database to monitor trends over time.

The technical note for this indicator is on page 235.

⊖ Soil Biological Condition

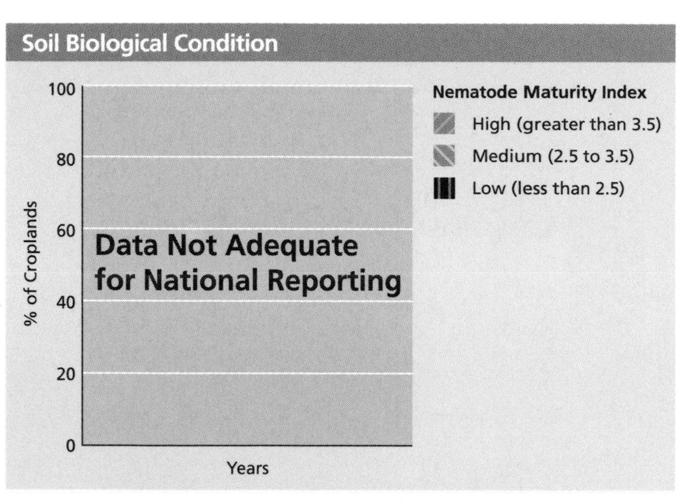

What Is This Indicator, and Why Is It Important? This indicator would report the percentage of croplands in three different ranges on the Nematode Maturity Index (NMI), an index that measures the types of roundworms, or nematodes, in the soil. A map showing the percentage of cropland in each major cropland region with low index values (indicating disturbed soils) would accompany the nationwide data.

Healthy soils contain many different microscopic animals. Agricultural practices often disturb the soil, and the amount of disturbance can be measured by changes in these microscopic animals. This indicator is based on the identification of various types of nematodes, each of which has a different tolerance for soil disturbance. Calculation of the NMI is based on the proportion of nematodes with different levels of tolerance for disturbance. Low NMI values (less than 2.5) are often found in soils subjected to intensive agricultural production methods, like monoculture and the use of high levels of nitrogen fertilizer and pesticides. Midrange values (from 2.5 to 3.5) suggest a more diverse soil community and often reflect such practices as crop mixtures and rotations and no-till farming. High NMI values (greater than 3.5) are rarely found on cultivated lands.

Soil biological condition, along with organic matter content (p. 99), erosion (p. 100), and soil salinity (p. 101) are key indicators of soil quality, reflecting the effect of agriculture on soils and the influence of changing crop and soil management practices

Why Can't This Indicator Be Reported at This Time? Measuring soil quality by measuring soil organisms has gained broad scientific acceptance. While the Nematode Maturity Index is a promising indicator, it has not yet been adopted by a nationwide monitoring program. However, NMI has been applied successfully in two statewide surveys (North Carolina and Nebraska) carried out in cooperation with the National Agricultural Statistics Service.

Reporting of soil quality based on nematode populations would require large-scale implementation of the indicator described here. This could be done through an existing national monitoring program, or state-based monitoring using consistent methods would allow the resulting information to be aggregated at the national level.

The technical note for this indicator is on page 236.

SYSTEM DIMENSIONS	CHEMICAL AND PHYSICAL	BIOLOGICAL COMPONENTS	HUMAN USES
Extent	Nutrients, Carbon, Oxygen	**Plants and Animals**	Food, Fiber, and Water
Pattern	Contaminants	Communities	Recreation and Other Services
	Physical	Ecological Productivity	

⑦ Status of Animal Species in Farmland Areas

What Is This Indicator, and Why Is It Important? This indicator would report on the status of wildlife in farmland areas.

Farmlands—including both croplands and the patches of natural lands that are intermingled with them—are home to many kinds of wildlife. Some species would be found in the forests, grasslands, or shrublands from which the farmlands were created. Such species may find fewer habitat opportunities in farmland areas, but may take advantage of remaining patches of habitat (see p. 94) and remain in the area, but at low population levels. However, there are many species that favor the kinds of conditions found in areas with extensive farmlands, and these species are often more common than they were before conversion to agriculture.

Status of Animal Species in Farmland Areas

Indicator Development Needed

Why Can't This Indicator be Reported at this Time? An index is needed that that would account for both types of species found on farmlands—those that favor the pre-agricultural landscape and those that favor landscapes dominated by agriculture. This approach must necessarily differ from that taken in reporting on marine, forest, grassland and shrubland, and freshwater species (see pp. 75, 124, 144, and 168), because it is not possible to define a set of "farmland" species in the same way that one can identify species that have evolved to depend on these other ecosystem types.

The technical note for this indicator is on page 237.

SYSTEM DIMENSIONS	CHEMICAL AND PHYSICAL	**BIOLOGICAL COMPONENTS**	HUMAN USES
Extent Pattern	Nutrients, Carbon, Oxygen Contaminants Physical	**Plants and Animals** Communities Ecological Productivity	Food, Fiber, and Water Recreation and Other Services

⑦ Native Vegetation in Areas Dominated by Croplands

Native Vegetation in Areas Dominated by Croplands

Indicator Development Needed

What Is This Indicator, and Why Is It Important? This indicator would report, for areas where croplands account for a large percentage of the land cover, how much of the remaining vegetation (outside of croplands) is native to the area.

Where croplands dominate the landscape, wildlife rely more heavily on the remaining areas for their habitat needs. Since vegetation dominated by non-native species often has much lower value as wildlife habitat, a high proportion of non-native plant species in the remaining non-cropland areas will have a harmful effect on wildlife populations. For example, when lands in the Conservation Reserve Program, which provides rental payments to farmers who retire lands important for conservation, are converted from non-native grasses to native prairie grass, upland bird populations increase significantly.

Why Can't This Indicator Be Reported at This Time? Several questions must be answered before this indicator can be implemented. These include the scale at which it should be reported (i.e., county, state, or region?); the threshold for including an area in the indicator (i.e., should the indicator include only areas with more than 50% croplands, or more than 75%?); and the proportion of non-native species that should be used to categorize areas as "dominated" by non-native rather than native species.

Once the indicator is clearly defined, obtaining data may also be difficult. The fraction of land in a county, state, or region that is cropland, and its location, are readily available from satellite data. Whether vegetation is dominated by native or non-native species cannot generally be determined using satellite imagery, but many state and federal agencies, nongovernmental organizations, and universities collect data on non-native plants. However, these data have never been brought together to provide consistent information over large areas. Many existing federal, state, and local government programs could contribute to reporting on the extent of non-native species, as could nongovernmental organizations and academic institutions.

The technical note for this indicator is on page 237.

⑦ Stream Habitat Quality

What Is This Indicator, and Why Is It Important? This indicator would describe the habitat quality of farmland streams by comparing a number of key attributes to those of relatively undisturbed streams in the same general area. The index would incorporate the presence of riffles and pools, the size of streambed sediments and the degree to which larger gravel and cobbles are buried in silt, the presence of branches, tree trunks, and other large woody pieces, and the stability of the bank. A companion indicator (p. 149) would report on all streams, not just on those in farmlands.

Stream Habitat Quality

Indicator Development Needed

Streams with higher condition ratings (those that resemble undisturbed streams) have a more natural and diverse array of underwater and bank habitats and are therefore capable of supporting diverse native species. These streams are also more likely to have relatively undisturbed flow patterns (see p. 142) and to have vegetation along their banks—features that help maintain the conditions necessary to support a healthy biological community over the long term.

Like their counterparts on land, stream-dwelling animals and plants require specific habitat conditions in order to survive and reproduce. Because each species has its own particular habitat requirements, a variety of habitats along a stream are needed to maintain the stream's natural complement of plants and animals.

Why Can't This Indicator Be Reported at This Time? Scientists generally agree on the key stream attributes that should be measured to evaluate stream habitat quality (riffles and pools, streambed sediments, and so on), and there is a considerable work under way by the Environmental Protection Agency, the U.S. Geological Survey, and state agencies to gather data and develop ranking methods. However, there is still no generally accepted method for combining data on individual attributes into a single index. In addition, habitat values for any particular stream must be evaluated in relation to the plants and animals in that region, so any stream habitat index would have to be tailored for different regions.

The technical note for this indicator is on page 237.

● Major Crop Yields

Crop Yields: Corn, Soybeans, and Wheat

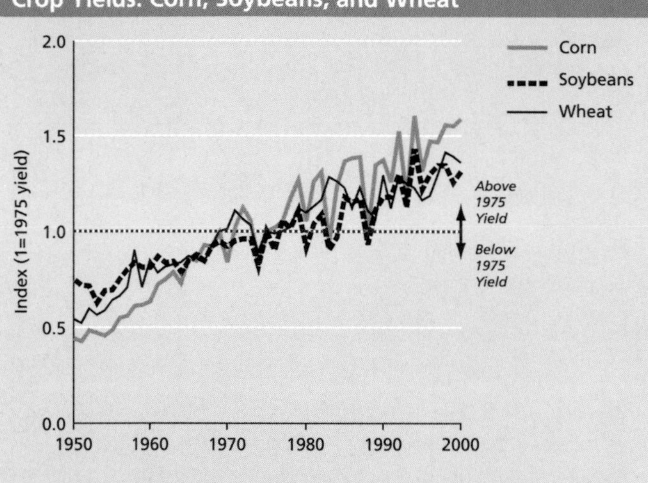

Data Source: USDA National Agricultural Statistics Service. Coverage: all 50 states.

Crop Yields: Hay and Cotton

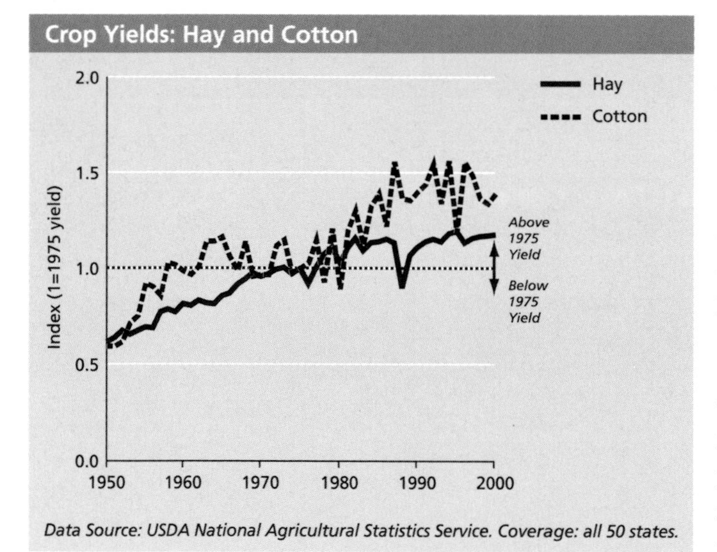

Data Source: USDA National Agricultural Statistics Service. Coverage: all 50 states.

What Is This Indicator, and Why Is It Important? This indicator reports the yield of corn, soybeans, wheat, hay, and cotton, as an index with 1975 as the base year. Values above 1.0 indicate higher yields, typically measured as tons or bushels per acre, than in 1975; values below 1.0 indicate lower yields than in 1975. These five crops account for about 90% of harvested acreage in the United States and more than half the monetary value of all crops (p. 108).

Increasing the amount of food grown per acre has allowed U.S. agriculture to produce more food and fiber without corresponding increases in farm acreage. The total acreage used for agricultural production has declined slightly over the past half-century (p. 91), and a significant increase in the acreage devoted to agriculture is generally considered unlikely.

What Do the Data Show? Per-acre yields of the major crops grown in the United States have increased dramatically over the past 50 years. Yields for three of the five major crops (corn, wheat, and cotton) more than doubled over this period, with corn yields increasing almost fourfold. Of these five major crops, soybean yields increased the least, but even they nearly doubled over the period.

Discussion Increases in crop yields are believed to result from a combination of factors. These include improvements in breeding, changes in cultivation practices, and increased use of a variety of inputs, including pesticides and fertilizers. More intensive use of farmland is thought to play an important role in improving yields, but it may also have negative effects, such as increased concentrations of nitrogen, phosphorus, or pesticides in streams, lakes, and coastal waters (see the farmland nitrogen and phosphorus indicators, pp. 95 and 96, the farmland pesticide indicator, p. 97, and the national nitrogen indicator, p. 46).

The technical note for this indicator is on page 238.

SYSTEM DIMENSIONS	CHEMICAL AND PHYSICAL	BIOLOGICAL COMPONENTS	HUMAN USES
Extent	Nutrients, Carbon, Oxygen	Plants and Animals	**Food, Fiber, and Water**
Pattern	Contaminants	Communities	Recreation and Other Services
	Physical	Ecological Productivity	

● Agricultural Inputs and Outputs

What Is This Indicator, and Why Is It Important? This indicator reports the amount of inputs used to produce one unit of output, with 1975 as the base year. So, for any input, the index value for a given year describes whether more or less of that input was used to produce a unit of output in that year than in 1975. The indicator also reports agricultural outputs over time, again compared to the output in 1975.

This is a very broad analysis. For example, all fertilizers used on U.S. farms were divided by all agricultural outputs—even if different amounts of fertilizer were used to produce each commodity. Agricultural production is driven by physical inputs and by the knowledge and skill of farmers, plant breeders, and others. A decreasing input index results because the input is used more efficiently by farmers (e.g., less fertilizer per ton of corn due to targeted application), or because of a series of advances (e.g., less labor required because of increased mechanization and more effective pesticides). Because inputs are often expensive and, like pesticides and fertilizers, may have environmental consequences, input trends are an important indicator of the long-term health of the agricultural enterprise and the level of its environmental impact.

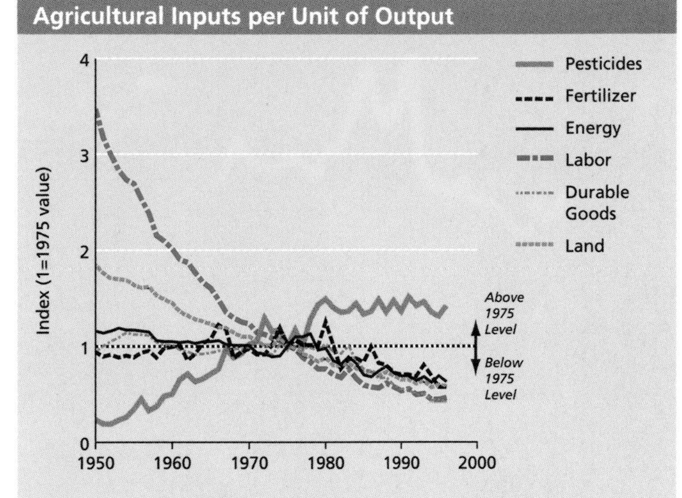

Data Source: USDA Economic Research Service. Coverage: all 50 states.

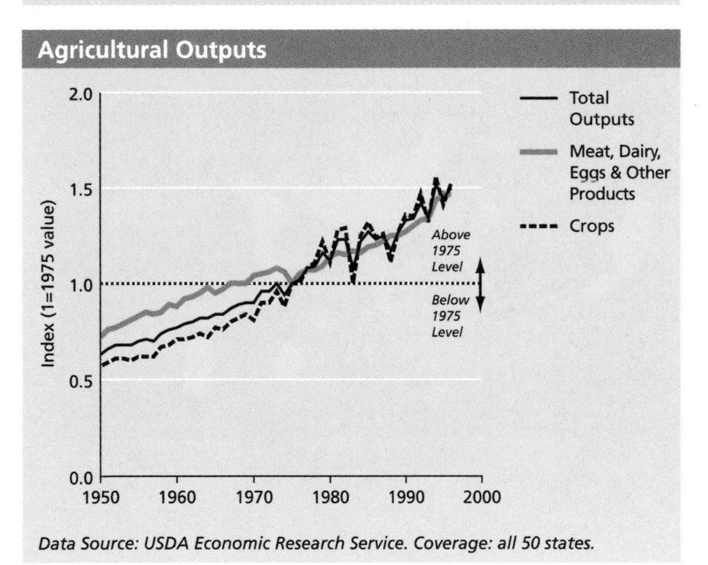

Data Source: USDA Economic Research Service. Coverage: all 50 states.

What Do the Data Show? U.S. agricultural output has been increasing steadily since 1950, while the major inputs required to produce a unit of that output—with the exception of pesticides—have decreased. The amounts of both labor and land needed to produce a unit of output have fallen steadily since 1950, although the decrease in labor has been much larger. Farmers have produced more output per unit of fertilizer, energy, and durable goods such as tractors since the mid-1970s. Pesticide use per unit of output, which showed steady increases from the 1950s, leveled off around 1980.

Discussion As technology and farming practices change, inputs can change considerably. For example, a pound of pesticides today provides far more pest control than did the same amount 30 years ago. For this reason, this indicator relies upon a complex analysis of the quantities and quality of inputs used (see the technical note). A similar analysis was used for outputs, because they cannot simply be added together (a pound of strawberries is not equal to a pound of corn).

The indicator focuses on a few major, quantifiable, inputs. This means that some factors, such as changes in plant breeding (including the introduction of genetically engineered crops), are not addressed at all, and some inputs, such as water, are addressed only indirectly (in this case, through the energy costs associated with irrigation).

The technical note for this indicator is on page 238.

Monetary Value of Agricultural Production

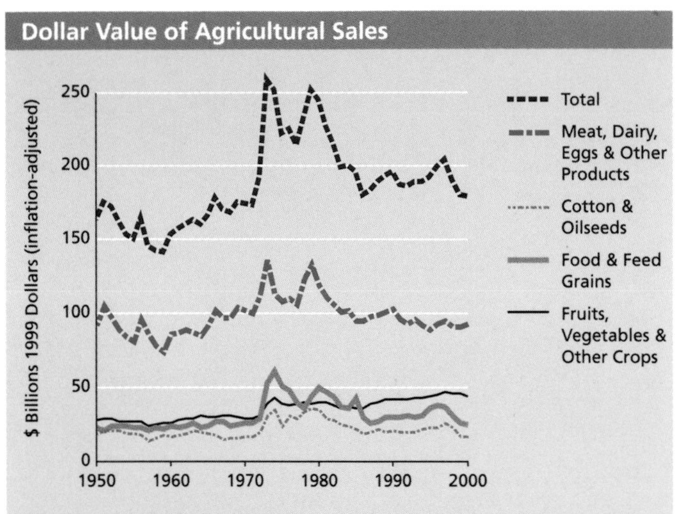

Data Source: USDA Economic Research Service. Coverage: all 50 states.

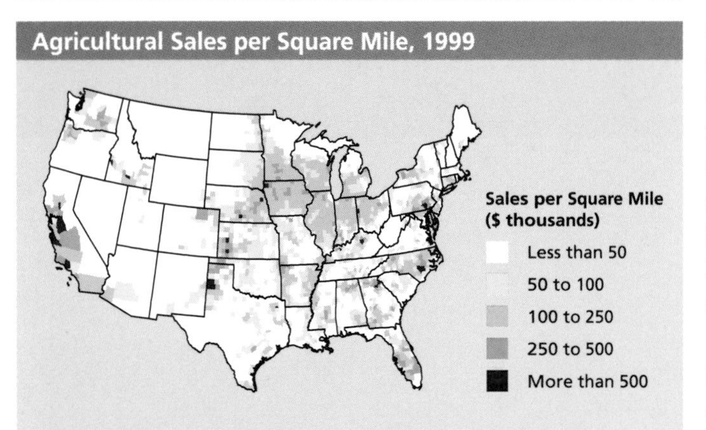

Data Source: Department of Commerce Bureau of Economic Analysis. Coverage: Lower 48 states

What Is This Indicator, and Why Is It Important? This indicator reports the dollar value of the annual output of major crops and livestock. The value is determined by multiplying the amount of output by the prices received by farmers (in 1999 dollars). The data are presented both nationally over time and by location for the most recent year available (in this case, 1999).

Farming is a business, and the monetary value of the goods produced is an indication of the importance to society of those goods. In addition, some areas have high concentrations of agriculture or produce high-value crops (or both). In these areas, farming is often a significant component of the local economy.

What Do the Data Show? The gross value of agricultural output (adjusted for inflation) was about $180 billion in 1999, or about 10% more than in 1950. Over this half-century, however, there were major fluctuations—from a low of about $140 billion in 1959 to a high of about $260 billion in 1973. Livestock products consistently account for about half of overall agricultural income. Agricultural production is concentrated in the Midwest, but there are concentrations of very high agricultural sales in many areas across the country.

Discussion Advances over the last 50 years have enabled farmers to produce more per acre of land (p. 106) and to increase total physical outputs, while requiring, in general, fewer inputs (p. 107). However, as shown here, these advances have not translated into steadily increasing farm sales. Note that the values reported here are gross revenues, meant to represent the value of the harvest from croplands—they reveal nothing about the profitability of American farming. This indicator also reports the money received by farmers, not the retail price of farm products.

The technical note for this indicator is on page 238.

⊖ Recreation

What Is This Indicator, and Why Is It Important? This indicator would report the number of days spent fishing, hunting, viewing wildlife, or engaged in other recreational activities on farmland.

A great deal of recreation takes place on our nation's farmlands, and those enjoying these recreational opportunities may be the farmers themselves, their friends, or visitors. In many areas, farmers supplement their income by charging to hunt or fish on their property, and they may even take steps to increase the abundance of wildlife in order to attract business. Wildlife-associated recreation is an important source of income for many small agricultural communities.

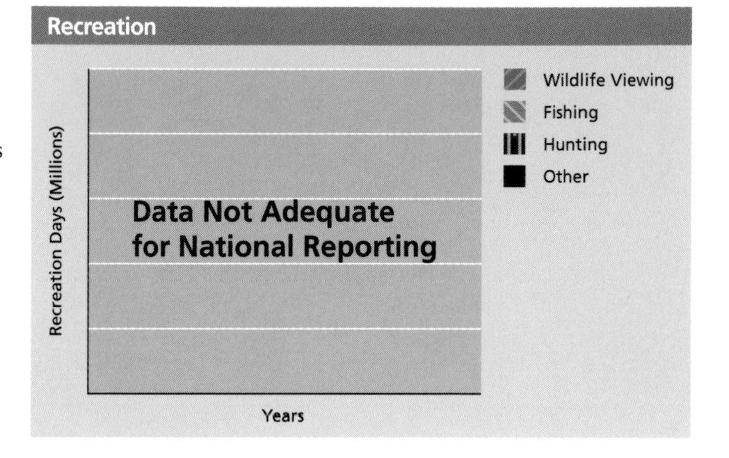

Why Can't This Indicator Be Reported at This Time? There are no national data sets that document the type and level of recreation on farmlands. The National Survey of Fishing, Hunting, and Wildlife-Associated Recreation (http://fa.r9.fws.gov/surveys/surveys.html) and the National Survey on Recreation and the Environment (http://www.srs.fs.fed.us/trends/nsre.html) both provide reliable data on these activities, but neither survey identifies activities that take place on farmlands. The Census of Agriculture (http://www.nass.usda.gov/census/), which provides information on a wide range of farm-related subjects, does not address recreational activities.

Adequate reporting would require modification of existing surveys to elicit information either on the location of recreational activities or on the amount of recreation on farms.

There is no technical note for this indicator.

Forests

What Indicators are used to describe Forests?			Can we report trends? Are there other useful reference points?
SYSTEM DIMENSIONS			
●	Forest Area and Ownership	How much forest land is there in the United States? How much is privately owned, and how much is publicly owned?	Trends, regional comparison
●	Forest Types	How is the area occupied by major forest types changing?	Trends, regional comparison
●	Forest Management Categories	How intensively managed are U.S. forest lands?	Trends, regional comparison
◐	Forest Pattern and Fragmentation	How fragmented are U.S. forests?	Regional comparison
CHEMICAL AND PHYSICAL CONDITIONS			
●	Nitrate in Forest Streams	How much nitrate is there in forest streams?	Current data only, federal standard, cross-ecosystem comparison
◐	Carbon Storage	How much carbon is stored in U.S forests?	Trends, regional comparison
BIOLOGICAL COMPONENTS			
◐	At-Risk Native Species	What are the percentages of forest-dwelling species that are at different levels of risk of extinction?	Current data only, regional comparison
⊖	Area Covered by Non-native Plants	What percentage of the plant cover in forests is not native to the region?	No data reported
◐	Forest Age	How much of the nation's forests is young, middle-aged, or old?	Regional comparison
●	Forest Disturbance: Fire, Insects, and Disease	How many acres are affected each year by fires, insects, disease, windstorms, and ice?	Trends
⊖	Fire Frequency	Are forest fires burning much more or less frequently than in presettlement times?	No data reported
⊖	Forest Community Types with Significantly Reduced Area	How much area is occupied by forest types that have significantly declined in area since presettlement times? Are these forest types increasing or decreasing in area at present?	No data reported
HUMAN USES			
●	Timber Harvest	How much timber is harvested each year, and what is it used for?	Trends, regional comparison
●	Timber Growth and Harvest	How much timber grows each year, compared to the amount that is cut?	Trends, regional comparison
⊖	Recreation in Forests	How much recreational activity takes place in the nation's forests?	No data reported

● All Necessary Data Available ◐ Partial Data Available ⊖ Data Not Adequate for National Reporting ? Indicator Development Needed

Chapter 7: Indicators of the Condition and Use of Forests

What makes a forest a forest is, obviously, the presence of trees: the giant redwoods of the West Coast; the maples, oaks, and hickories that give New England its flaming fall foliage; the trackless wilderness of interior Alaska; even the Pine Barrens of New Jersey—all these fit into the picture conjured up by the word "forest." But forests also include the coastal live oak woodlands of California, the cypress swamps and savannahs of the South, and the pinyon–juniper woodlands of the arid Southwest. Highly managed timberlands are also forests. Many of these "forest" types overlap with, or even occupy the same space as, other ecosystem types (wetlands, grasslands and shrublands, farmlands, urban and suburban areas). These varied forest lands provide Americans with timber and other wood products, but they also offer the opportunity for solitude, hunting, hiking, birdwatching, and camping.

What can we say about the conditions and use of U.S. forests?

Fifteen indicators describe the condition and use of forests in the United States. Partial or complete data are available for eleven of these fifteen indicators; data are available for a higher percentage of forest indicators than for any other ecosystem. Seven indicators have a data record that is long enough to judge trends, and for one there is a regulatory standard for comparison. For four indicators, data are not adequate for national reporting. Nine of the eleven indicators for which data are available are presented by region, allowing comparison of conditions in different regions.

After the following brief summaries of the findings and data availability for each indicator, the remainder of this chapter consists of the indicators themselves. Each indicator page offers a graphic representation of the available data, defines the indicator and explains why it is important, and describes either the available data or the gaps in those data.

System Dimensions

Three of the four indicators of forest system dimensions track forest acreage, each in a different way. These are total forest acreage, including the split between public and private ownership; the acreage of about 20 major forest types, each of which provides habitat for a different mix of plants and animals; and the percentage of forest under various management regimes, from planted timberland to national parks and wilderness areas, where timber harvesting is prohibited. Both the type of forest and the degree to which the forest is broken into smaller patches and intermingled with nonforest areas are important for many forest species, so the fourth indicator reports the percentage of forest surrounded by small, medium, and large expanses of more-or-less complete forest cover.

- **How much forest land is there in the United States? How much is privately owned, and how much is publicly owned?** Forests cover 747 million acres, or about a third of the land area of the United States, down from about 1 billion acres at the time of European settlement. The area of forest is divided about evenly between East and West, but most historic conversion of forest to other uses has taken place in the East. In recent decades, the overall acreage of forest has been relatively stable. In the East, more than 80% of forest lands are privately owned; in the West, about two-thirds are publicly owned.
- **How is the area occupied by major forest types changing?** Over the past several decades, the major forest types with the largest increases in acreage were oak–hickory and maple–beech–birch in the East, and fir–spruce in the West. Forest types declining in area included elm–ash–cottonwood and oak–gum–cypress in the East and hemlock–sitka spruce, ponderosa pine, and lodgepole pine in the

West. Over this period, overall forest area changed very little, so reductions in area occupied by one forest type were generally balanced by increases in area by other types.

- **How intensively managed are U.S. forest lands?** In 1997, 11% of western forests and 3% of eastern forests were in federal wilderness areas and national parks (reserved forest), while 10% of eastern forests and 4% of western forests resulted from replanting with seedlings in anticipation of future timber harvest. Nationwide, reserved forest land has doubled since 1953, to 14 million acres; during the same period, planted timberland increased tenfold, also to 14 million acres. Most forest lands in the United States—including those used for timber production—are neither national parks or wilderness areas nor planted forests.
- **How fragmented are U.S. forests?** One way to report on forest fragmentation pattern is to describe the degree to which any forested point is surrounded by land that is mostly forested (at least 90% forest cover). About two-thirds of all points in both eastern and western forests are surrounded by mostly forest cover within a radius of about 250 feet. About a quarter of all forest points have mostly forest cover within about a 2½-mile radius. Tracking this indicator over time will make it possible to distinguish between natural forest patterns and changes caused by human activity. In addition, methods available in the future may allow identification of smaller features (for example, forest roads and small clearings with houses) than can readily be mapped using the current satellite data that this indicator relies upon.

Chemical and Physical Conditions

Two indicators describe the chemical and physical condition of forests. We track nitrate in forest streams because elevated concentrations of this nutrient can be a sign of plants under stress or of increased inputs from sources such as atmospheric deposition and conversion of forest to other land use. We track carbon storage because carbon is the major building block of forest systems and because increased carbon storage in forests can offset emissions of carbon dioxide from burning fossil fuels.

- **How much nitrate is there in forest streams?** Almost all forest streams had nitrate concentrations below 1 part per million; more than half had concentrations below 0.1 part per million. The federal drinking water standard is 10 parts per million. No trend data are available, but streams in forested regions have the lowest nitrogen concentrations, farmland streams the highest.
- **How much carbon is stored in U.S. forests?** Increased carbon storage by forests and other ecosystems can offset emissions of carbon dioxide from the burning of fossil fuels, of concern because of climate change. The amount of carbon stored in trees on timberlands (a USDA Forest Service designation for areas with trees that grow fast enough to support timber harvests and on which harvest is not prohibited by law) has gone up steadily in the East and remained stable in the West. Data are not adequate for national reporting on carbon stored in roots, forest floor litter, and soil.

Biological Components

Six indicators describe the biological condition of the forests. One tracks the fraction of native forest species according to their relative risk of extinction. A related measure tracks changes in forest plant communities—specific plant groupings—whose area has declined by at least 70% since presettlement times. Because forests of different age structures provide different goods, services, and values, a third measure tallies the age distribution of forest trees. The remaining three indicators focus on several key forest disturbances. The first focuses on non-native plants, which can be ecologically disruptive; the second tracks acres of forest affected by fire, insects, and disease; and the third indicator will focus on fire frequency, a key determinant of forest composition, once adequate data become available.

- **What are the percentages of forest-dwelling species that are at different levels of risk of extinction?** About 9% of 1,700 native animal species that depend on forests are considered critically imperiled or imperiled, and about 1.5% of forest species may already be extinct. When "vulnerable" species are counted, a total of 20% of forest species are considered to be at risk. Interpretation of these data will be greatly enhanced when it is possible to present information on population trends for these species.

- **What percentage of the plant cover in forests is not native to the region?** Non-native plants can crowd out native plants and may provide poorer quality habitat for wildlife. Data are not adequate for national reporting on the amount of non-native cover in forests.
- **How much of the nation's forests is young, middle-aged, or old?** Data are currently available only for timberlands, a USDA Forest Service designation for areas with trees that grow fast enough to support timber harvests and on which harvest is not prohibited by law. Sixty-five percent of eastern timberlands are less than 60 years old, and 90% are less than 100 years old. About 35% of western timberlands are more than 100 years old. No trend data are available. Forest age is affected by historical and management factors, as well as by the difference in life spans of different species.
- **How many acres are affected each year by fires, insects, and tree disease?** Since 1980, between 2 million and 7 million acres were burned by wildfire per year, down from a high of 52 million acres in 1930 (note that these figures include some grassland and shrubland fire acreage). Insect damage decreased overall from 1979 to1999, but there are dramatic year-to-year variations (over these two decades, damage ranged from 8 million acres to 46 million acres).
- **Are forest fires burning much more or less frequently than in presettlement times?** The frequency with which forests burn is an important factor in shaping the composition of the forest. Data are not adequate for national reporting on this indicator.
- **How much area is occupied by forest types that have declined in area significantly since presettlement times?** Are these forest types increasing or decreasing in area now? Many forest community types now occupy a small fraction of their former area. Data are not adequate for national reporting on this indicator.

Human Use

Two of the indicators of human use of forests focus on timber: the first tracks timber harvest and the products into which it is made (for example, sawlogs or pulpwood). The second reports whether each year's harvest is greater or less than that year's growth. A third measure focuses on recreational use of forests.

- **How much timber is harvested each year, and what is it used for?** Nationally, timber harvest grew by 40% from 1952 to 1996. There was slow, steady growth through 1976, followed by a sharp increase from 1976 to 1986, and a subsequent decline. Pulpwood and sawlogs account for more than half of all harvest; other uses include fuelwood and veneer logs.
- **How much timber grows each year, compared to the amount that is cut?** Growth exceeds harvest on both public and private timberlands in the East and West; this has been true for most of the past 50 years. In 1997, growth was higher than it was in the 1950s on all categories of land, although growth on eastern forest lands (both public and private) was lower than it was at its peak in the 1970s. Nationally, almost 90% of harvest occurs on private lands.
- **How much recreational activity takes place in the nation's forests?** People use forest lands for hunting, fishing, hiking, skiing, and many other recreational activities. Data are not adequate for national reporting on this indicator.

What do we mean by "forests"?

Many of the data reported here are based on the USDA Forest Service definition of forest: any lands at least 10% covered by trees of any size, at least one acre in extent. This includes both heavily treed areas and areas where trees are intermingled with other cover, such as the chaparral and pinyon–juniper areas of the Southwest. This definition includes both naturally regenerating forests and areas planted for future harvest (plantations or "tree farms")—that is, areas that may not have mature trees now, but that will in the future, are classified as forest.

This definition overlaps to some degree with the definition of "grasslands and shrublands" in this report, largely because some areas share characteristics of both forests and grasslands and shrublands. For example, the Forest Service classifies pinyon–juniper and chaparral as forest; in this report, these lands are considered "grasslands and shrublands" as well.

Other approaches to defining and delineating forests, which did not meet our needs, include restricting the definition to "closed canopy" areas—this would eliminate consideration of woodland areas with less complete tree cover, such as chaparral and pinyon–juniper—and excluding areas that are stocked with nursery seedlings for later harvest.

Forest Service estimates reported here are part of an extensive monitoring program that provides information on many aspects of forest extent, use, and condition, and the program's breadth of coverage, historical trends, and internal consistency are quite useful. However, since this program does not produce comparable information about other ecosystem types (grasslands and shrublands, farmlands, etc.), the data cannot be used for reporting on ecosystem extent and change nationwide.

One method that does produce consistent nationwide estimates of ecosystem extent uses satellite remote sensing information (see Map 4.2, p. 40). For forests, the remote sensing method produces estimates that are about 55 million acres (9%) lower than the Forest Service estimates reported in this chapter. Differences between the methods include the scale of measurement (the satellite data include areas as small as about 100 feet on a side, or just over one-fifth of an acre) and the fact that the Forest Service approach considers as forest any areas that *will* become or return to forest cover—including areas on which timber harvest has occurred and that are either replanted or are being reseeded naturally, even if they are currently covered with grass, shrubs, or other nonwoody vegetation.

Map 7.1. Eastern and Western Regions, Used for the Majority of Indicators in this Chapter

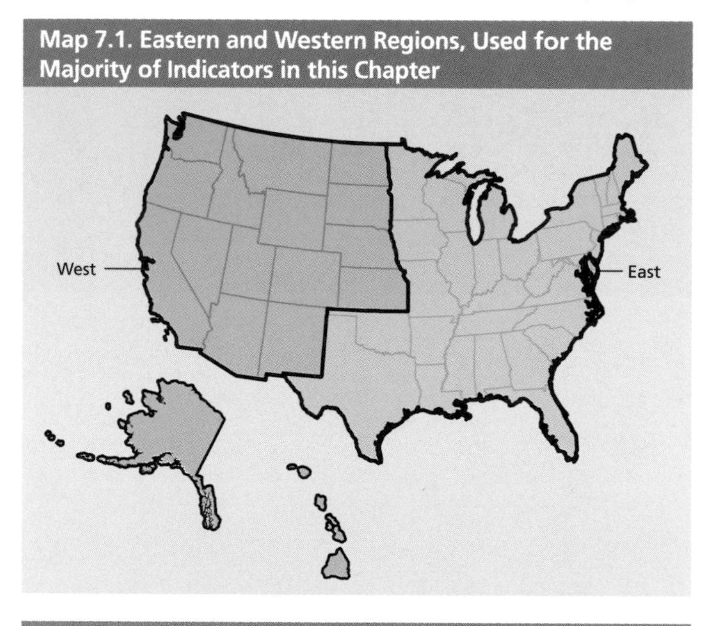

Map 7.2. Regional Boundaries for the At-Risk Species Indicator, p. 124

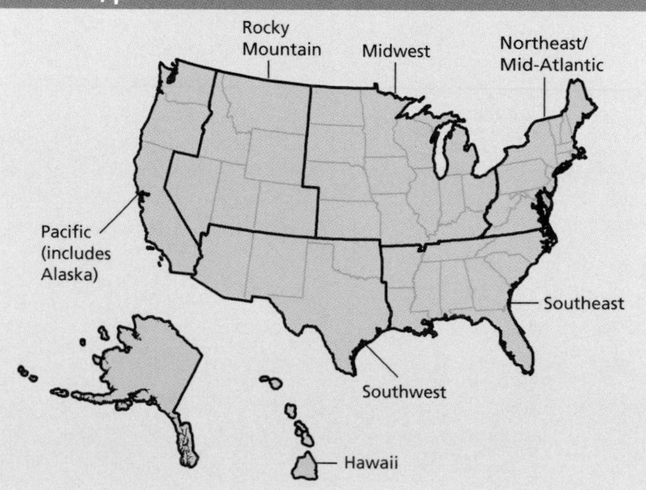

A Note about Regions

The eastern and western regions used to present data on eight of the indicators in this chapter (Map 7.1) are aggregates of the USDA Forest Service regional structure and mirror the distribution of forest lands in the United States, which is interrupted by the major expanse of farmland and grassland and shrubland that occupies the Midwest and the Great Plains. This approach was selected to enable the report to focus on very broad regional trends. Data from Alaska, when available, are included in the western region.

One indicator (At-Risk Species) is presented on the basis of seven regions (see Map 7.2), consistent with the data presented in other at-risk species indicators (pp. 52, 144, and 168).

SYSTEM DIMENSIONS	CHEMICAL AND PHYSICAL	BIOLOGICAL COMPONENTS	HUMAN USES
Extent	Nutrients, Carbon, Oxygen	Plants and Animals	Food, Fiber, and Water
Pattern	Contaminants	Communities	Recreation and Other Services
	Physical	Ecological Productivity	

● Forest Area and Ownership

What Is This Indicator, and Why Is It Important? This indicator reports how much forest land there is in the United States and who owns it.

Knowing how much land is forested and who owns that land is vital to making informed decisions about forests. Gains and losses in forest area directly affect the public's continued enjoyment of the goods and services that forests provide—recreation, lumber, watershed protection, and many other things. Public and private owners often have very different goals and assumptions, differences that are reflected in management priorities and practices.

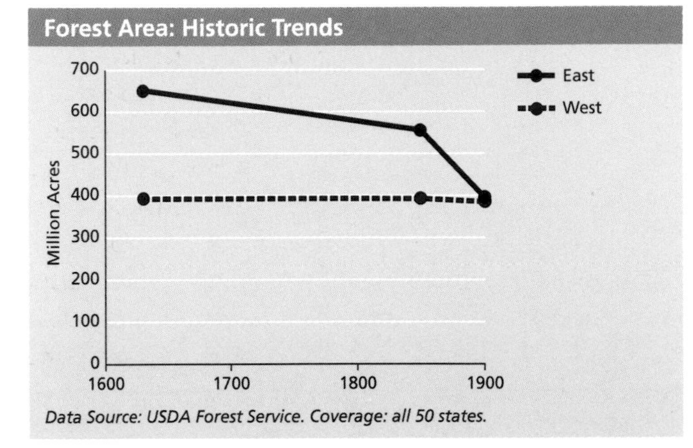

Data Source: USDA Forest Service. Coverage: all 50 states.

What Do the Data Show? Forests today cover about 747 million acres, or about a third of the total land area of the United States, down from about 1 billion acres at the time of European settlement. Most forest clearing occurred in the East, ending by 1900. In recent decades, the amount of forest land has been nearly stable, with an increase of about 1%, or 8 million acres, from 1987 to 1997.

There are striking regional differences in patterns of ownership: in the East, more than 80% of forest land is privately owned, while in the West, about two-thirds is publicly owned. Forest industry ownership accounts for about 13% of eastern forest land and 4% of western forest land; a wide variety of individuals and corporations own the rest.

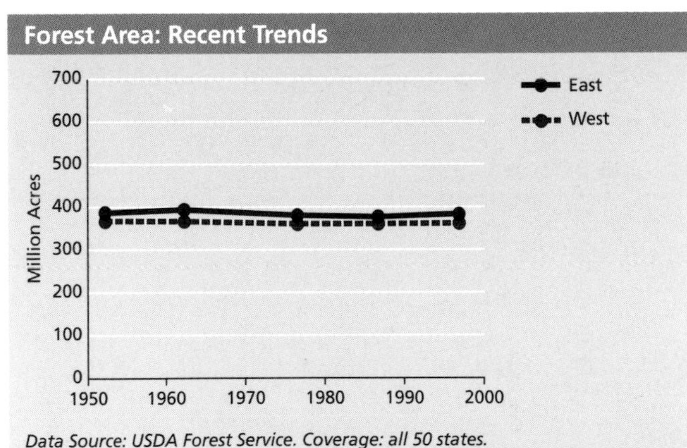

Data Source: USDA Forest Service. Coverage: all 50 states.

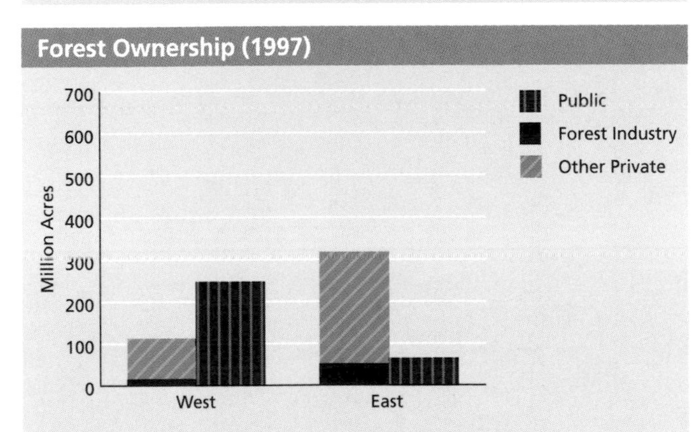

Data Source: USDA Forest Service. Coverage: all 50 states.

The technical note for this indicator is on page 239.

SYSTEM DIMENSIONS	CHEMICAL AND PHYSICAL	BIOLOGICAL COMPONENTS	HUMAN USES
Extent	Nutrients, Carbon, Oxygen	Plants and Animals	Food, Fiber, and Water
Pattern	Contaminants	Communities	Recreation and Other Services
	Physical	Ecological Productivity	

Forest Types

Western Forest Cover Types Increasing in Area

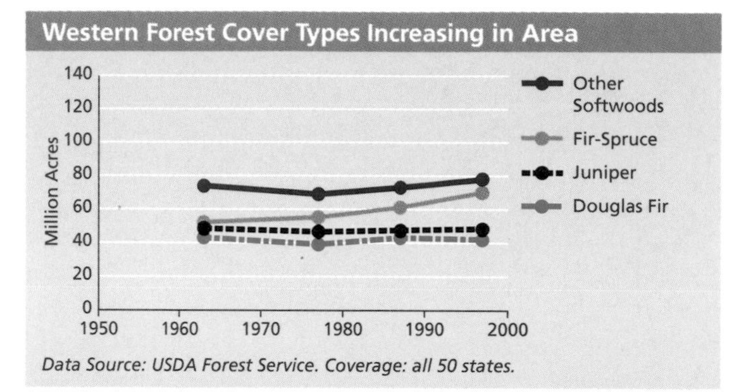

Data Source: USDA Forest Service. Coverage: all 50 states.

Western Forest Cover Types Decreasing in Area

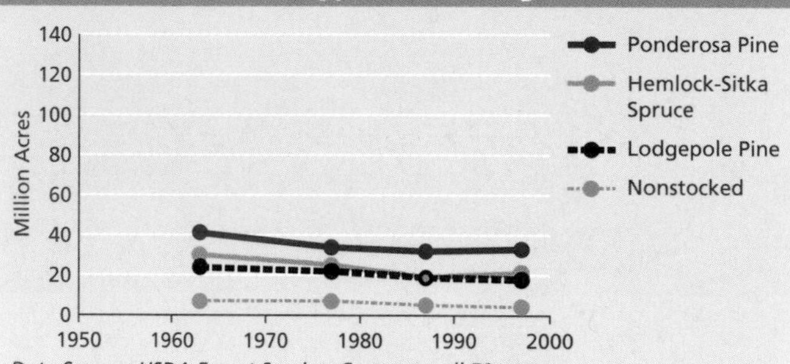

Data Source: USDA Forest Service. Coverage: all 50 states.

Eastern Forest Cover Types Increasing in Area

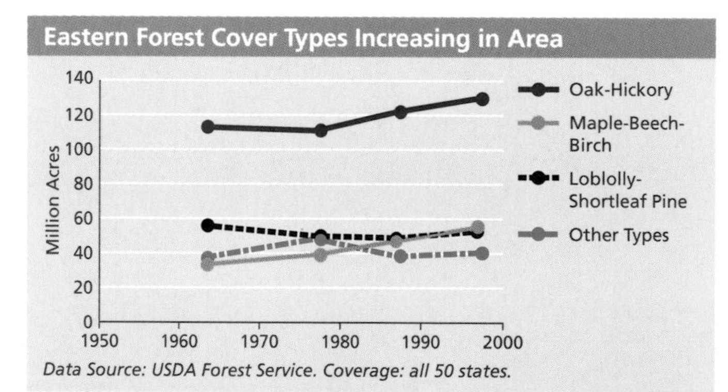

Data Source: USDA Forest Service. Coverage: all 50 states.

Eastern Forest Cover Types Decreasing in Area

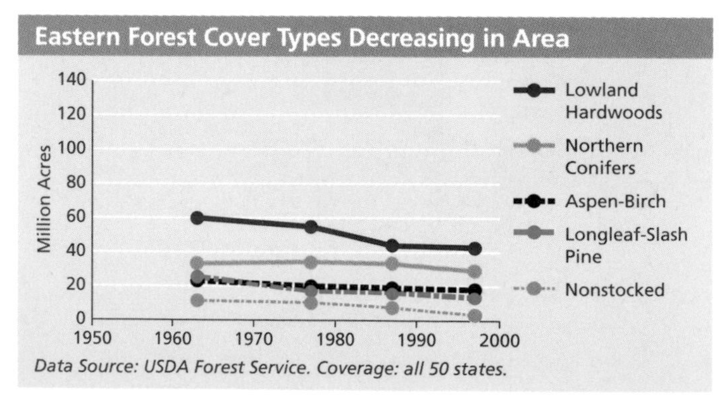

Data Source: USDA Forest Service. Coverage: all 50 states.

What Is This Indicator, and Why Is It Important? This indicator reports the acreage of a variety of forest "cover types." Cover types describe the dominant species of trees found in the forests (e.g., oak–hickory forests are dominated by oaks and hickories, but include other kinds of trees as well).

Forest type may change as a result of direct human intervention (fire suppression, planting and harvesting, development, and grazing) or because of natural succession. Changes in climate may also affect the range of different forest types.

Different plants and animals live in different types of forests. In addition, the types of forest available influence the way people use them for recreation and other purposes.

What Do the Data Show? From 1963 to 1997, oak–hickory and maple–beech–birch in the East and fir–spruce in the West increased the most (by 18 million, 22 million, and 18 million acres, respectively).

In the East, longleaf–slash pine and lowland hardwoods (elm–ash–cottonwood and oak–gum–cypress) had the largest decreases in acreage (12 million and 17 million acres, respectively). In the West, hemlock–sitka spruce, ponderosa pine, and lodgepole pine decreased the most (by 9 million, 8 million, and 6 million acres, respectively).

In both regions, "nonstocked" land (land where trees have been cut but that has not yet regrown as forest) has declined steadily.

It is important to note that total forest area changed very little over this period. In general, the increases or reductions described here represent shifts from one forest type to another.

The technical note for this indicator is on page 240.

SYSTEM DIMENSIONS	CHEMICAL AND PHYSICAL	BIOLOGICAL COMPONENTS	HUMAN USES
Extent	Nutrients, Carbon, Oxygen	Plants and Animals	Food, Fiber, and Water
Pattern	Contaminants	Communities	Recreation and Other Services
	Physical	Ecological Productivity	

● Forest Management Categories

What Is This Indicator, and Why Is It Important? This indicator reports the percentage of forest area in several different management categories. These range from "reserved lands" (forests in national parks, wilderness areas, and other similar areas) to forests under intensive management involving replanting after harvest. Other forest lands are subject to a wide variety of both management practices and restrictions on use.

How a forest is managed influences the goods and services that it provides. Heavily managed areas produce fiber and other wood products, while the value of reserved areas may lie in the solitude they offer, the rare plants and animals they shelter, or the watersheds they protect.

What Do the Data Show? In 1997, 10% of eastern forests and 4% of western forests were in intensively managed plantations (planted timberlands). Eleven percent of western forests and 3% of eastern forests were in reserved forest lands (federally designated wilderness areas or national parks). (Data are not currently available to support national reporting on reserved lands in private or other public ownership.) Nationwide (East plus West), reserved forest land doubled between 1953 and 1997, while planted timberland increased tenfold.

Other forest lands receive less-intensive management activity, which may include periodic timber harvest. Nineteen percent of forests in the West grow too slowly to support timber harvest under current economic conditions; these forests are identified as "other natural or semi-natural forest lands" in the graph above.

The technical note for this indicator is on page 240.

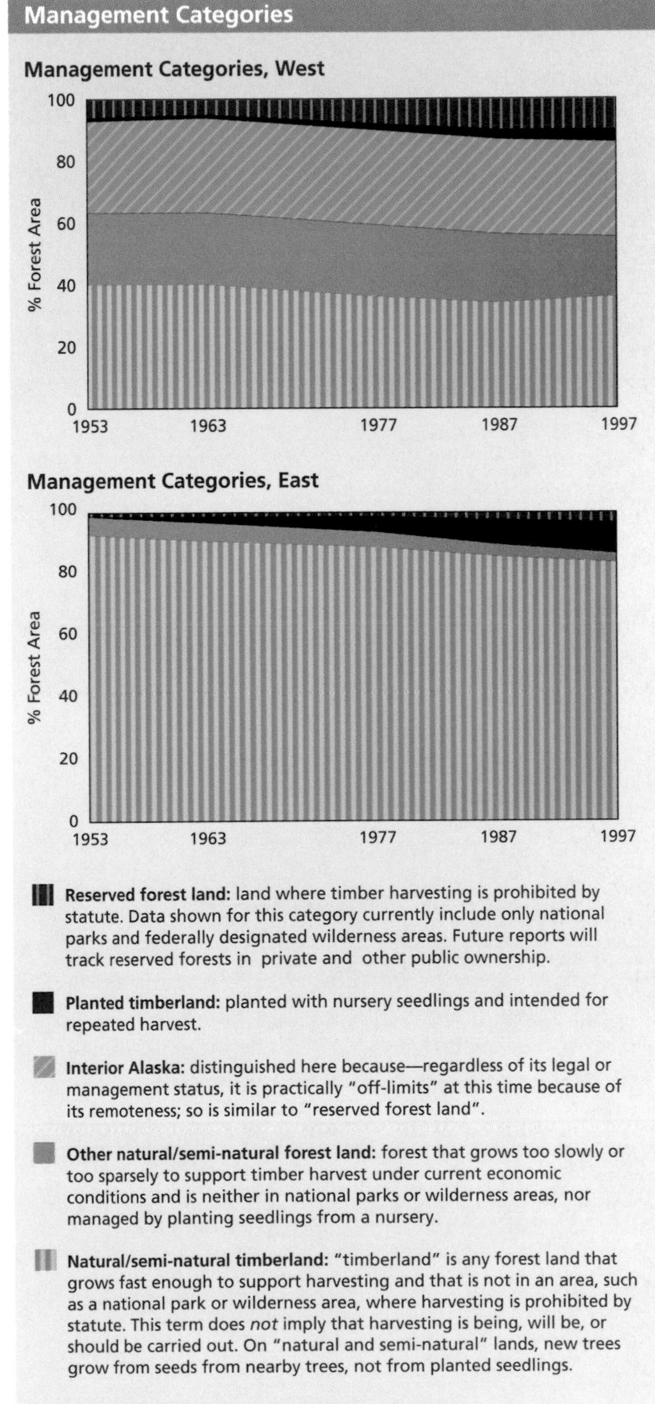

SYSTEM DIMENSIONS	CHEMICAL AND PHYSICAL	BIOLOGICAL COMPONENTS	HUMAN USES
Extent	Nutrients, Carbon, Oxygen	Plants and Animals	Food, Fiber, and Water
Pattern	Contaminants	Communities	Recreation and Other Services
	Physical	Ecological Productivity	

◐ Forest Pattern and Fragmentation

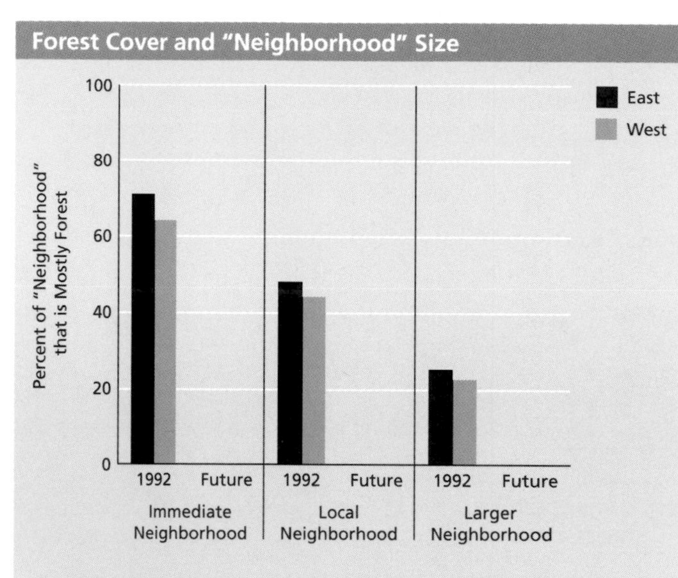

Immediate neighborhood: land within a radius of about 250 ft from each forest point

Local neighborhood: land within a radius of about 1/4 mile from each forest point

Larger neighborhood: land within a radius of about 21/2 miles from each forest point

Mostly forest: land that is at least 90% forested (less than 10% nonforest)

Data Source: Multi-Resolution Land Characterization Consortium; USDA Forest Service. Coverage: lower 48 states.

What Is This Indicator, and Why Is It Important? Imagine that it was possible to measure, for each tree in the nation's forests, whether that tree was surrounded by more-or-less complete forest cover, or whether its "neighborhood" contained a substantial amount of nonforest cover (agriculture, developed areas, recently cleared land, roads, railroads, powerline rights-of way, etc.). Some trees have only small forested neighborhoods, while others are surrounded by larger forested regions.

This indicator describes a tree's forest neighborhood according to the degree of forest cover within various distances. Thus, the "immediate neighborhood" of a particular tree is everything within about 250 feet in all directions. This "immediate neighborhood" is "mostly forest" if the land is at least 90% forested. A tree's "local neighborhood" extends about ¼ mile in all directions, and its "larger neighborhood" extends about 2½ miles. This analysis relies upon computer analyses of satellite data on millions of individual forest points. While these points (called "pixels") are not individual trees—they are squares about 100 feet on a side—they serve much the same purpose.

"Forest fragmentation" describes the degree to which forested areas are being broken into smaller patches and interspersed with nonforest areas. Research has shown that forest close to nonforest cover is often warmer and drier, more likely to be affected by wind, and more likely to be invaded by non-native species. In addition, forest animals that live near developed areas, farmlands, or roads are more likely to be affected by collisions with cars, increased hunting pressure, noise, lights, predation by cats and dogs, etc.

These effects may be felt at different distances from the nonforest edge. In addition, some species are quite sensitive to these effects, while others are less affected. Because these variations in both effect and response by species mean that there is no single distance threshold for the extent of such effects, this indicator presents a range of different neighborhood sizes.

What Do the Data Show? About two-thirds of all points in both eastern and western forests have land cover that is mostly forest—that is, 10% or less of the area is nonforest—within their immediate neighborhood (roughly 250-foot radius). About a quarter of all forest points are surrounded by larger (roughly 2½-mile radius) neighborhoods that are "mostly forest." Tracking this indicator through time is important, because it will help distinguish between natural forest patterns and changes caused by human activity.

SYSTEM DIMENSIONS	CHEMICAL AND PHYSICAL	BIOLOGICAL COMPONENTS	HUMAN USES
Extent	Nutrients, Carbon, Oxygen	Plants and Animals	Food, Fiber, and Water
Pattern	Contaminants	Communities	Recreation and Other Services
	Physical	Ecological Productivity	

◐ Forest Pattern and Fragmentation *(continued)*

Discussion This analysis does not distinguish between fragmentation caused by human activity (development, agriculture, etc.) and natural patchworks of forest and nonforest cover. Many western forests in particular are characterized by natural intermingling of forest and grasslands or shrublands.

The satellite remote sensing data presented here can generally identify forest features that are at least 10,000 feet square, or about 100 feet on a side. Thus, features that are narrower than about 100 feet—for example, some roads, powerlines, residential development within otherwise-wooded areas, or other small nonforest land uses—are missing. (Somewhat larger features may also be missed if they are split between multiple pixels.) Future analyses could include these smaller features by using satellite data that can discern smaller nonforest areas, or using ancillary information, such as mapped databases showing the location of these smaller features.

This analysis treats all nonforest land uses similarly, whether they are clumped together, spread evenly across a landscape, or strung together in a line (e.g., as a road or powerline). Different types of breaks in forest cover may affect forests in different ways—concentrations of nonforest cover may have major impacts on local habitat suitability, while linear features such as roads can serve as barriers to species movement. Future analyses might weight some nonforest areas or patterns more than others.

The "neighborhoods" used in this analysis are intended to provide a perspective on forest pattern, not to represent the habitat needs of particular species.

The technical note for this indicator is on page 240.

● Nitrate in Streams

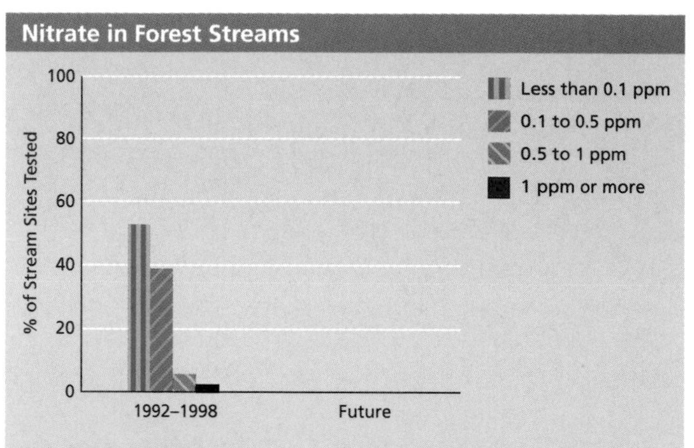

Data Source: USGS National Water Quality Assessment. Coverage: lower 48 states. Each sampling area was sampled intensively for approximately 2 years during 1992–1998.

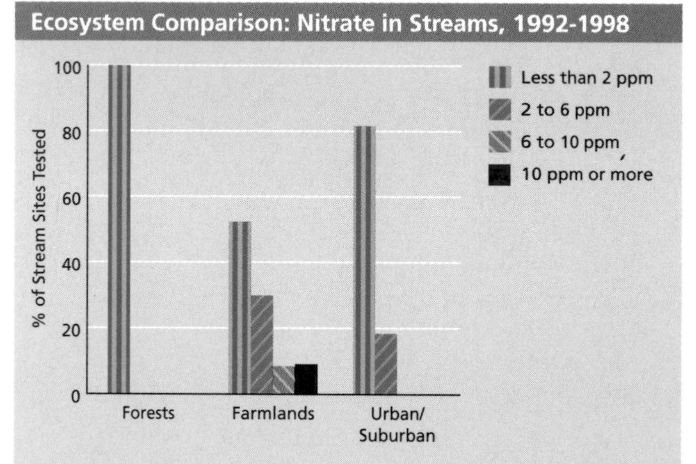

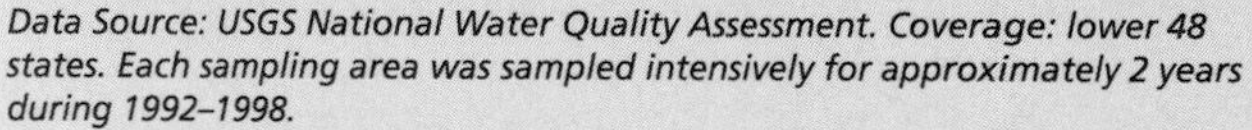
Data Source: USGS National Water Quality Assessment. Coverage: lower 48 states. Each sampling area was sampled intensively for approximately 2 years during 1992–1998.

What Is This Indicator, and Why Is It Important? This indicator reports on the concentration of nitrate in representative streams in forested areas. Specifically, the indicator reports the percentage of streams with average nitrate concentrations in one of four ranges, for streams draining watersheds that are primarily forested.

Nitrate is a naturally occurring form of nitrogen and an important plant nutrient; it is often the most abundant of the forms of nitrogen that are readily usable by plants, including algae. Increased nitrate in streams that ultimately empty into coastal waters can lead to algal blooms in those waters, which can decrease recreational and aesthetic values and help deplete oxygen needed by fish and other animals (see the national nitrogen indicator and the hypoxia indicator, pp. 46 and 71). Elevated nitrate in drinking water can also cause human health problems.

Elevated amounts of nitrate in streams are a sign that inputs from human sources have increased or that that plants in the system are under stress. Nitrogen is a critical plant nutrient, and most nitrogen, including nitrate, is used and reused by plants within an ecosystem. Thus, in undisturbed forested ecosystems, there is relatively little "leakage" into either surface runoff or groundwater, and concentrations are very low. Elevated stream nitrate might come from land clearing, the use of fertilizer in the watershed, or from rain and snowfall (in the form of acid rain).

What Do the Data Show? Almost all forest stream sites (97%) had nitrate concentrations below 1 part per million (ppm), more than three-fourths had concentrations of less than 0.5 ppm, and more than half had concentrations of less than 0.1 ppm.

Most streams in urban/suburban areas also have low average nitrate concentrations (less than 2 ppm), while farmland streams have the highest nitrate concentrations (see pp. 95 and 186). There is also a core national indicator for nitrogen (p. 46).

The federal drinking water standard for the protection of human health is 10 ppm of nitrate, which is exceeded only in agricultural areas (see p. 95).

The technical note for this indicator is on page 232.

SYSTEM DIMENSIONS	CHEMICAL AND PHYSICAL	BIOLOGICAL COMPONENTS	HUMAN USES
Extent	**Nutrients, Carbon, Oxygen**	Plants and Animals	Food, Fiber, and Water
Pattern	Contaminants	Communities	Recreation and Other Services
	Physical	Ecological Productivity	

◐ Carbon Storage

What Is This Indicator, and Why Is It Important? This indicator reports how much carbon—an essential component of all organisms—is stored in forests, including trees, soil, and plant litter on the forest floor, and in wood products.

Carbon storage has become important in international negotiations on the management of greenhouse gas emissions, because increased carbon storage can be useful in offsetting emissions of carbon from fossil fuel burning and other sources. The amount of carbon stored in forests can change through the adoption of forest management practices that allow the incorporation of more plant materials into forest soils, changes in age structure (see Forest Age, p. 126), and increases in the extent of forested areas (see Forest Area and Ownership, p. 117).

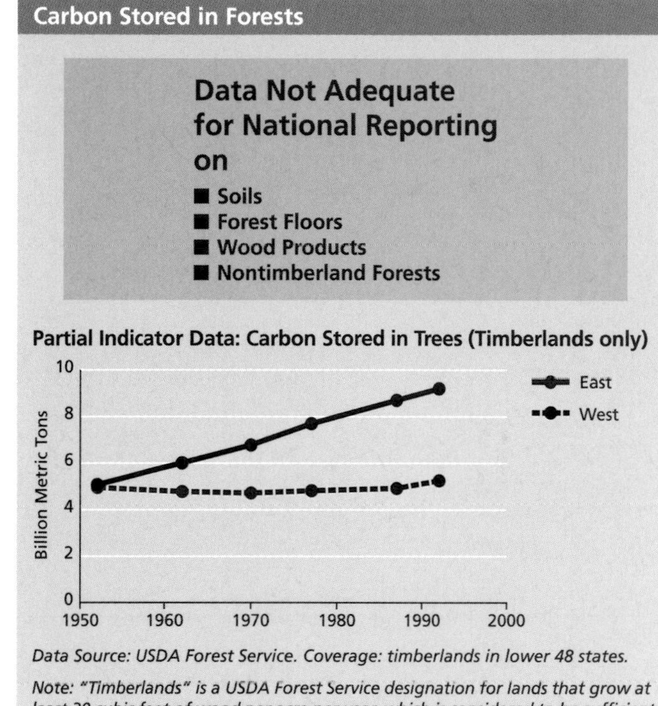

Data Source: USDA Forest Service. Coverage: timberlands in lower 48 states.

Note: "Timberlands" is a USDA Forest Service designation for lands that grow at least 20 cubic feet of wood per acre per year, which is considered to be sufficient to support commercial harvest under current economic conditions. Lands on which harvest is prohibited by statute are not included as "timberlands."

What Do the Data Show? The amount of carbon stored in trees in the East increased by 80% from1950 to 1992, despite relatively modest changes in forest area. This increase has been attributed to growth on farmlands allowed to revert to forests, maturing of second-growth forests, and to increased growth in some southeastern forests. In western forests, the addition of new carbon through forest growth was offset by harvest, resulting in little change in the overall amount stored. Note that, unlike many other forest indicators, these data do not reflect changes that occurred after 1992.

Why Can't This Entire Indicator Be Reported at This Time? More data are needed to report on the amount of carbon stored in forest soils, in leaf litter and other decomposing matter on the ground in forests, and in forest products in use or slowly decomposing in landfills.

In addition, available data are limited to timberlands, but data collection will be more comprehensive in future.

The technical note for this indicator is on page 241.

◐ At-Risk Native Forest Species

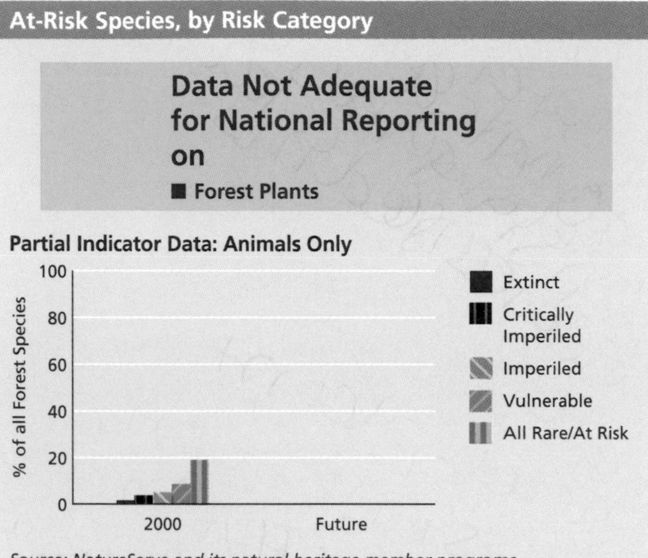

Source: NatureServe and its natural heritage member programs.
Coverage: all 50 states.

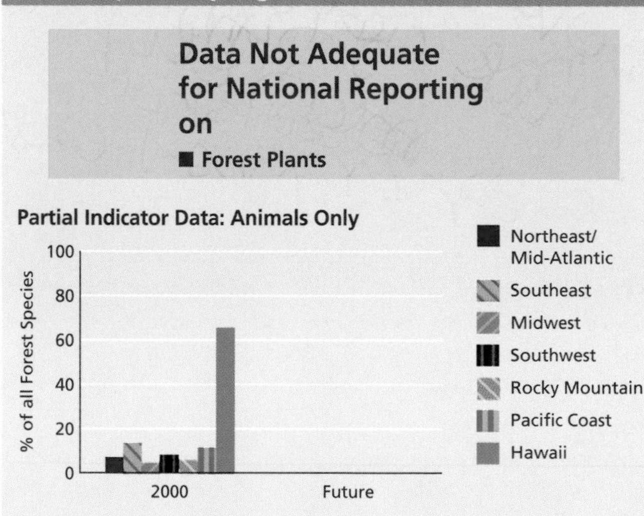

Source: NatureServe and its natural heritage member programs.
Coverage: all 50 states.

What Is This Indicator, and Why Is It Important? This indicator reports on the relative risk of extinction of native forest species. The risk categories are based on such factors as the number and condition of individuals and populations, the area occupied by the species, population trends, and known threats. Degrees of risk reported here range from very high ("critically imperiled" species are often found in five or fewer places or have experienced very steep declines) to moderate ("vulnerable" species are often found in fewer than 80 places or have recently experienced widespread declines). In all cases, a wide variety of factors contribute to the overall ratings. "Forest species" live in forests during at least part of their life and depend on forest habitats for survival.

Species are valued for a variety of reasons: they provide products, including food, fiber, and, more recently, genetic materials; they are key elements of ecosystems, which themselves provide valuable goods and services; and many people value them for their intrinsic worth or beauty.

What Do the Data Show? About 3.5% of native forest animal species are critically imperiled, about 5% are imperiled, and 1.5% are or may be extinct. When vulnerable species (9%) are counted, a total of about 20% of forest animals are considered to be "at risk." Hawaii has a much larger percentage of at-risk forest species than any other region.

Interpreting these figures is complicated, however, because some species are naturally rare. Thus, the rankings are influenced by differences among regions and species groups in the number of naturally rare species, as well as by different types and levels of human activities that can cause species declines. Interpretation of these data will be greatly enhanced when information on population trends for these at-risk species becomes available.

Why Can't This Entire Indicator Be Reported? This indicator reports on mammals, birds, reptiles, amphibians, grasshoppers, and butterflies. Data on other groups have not been included either because too little is known to assign risk categories or, as with most plants, because determinations of which species are associated with forests, grasslands, or other habitats have not been completed.

See also the national at-risk species indicator (p. 52) and the indicators for at-risk coastal, freshwater, and grassland and shrubland species (pp. 75, 144, and 168), as well as the indicators for species in farmlands (p. 103) and urban and suburban areas (p. 191).

The technical note for this indicator is on page 214.

⊖ Area Covered by Non-native Plants

What Is This Indicator, and Why Is It Important? This indicator describes the degree to which non-native plants are found in U.S. forests. It will report the percentage of the total area covered by overstory (large trees that form the canopy) and understory (shrubs, ground plants, and smaller trees) that is made up of non-native plants.

Species are generally considered to be non-native if their natural range does not include North America, although there is growing recognition that species that are native to one part of the United States may cause problems if they spread to other areas. Non-native species may spread aggressively and crowd out species that are native to a region; they may also alter essential habitat of native species, by shading native plants or by consuming large quantities of water, for example.

Well-known non-native species in the East include kudzu, melaleuca, and ailanthus, while western species include eucalyptus and Russian olive. Some non-native plants were introduced accidentally; others were originally planted for landscaping (e.g., Norway maple, multiflora rose) or for purposes such as erosion control (Russian olive). In general, forests with greater coverage by non-native species are subject to higher levels of ecological disruption, which may in turn have economic consequences.

Forest Area Covered by Non-native Plants, West

% Non-native Cover
100
80
60
40
20
0
Data Not Adequate for National Reporting
Years
Understory
Overstory

Forest Area Covered by Non-native Plants, East

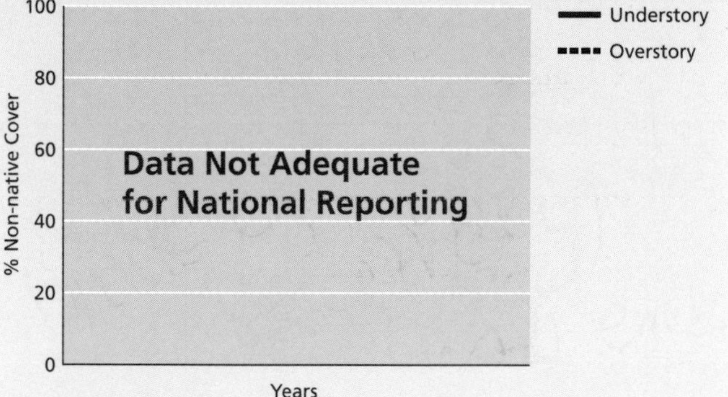

Why Can't This Indicator Be Reported at This Time? The USDA Forest Service Forest Inventory and Analysis (FIA) program is developing and testing protocols for reporting non-native plant cover. Data from this program will be included in future reports.

For other non-native species indicators, see pp. 76, 104, 145, and 169.

The technical note for this indicator is on page 242.

SYSTEM DIMENSIONS | CHEMICAL AND PHYSICAL | **BIOLOGICAL COMPONENTS** | HUMAN USES
Extent, Pattern | Nutrients, Carbon, Oxygen; Contaminants; Physical | Plants and Animals; **Communities**; Ecological Productivity | Food, Fiber, and Water; Recreation and Other Services

◐ Forest Age

Forest Age

Data Not Adequate for National Reporting on

- Forest Lands Other Than Those Classified As Timberlands

Partial Indicator Data: West (Timberlands Only)

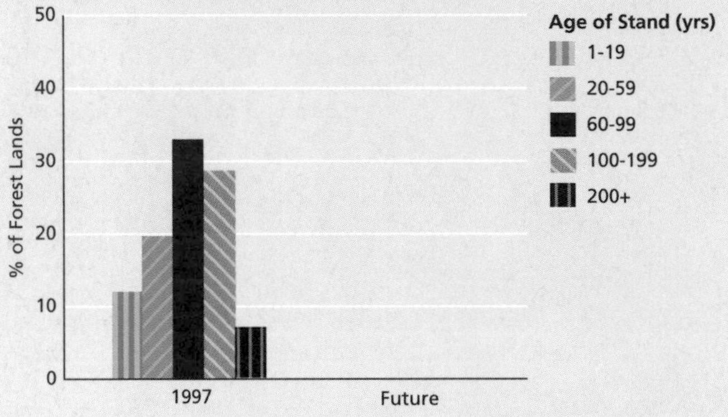

Partial Indicator Data: East (Timberlands Only)

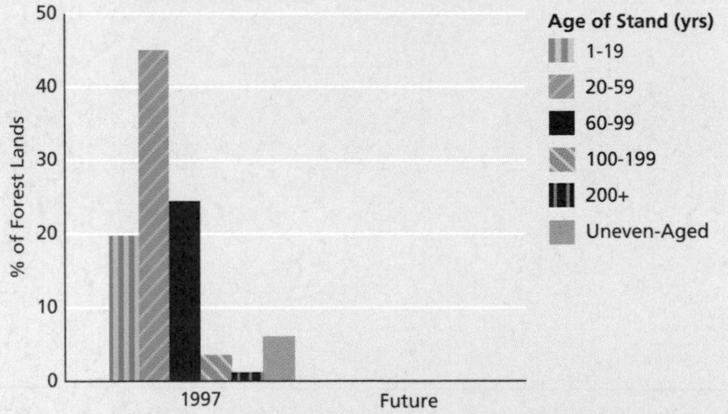

Data Source: USDA Forest Service. Coverage: all 50 states (timberlands only.)

Note: "Timberlands" is a USDA Forest Service designation for lands that grow at least 20 cubic feet of wood per acre per year, which is considered be sufficient to support commercial harvest under current economic conditions. Lands on which harvest is prohibited by statute are not included as "timberlands." Note also that the term "uneven-aged" is being phased out; such stands are composed of intermingled trees that differ considerably in age.

What Is This Indicator, and Why Is It Important? This indicator reports the percentage of forest lands with stands in several age classes. Forests of different ages often provide different goods, services, and values. For example, woodpeckers and species that need trunk cavities for nesting find older forests, with their dead trees, a suitable habitat. Younger forests, with their rapid growth and smaller trees, provide habitat for species such as the Kirtland's warbler, which can only live in forests recently regrown after fire.

What Do the Data Show? Sixty-five percent of eastern timberlands, where most of the nation's timber products are produced (see Timber Harvest, p. 130), are less than 60 years old, and 90% are less than 100 years old. Most of the nation's older timberland stands are in the West—about 35% of western timberlands are more than 100 years old, and about 30% are less than 60 years old. Although not included in this indicator, most of the nation's forests in wilderness areas and national parks, which contain many old stands, are also in the West (see Forest Management Categories, page 119).

Why Can't This Entire Indicator Be Reported at This Time? Data are currently available only for timberlands. Data on the age class of forest trees are not available for national parks and wilderness areas and other forest land not classified as timberlands. These data will be available for future reports.

Discussion Forest age structure reflects historic and current management as well as natural factors. For example, the high percentage of younger forests in the eastern United States reflects such factors as the reforestation of former agricultural land, the management of many private landholdings for commercial harvesting, and the fact that very old stands are much less common in the East.

The technical note for this indicator is on page 242.

Forest Disturbance: Fire, Insects, Disease

What Is This Indicator, and Why Is It Important? This indicator reports the acreage of forest affected each year by several important types of disturbance: forest fires, insects, and diseases of trees.

Fires, insects, and diseases are, for the most part, natural influences on forests. However, at times, such influences can exceed or otherwise differ from what might be encountered in an undisturbed ecosystem. For example, fire suppression may foster the conditions necessary for catastrophic fires, and introduced pests like gypsy moths and Dutch elm disease can devastate large areas.

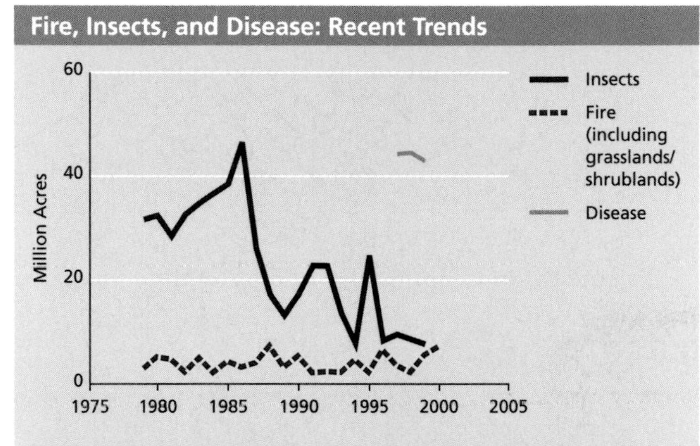

Insects: gypsy moth, spruce budworm, southern pine beetle, mountain pine beetle, western spruce budworm (all but the gypsy moth are native to the United States.)

Diseases: fusiform rust, dwarf mistletoe

Data Source: USDA Forest Service Forest Health Protection/Forest Health Monitoring Program (insects, disease), National Forest System (fire); note that these data are not limited to national forests. Coverage: all 50 states

What Do the Data Show? Fewer acres have been burned by wildfire in recent decades than in the first half of the 20th century. Since 1980, between 2 million and 7 million acres burned per year, down from a high of 52 million acres in 1930; the decline is largely due to fire suppression policies and practices. Note that the data reported here describe all wildfires, including both forest fires and grassland/shrubland fires. Although nationwide data do not show an increase in recent decades, data (not shown here) from national forests, which are mostly in the West, do show a significant increase.

Insect damage varies dramatically from year to year. Five insect species together affected between 8 million and 46 million acres per year from 1979 to 1999, with a clear trend toward fewer acres over that time. Many insect populations go through major cycles of year-to-year variation. For example, much of the variation over the past 20 years results from such cycles for gypsy moth and southern pine beetle.

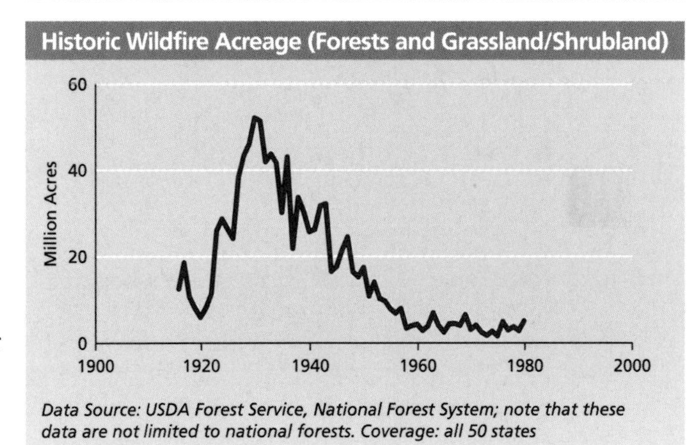

Data Source: USDA Forest Service, National Forest System; note that these data are not limited to national forests. Coverage: all 50 states

In recent years, 43–44 million acres have been affected by two major diseases/parasites (fusiform rust and dwarf mistletoe).

Discussion It would be desirable to be able to report on acreage affected by forest fires (as distinct from other wildfires), on the acreage subject to different levels of fire intensity, and on the acreage of prescribed fire (fires that are intentionally set as a management tool). In addition, data on the acreage affected by other diseases is not available. Finally, some non-native insects, such as the hemlock woolly adelgid, which affects half of all eastern hemlock forests, may spread widely before it causes damage that is apparent from aerial surveys.

See also Fire Frequency (p. 128).

The technical note for this indicator is on page 242.

⊖ Fire Frequency

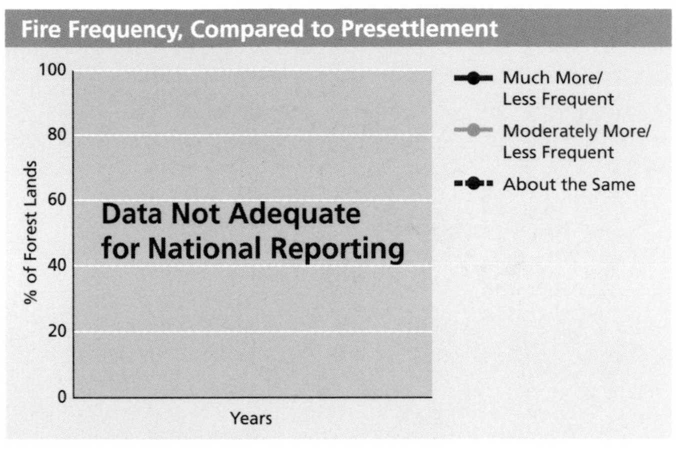

What Is This Indicator, and Why Is It Important? This indicator describes the frequency with which forests are burned by wildfire. It would report the fraction of forest lands that experience wildfire much more or less frequently, moderately more or less frequently, or with about the same frequency as in presettlement times. Thus, a forest that, historically, burned every 50 years on average will be considered moderately altered if it burns every 100 years, and significantly altered if it burns only every 150 years, and about the same if it burns once every 50 years.

Fire has always been an important influence on most forest types—indeed, it is necessary for the maintenance of some forest types—and it will continue to be important in the future. Periodic fires shape forest composition by allowing certain fire-adapted species to thrive while removing other, less tolerant, trees. For most of the past 10,000 years (since the last Ice Age), most forests in the lower 48 states burned regularly, with fires started by lightning or by American Indians, who used fire to manage forests and grasslands. There is increasing interest in forest management practices that incorporate fire and other disturbances in ways that mimic historic patterns.

Why Can't This Indicator Be Reported at This Time? This indicator requires information on both the historic and current fire frequency. While current fire frequency data are not difficult to collect, it is not simple to determine the historic fire frequency of an area or forest type. Researchers have estimated historic fire frequencies, but at this time, fire frequency data has been measured (from tree ring scars and similar evidence) at only a few sites.

Discussion Active suppression of forest fires dramatically changes forest composition, structure, and ecology. In suppressed areas, there are often more trees per acre and a higher frequency of certain species whose spread was formerly controlled by fire. In the East, for example, red maple has increased in eastern oak and pine forests, and in the West, white fir and incense cedar are now more common in ponderosa pine and giant sequoia forests. In some forests, like ponderosa pine, the denser forests produced by fire suppression are subject to hotter fires, which kill more trees. In other areas, such as eastern oak forests, fire suppression favors trees like maples, birches, and beech, with a corresponding decrease in both flammability and the number of oaks.

See page 171 for an indicator of fire frequency in grasslands and shrublands.

The technical note for this indicator is on page 243.

⊖ Forest Community Types with Significantly Reduced Area

What Is This Indicator, and Why Is It Important? This indicator would report whether those forest community types that cover significantly fewer acres than they did in presettlement times are increasing or decreasing in area, and by how much. It would also report the total area occupied by these much-reduced forest community types—those that have been reduced by 70% or more in area.

Forest community types, such as Virginia pine–oak, American beech–southern magnolia, Douglas fir, and longleaf pine–oak, are characterized by certain plant and animal species that depend on the particular habitat provided by that forest type. When the area occupied by a forest community declines, populations of animals and plants that are highly dependent upon that community type may also decrease.

Some forest community types occupy much less area than they did at the time of European settlement. For example, redwood forest, which occupied an estimated 2.19 million acres before European settlement, now occupies 1.32 million acres, a decline of 40%. Similarly, Great Lakes pine forest, which occupied an estimated 18 million acres before European settlement, now occupies 4.1 million acres, a decline of 77%, and oak savanna, which covered about 30 million acres of the Midwest at the time of European settlement, covered only about 7000 acres, or about 0.02% of its historic area, in 1985.

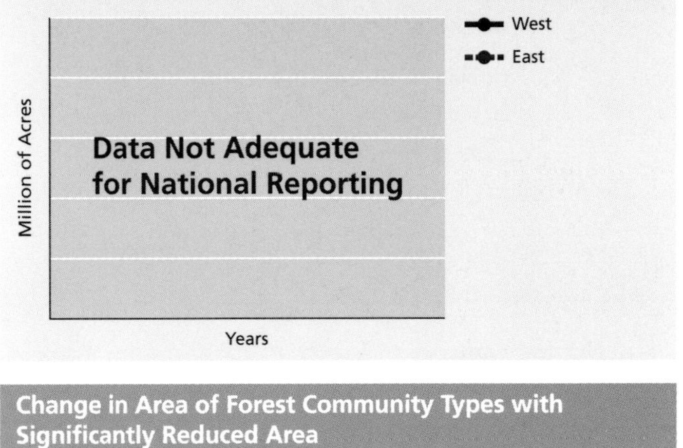

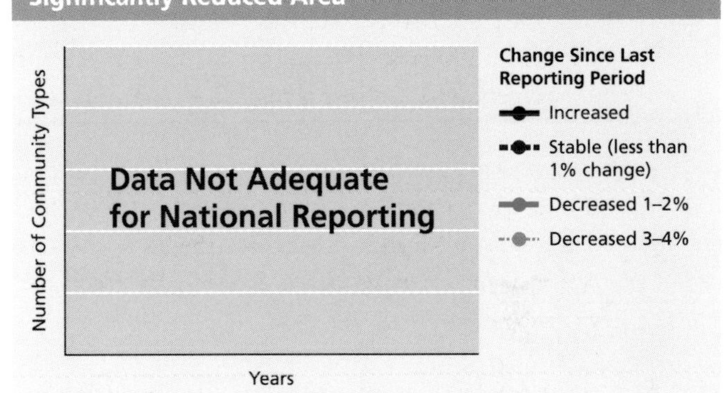

These declines may result from outright conversion, such as the clearing of forests for agriculture, or they may result from less direct changes: when fire is suppressed for long periods, different species thrive, creating a different community type.

Why Can't This Indicator Be Reported at This Time? Data on historic and current area of many forest types are not available. Methods are being developed to obtain estimates of current area from existing USDA Forest Service data. It is also possible to estimate historic area, but this has not been done on a comprehensive basis.

The technical note for this indicator is on page 243.

SYSTEM DIMENSIONS	CHEMICAL AND PHYSICAL	BIOLOGICAL COMPONENTS	**HUMAN USES**
Extent Pattern	Nutrients, Carbon, Oxygen Contaminants Physical	Plants and Animals Communities Ecological Productivity	**Food, Fiber, and Water** Recreation and Other Services

Timber Harvest

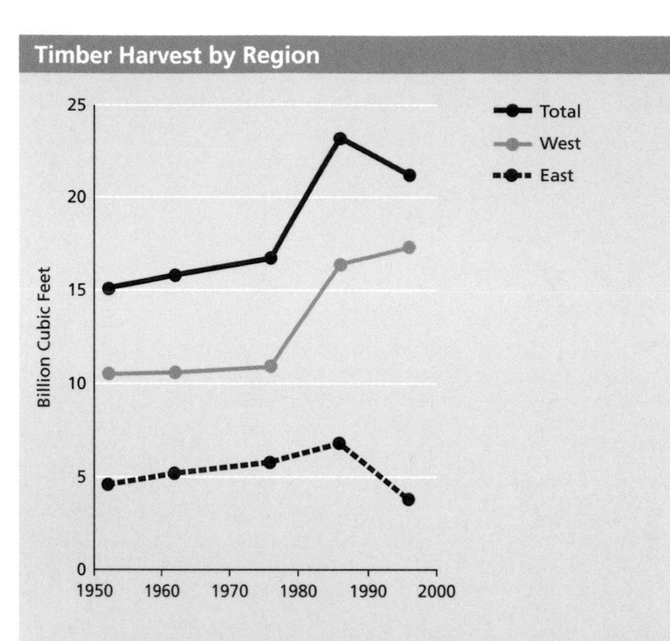

Data Source: USDA Forest Service. Coverage: all 50 states.

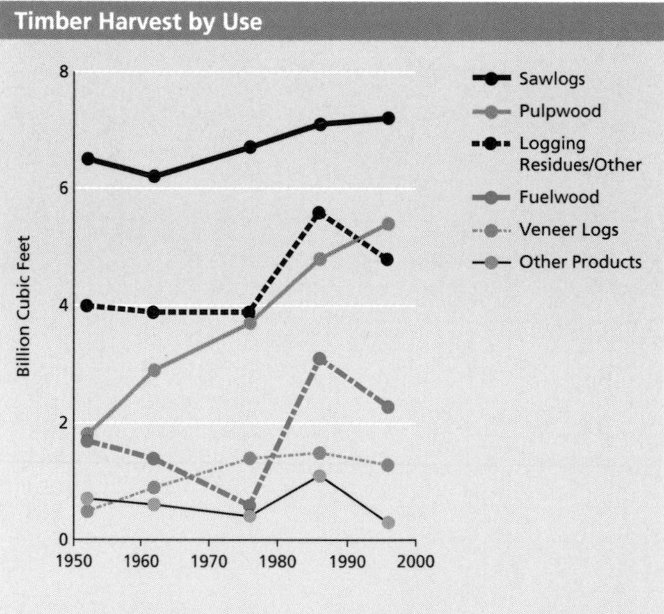

Data Source: USDA Forest Service. Coverage: all 50 states.

What Is This Indicator, and Why Is It Important? This indicator reports trends in timber harvest, by region and by primary product category (sawlogs, pulpwood, etc.)

The production of wood products provides employment, generates economic benefits, and meets society's needs for wood, paper and other products. Demand for these products drives harvesting and other forest management activities.

What Do the Data Show? Nationally, timber harvest increased by about 40% from 1952 to 1996. There was slow, steady growth through 1976, followed by a sharp increase from 1976 to 1986, and a subsequent decline. After 1986, harvest continued to rise in the East, but this increase was more than offset by decreases in harvest in the West.

Pulpwood production tripled from 1952 to 1996, increasing to 25% of total harvest (pulpwood is used for paper and similar products). One-third of the total harvest is used to produce sawlogs; this fraction is down slightly from 1952, despite a 20% increase in harvest for this purpose. Harvest for all uses other than pulpwood and sawlogs declined in 1996 compared to 1986.

See Growth and Harvest (opposite page) for a discussion of harvest trends on public and private lands.

The technical note for this indicator is on page 244.

SYSTEM DIMENSIONS	CHEMICAL AND PHYSICAL	BIOLOGICAL COMPONENTS	HUMAN USES
Extent	Nutrients, Carbon, Oxygen	Plants and Animals	Food, Fiber, and Water
Pattern	Contaminants	Communities	Recreation and Other Services
	Physical	Ecological Productivity	

Timber Growth and Harvest

What Is This Indicator, and Why Is It Important? This indicator reports the amount of new wood grown and the amount of wood harvested each year on public and private timberlands, by region. The balance between growth and harvest tells us whether the amount of wood potentially available for harvest is increasing or decreasing.

What Do the Data Show? Growth exceeds harvest on both public and private timberlands in both the East and West; with the exception of private timberlands in the West, this has been true for the past 50 years. This difference is increasing in the West (particularly on public lands) and decreasing in the East.

Growth is higher on public and private lands in both East and West than it was in the 1950s, although growth in eastern forests (both public and private) is lower than it was at its peak in the mid-1970s.

Following a peak in the mid-1980s, harvest decreased on public lands in the West; harvest levels in the 1990s on both public and private lands were below those of the 1950s. Harvest on public and private lands in the East increased from the mid-1980s to the mid-1990s, with private lands accounting for the vast majority of both overall and increased production. Nationally, private lands account for almost 90% of total harvest, a figure that has grown only slightly since the 1950s.

Although not shown, there may be substantial differences between northern and southern areas within the eastern and western regions shown here.

See also Timber Harvest (opposite page).

The technical note for this indicator is on page 245.

Timber Growth and Harvest

Timber Growth and Harvest, West, Public Timberlands

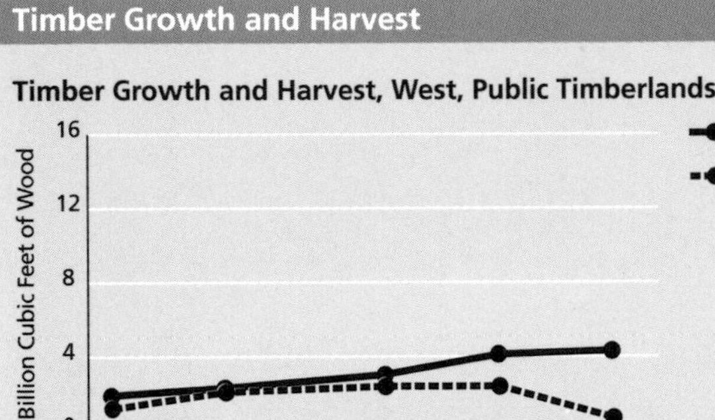
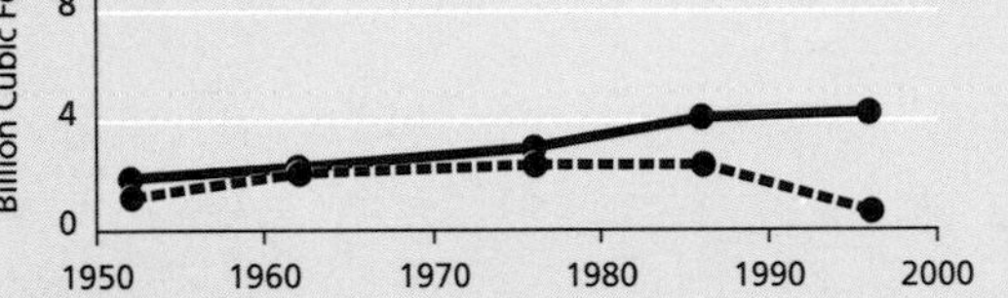

Timber Growth and Harvest, West, Private Timberlands

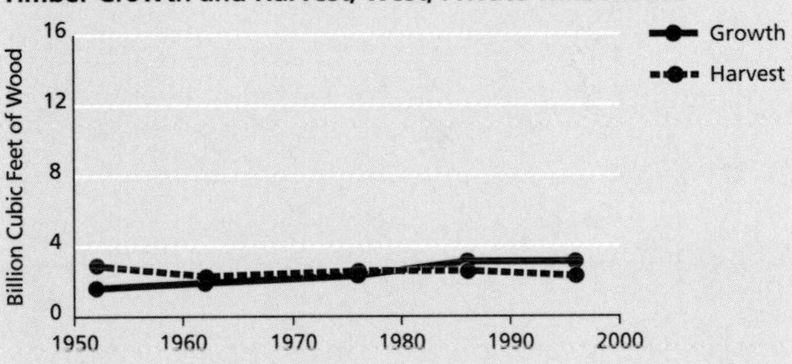

Timber Growth and Harvest, East, Public Timberlands

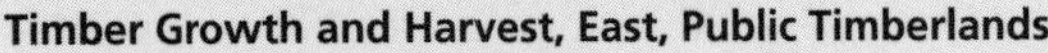
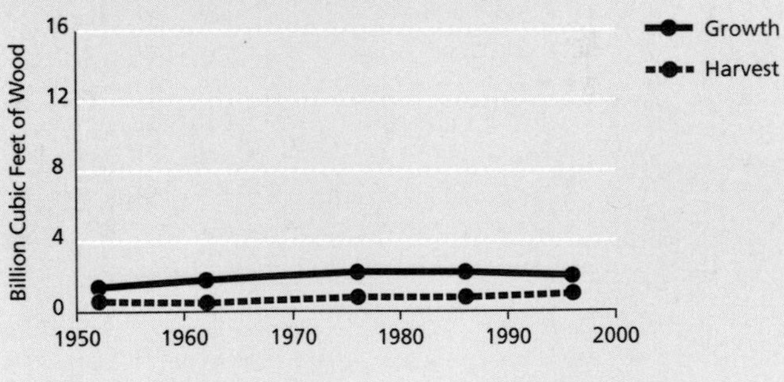

Timber Growth and Harvest, East, Private Timberlands

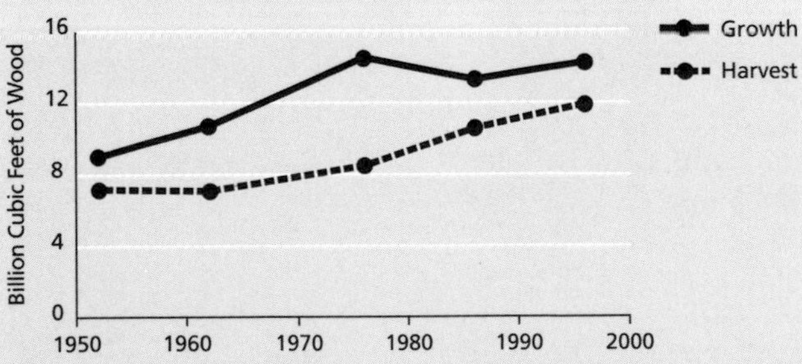

Data Source: USDA Forest Service. Coverage: all 50 states.

Note: "Timberlands" is a USDA Forest Service designation for lands that grow at least 20 cubic feet of wood per acre per year, which is considered to be sufficient to support commercial harvest under current economic conditions. Lands on which harvest is prohibited by statute are not included as "timberlands."

SYSTEM DIMENSIONS	CHEMICAL AND PHYSICAL	BIOLOGICAL COMPONENTS	HUMAN USES
Extent	Nutrients, Carbon, Oxygen	Plants and Animals	Food, Fiber, and Water
Pattern	Contaminants	Communities	**Recreation and Other Services**
	Physical	Ecological Productivity	

⊖ Recreation in Forests

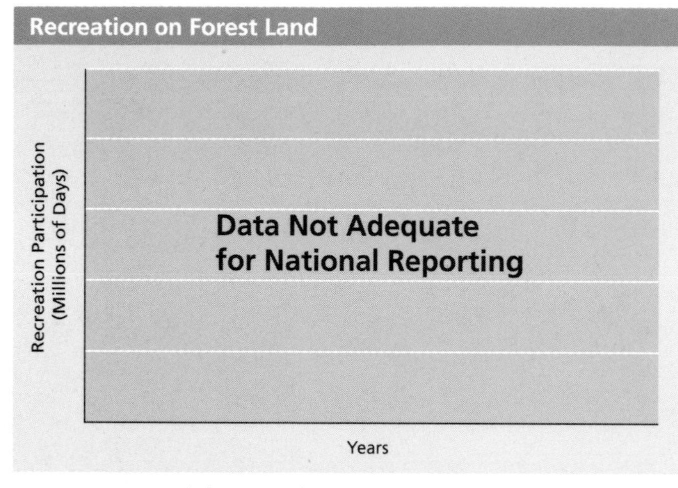

What Is This Indicator, and Why Is It Important? This indicator would report the number of days per year that people engage in a variety of recreational activities in forests. Activities such as walking, hiking and backpacking, fishing and hunting, wildlife viewing, cross-country and downhill skiing, and snowmobiling would be included.

A great deal of recreational activities takes place within our nation's forests. Recreation is a benefit that is derived from forests in much the same way that we derive products such as timber.

Why Can't This Indicator Be Reported at This Time? There are no national data sets that document the type and amount of recreation in forests. The National Survey of Fishing, Hunting, and Wildlife-Associated Recreation (http://fa.r9.fws.gov/surveys/surveys.html) and the National Survey on Recreation and the Environment (http://www.srs.fs.fed.us/trends/nsre.html) both provide reliable data on these activities, but neither survey identifies whether these activities take place in forests, on grasslands or shrublands, on farmlands, or elsewhere.

Adequate reporting would require modification of existing surveys to elicit information either on the location of recreational activities or on the amount of recreation in forested areas.

This report also includes other indicators of recreational activity. See pp. 60, 109, 153, and 174.

There is no technical note for this indicator.

Fresh Waters

	What Indicators are used to describe Fresh Waters?		Can we report trends? Are there other useful reference points?
SYSTEM DIMENSIONS			
◐	Extent of Freshwater Ecosystems	What is the area of lakes and wetlands, and the length of streams, rivers, and their stream bank (riparian) areas?	Some trends
◐	Altered Freshwater Ecosystems	How much of the nation's lakes, wetlands, streams, and riparian areas has been significantly altered?	Current data only
CHEMICAL AND PHYSICAL CONDITIONS			
◐	Phosphorus in Lakes, Reservoirs, and Large Rivers	How much phosphorus is there in lakes, reservoirs, and large rivers?	Current data only, federal limit
Nutrients in Fresh Waters: Additional indicators may be found in the core national indicators chapter, as well as in the forest, farmlands, grassland/shrubland, and urban/suburban chapters.			
Chemical Contamination in Fresh Waters: Indicators dealing with chemical contamination in fresh waters may be found in the core national indicators chapter and in the farmland and urban/suburban chapters.			
●	Changing Stream Flows	How many streams have had major changes in the size or timing of their lowest or highest flows since the 1930s and 1940s?	Trends
⊖	Water Clarity	How clear are lakes in the United States?	No data reported
BIOLOGICAL COMPONENTS			
◐	At-Risk Native Species	How many freshwater species are at different levels of risk of extinction?	Current data only, regional comparison
◐	Non-native Species	How many non-native species are found in watersheds throughout the United States?	Current data only, regional comparison
◐	Animal Deaths and Deformities	How many die-offs of waterfowl, fish, mammals, and amphibians occur? How common are amphibian deformities?	Trends, regional comparison
⊖	Status of Freshwater Animal Communities	What is the condition of communities of fish and bottom-dwelling animals in the nation's streams?	No data reported
◐	At-Risk Freshwater Plant Communities	How many wetland and stream bank plant communities are rare and thus potentially at risk?	Current data only, regional comparison
?	Stream Habitat Quality	What is the quality of the habitat in the nation's streams?	No data reported
HUMAN USES			
●	Water Withdrawals	How much fresh water do people withdraw, and what do they use it for?	Trends
⊖	Groundwater Levels	Are groundwater levels changing? Are they increasing or decreasing and at what rate?	No data reported
●	Waterborne Human Disease Outbreaks	How often do people get sick from drinking or swimming in contaminated water?	Trends
⊖	Freshwater Recreation Activities	How much recreation takes place in the nation's fresh waters?	No data reported

● All Necessary Data Available ◐ Partial Data Available ⊖ Data Not Adequate for National Reporting ? Indicator Development Needed

Chapter 8: Indicators of the Condition and Use of Fresh Waters

The nation's freshwater ecosystems are amazingly diverse, yet together they form an interconnected whole. They include streams and rivers, lakes and ponds, reservoirs, freshwater wetlands, groundwater, and riparian areas—the narrow strips of land along the edge of many of these bodies of water. From the Mississippi to a seasonal desert stream, from the Great Lakes to a farm pond, and from the Everglades to a prairie pothole, the nation's fresh waters provide Americans with drinking water, food, recreational opportunities, and energy, among many other goods and services. Besides being ecosystems in their own right, freshwater systems are an essential part of every one of the other terrestrial ecosystems. Because the state of America's waters reflects and affects the health of all other ecosystems, freshwater indicators are found throughout this report.

What can we say about the condition and use of fresh waters?

Fifteen indicators describe the condition and use of freshwater ecosystems in the United States. Partial or complete data are available for ten of these indicators. Five of these have a long enough data record from which to judge trends, and one has a federally adopted goal to use in judging current conditions. For four indicators, data are not adequate for national reporting, and one indicator requires additional development before it will be possible to assess the availability of data. In addition, indicators of nutrients and chemical contamination in fresh waters are included in every indicator chapter except Coasts and Oceans.

After the following brief summaries of the findings and data availability for each indicator, the remainder of this chapter consists of the indicators themselves. Each indicator page offers a graphic representation of the available data, defines the indicator and explains why it is important, and describes either the available data or the gaps in those data.

System Dimensions

As in each of the other systems, tracking changes in the size of the many types of freshwater ecosystems is the most basic way of describing the condition of the nation's fresh waters. Thus our first indicator of freshwater system dimension tallies the area of lakes and wetlands and the length of streams, rivers, and riparian areas along stream banks. The second tracks the alteration of many of the elements of this complex system.

- **What is the area of lakes and wetlands, and the length of streams, rivers, and their stream bank (riparian) areas?** About half of all Colonial-era wetland acreage in the lower 48 states has been converted to agriculture, development, or other land uses. By the 1990s, about 10% of wetlands that had existed in the 1950s had been lost, with the rate of loss considerably lower after 1985. Lakes, reservoirs, and ponds cover about 21 million acres, and wetlands cover 94 million acres. The area of ponds (usually less than 20 acres) has increased by over 100% since the mid-1950s. This is believed to reflect the construction of small ponds, but the data do not distinguish natural from constructed ponds. More than three-fourths of streams and rivers have forests or other natural vegetation along their banks and riparian area. Data are not adequate for national reporting on the miles of streams of different sizes.
- **How much of the nation's lakes, wetlands, streams, and riparian areas has been significantly altered?** Freshwater systems can be altered in many ways—by damming or channelizing rivers and

streams, by excavating or impounding wetlands, or by converting the edge of a lake or river to a different land use, such as urban/suburban or agriculture. About one-fourth of streams and rivers have either farmlands or urban development in the narrow (about 100-foot-wide) area immediately adjacent to the water's edge. Data are not adequate for national reporting on alterations to lakeshores or wetlands, or on streams and rivers that have been leveed, channelized, or impounded.

Chemical and Physical Conditions

Three indicators describe the chemical and physical condition of fresh waters; these are complemented by several related measures included in other chapters (see below). Two indicators focus on water quality: the concentration of phosphorus, a vital plant nutrient that can lead to problems if present in excess, and the clarity of lake and reservoir waters. Water quantity is also important, so a third measure tracks changes in key flow characteristics of streams and rivers.

Because it is important to all ecosystems, many additional indicators of water quality are presented elsewhere in this report. These include core national indicators dealing with nitrogen and contaminants such as pesticides, PCBs, and heavy metals in streams, sediment, groundwater, and fish. There are also measures dealing with nitrogen or phosphorus or both in the farmlands, forest, grasslands and shrublands, and urban and suburban chapters and indicators dealing with contaminants in the farmlands and urban and suburban chapters.

- **How much phosphorus is there in lakes, reservoirs, and large rivers?** About half of all river sites tested had phosphorus concentrations that exceeded the Environmental Protection Agency's recommended level for preventing excess algae growth. Data are not adequate for national reporting on phosphorus in lakes. See pp. 96 and 187 for data on phosphorus in smaller streams.
- **How many streams have had major changes in the size or timing of their lowest or highest flows since the 1930s–1940s?** Changes in these key flow characteristics can disrupt the plants and animals that live in or near streams. The percentage of streams with major changes in the size of their highest or lowest flow, or in the timing of these flows, increased slightly (from 55% to 60%) from the 1970s to the 1990s. The number of streams with high flows that were well above the 1930–1949 reference period increased markedly from the 1980s to 1990s.
- **How clear are lakes in the United States?** Lakes in some regions are normally quite clear; in other places, lakes are less clear because of naturally occurring algae, sediment, and other suspended materials. Decreases in lake clarity can harm fish and aquatic plants, reduce recreational values, and increase water supply costs. Data are not adequate for national reporting on this indicator.

Biological Components

Six indicators describe the biological condition of the freshwater system. As in other systems, one tracks the native freshwater plant and animal species that are at varying levels of risk of extinction. A parallel indicator tracks the fraction of wetland and riparian (stream bank) communities—specific plant groupings—that are at risk of elimination. A third indicator tracks often unwanted non-native species, reporting now on non-native fish breeding in major watersheds but eventually including amphibians, mollusks, and plants. A fourth indicator would measure how closely fish and bottom-dwelling animal communities resemble those in relatively undisturbed lakes and streams in each region. Because abnormal environmental conditions sometimes lead to unusual animal mortality events, a fifth indicator tallies unusual mortality events among birds, fish, mammals, and amphibians (so far, data are available only for waterfowl) The final indicator will focus on measures of stream habitat quality; a companion indicator is included in the farmlands chapter.

- **How many freshwater species are at different levels of risk of extinction?** About 20% of more than 4000 native animal species that depend on streams, lakes, wetlands, or riparian areas are considered "imperiled" or "critically imperiled," and 4% may already be extinct. When "vulnerable" species are counted, about a third of freshwater species are considered to be "at risk." Hawaii and the Southeast have a much higher percentage of at-risk freshwater species than any other region. Interpretation of these data will be greatly enhanced when information on population trends for these species becomes available.

- **How many non-native species are found in watersheds throughout the United States?** Some non-native species can outcompete native species for food or habitat, and others may act as predators of native species. At least one species of non-native fish has established a breeding population in 99% of the 350 major watersheds in the United States. About 60% of major watersheds have 1–10 non-native species, and two watersheds have 41–50. Watersheds in the central United States—including those on the Gulf Coast—have, in general, the fewest established non-native fish species. Data are not adequate to report nationally on non-native species other than fish.
- **How many die-offs of waterfowl, fish, mammals, and amphibians occur? How common are amphibian deformities?** Such mortalities are typically quite visible and can lead to considerable public concern. While causes are not always known, many scientists believe that increased numbers of mortality events signal serious problems in an ecosystem. The total number of waterfowl die-offs—about 500—was about 20% less in 1995–1999 than it was in either of the two preceding five-year periods. In general, die-offs are more frequent in the Pacific Coast and Midwest regions. Data are not adequate for national reporting on die-offs of fish, amphibians, or mammals or on amphibian deformities.
- **What is the condition of communities of fish and bottom-dwelling animals in the nation's streams?** Modifying a stream—through pollution, changes to the streambed or bank, flow modification, or other means—can change the number and diversity of fish and bottom-dwelling animals. Data are not adequate for national reporting on this indicator.
- **How many wetland and riparian plant communities are at risk?** About 60% of the 1560 wetland communities whose status is known are considered to be at risk, including 12% that are critically imperiled, 24% that are imperiled, and 25% that are vulnerable. Hawaii and the Southeast have a higher percentage of at-risk wetland communities, but in all regions except the Northeast more than 50% of wetland communities are at risk. Interpreting these figures is complicated, however, because some of these wetland community types have never been widely distributed, while others once covered much larger areas and have been reduced in area by conversion to other uses. Data are not adequate to report on riparian (stream bank) communities.
- **What is the quality of the habitat in the nation's streams?** Habitat quality, which varies naturally from stream to stream, directly affects a stream's ability to support native species. This indicator requires further development.

Human Use

Four indicators describe the human uses of fresh waters, two related to water withdrawals, one on recreation, and one on waterborne disease. The first indicator tracks withdrawals by use (e.g., for irrigation, electricity generation, or municipal use) and by source (surface or groundwater). The second tracks whether groundwater levels are changing, in part as a result of withdrawals exceeding recharge. A third indicator tracks a human-focused measure of water quality—the frequency of waterborne disease outbreaks attributed to both drinking and swimming in contaminated water. The fourth tracks such recreational activities as swimming and fishing.

- **How much fresh water do people withdraw, and what do they use it for?** Groundwater and surface water withdrawals increased from 1960 to 1980, and these increases were attributed to growing demand from all major types of use. Total water withdrawals declined about 10% between 1980 and 1985, then grew slightly from 1985 to 1995. Reduced demand for irrigation, thermoelectric power generation, and self-supplied industrial use was responsible for the decline in total withdrawals between 1980 and 1985; demand for these three uses was nearly flat from 1985 to 1995.

- **Are groundwater levels changing?** Are they increasing or decreasing, and at what rate? Groundwater is a major source of water for drinking, irrigation, and other uses and it provides water to many streams, rivers, and wetlands. Deeper water levels mean higher pumping costs and reduced contributions to surface waters. Data are not adequate for national reporting on this indicator.
- **How often do people get sick from drinking or swimming in contaminated water?** The number of disease outbreaks attributable to contaminated drinking water declined markedly from 1973 to 1998. Over approximately the same period, outbreaks attributed to recreational contact increased.
- **How much recreation takes place in the nation's fresh waters?** Americans frequently take part in recreational activities in and around fresh waters; however, data are not adequate for national reporting on this indicator.

What do we mean by "fresh waters"?

Freshwater ecosystems include

- Rivers and streams, including those that flow only intermittently
- Lakes, ponds, and reservoirs, from the Great Lakes to small farm ponds
- Groundwater
- Freshwater wetlands, including forested, shrub, and emergent (marsh) wetlands
- Riparian areas: the strip of land, usually vegetated, along the edge of streams and rivers (this term can also apply to lake edges)

Obviously, there are overlaps and gradations between these areas. Wetlands often occur at the margins of streams and rivers, in what is also considered the riparian area. Some ponds are shallow and thus may also be classified as wetlands. Reservoirs created when rivers are dammed may be classified as rivers, reservoirs, or both, and groundwater often connects all these systems.

Data sources currently available may not identify the smallest ponds, streams, and wetlands. For example, the U.S. Geological Survey dataset (the National Hydrography Dataset, http://nhd.usgs.gov/) that is used to identify lakes generally records neither lakes of less than 6 acres nor very small streams. Since the number and area of these smaller features may be subject to greater change than larger bodies of water—both because they are more numerous and because they are easier to affect—it is important to improve the resolution of these datasets in the future.

Map 8.1. Regional Boundaries for the Following Indicators: At-Risk Native Freshwater Species, Animal Deaths and Deformities, and At-Risk Freshwater Plant Communities

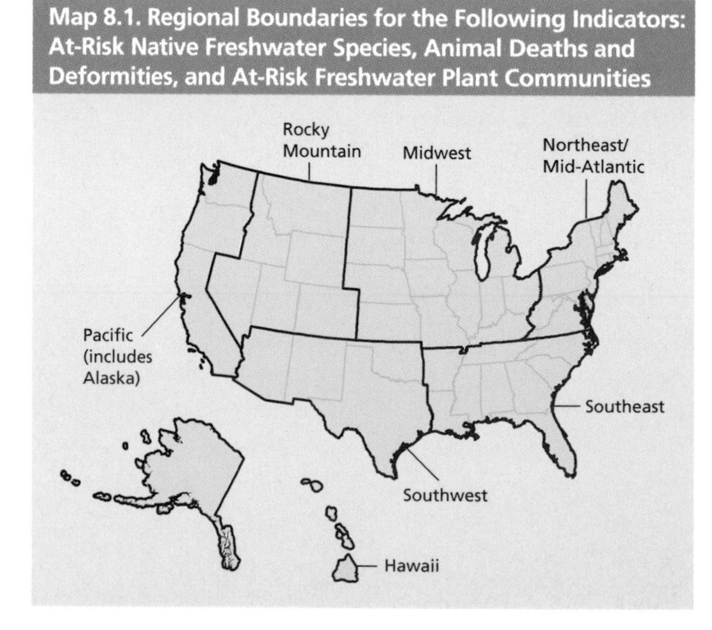

A Note about Regions

This chapter uses a regional approach (see Map 8.1) to providing data for three indicators: At-Risk Native Freshwater Species; Animal Deaths and Deformities; and At-Risk Freshwater Plant Communities. This regional scheme was developed for this report, and is used for the national plant growth and at-risk species indicators, as well as for the at-risk species indicators for forests and grasslands and shrublands.

◐ Extent of Freshwater Ecosystems

What Is This Indicator, and Why Is It Important? This indicator reports the area of wetlands and lakes, reservoirs, and ponds and the length of small, medium, and large streams and rivers. For streams and rivers, the indicator also reports on the type of land cover on their shorelines and adjacent areas ("riparian" areas): forest; grasslands, shrublands, or wetlands; and urban/suburban or agricultural land.

America's fresh waters provide critical fish and wildlife habitat and are an important component of most other ecosystems. They also provide people with many goods and services, including drinking water; water for industrial use, livestock, and irrigation; and opportunities for recreation. In addition, wetlands and riparian areas help filter runoff and reduce flooding, and rivers and lakes receive a variety of discharged wastes.

Why Can't This Entire Indicator Be Reported at This Time? Several methods are used to classify streams—by discharge, by drainage area, or by the number of tributaries a stream has. Since no single method has been agreed upon for general use, there are no national data sets for reporting on stream size.

What Do the Data Show? About half of all Colonial-era wetland acreage in the lower 48 states has been converted to agriculture, development, or other land uses. By the 1990s, about 10% of the wetlands that existed in the1950s had been lost, although the rate of loss slowed after 1985.

Lakes, ponds, and reservoirs occupy about 21 million acres, or one-fifth as much area as is occupied by wetlands. The area of ponds (usually less than 20 acres) has increased by over 100% since the mid-1950s. This is believed to reflect the construction of small ponds, but the data do not distinguish natural from constructed ponds.

For more than three-fourths of their length, the riparian areas of streams and rivers are forested or covered with other natural vegetation.

The technical note for this indicator is on page 246.

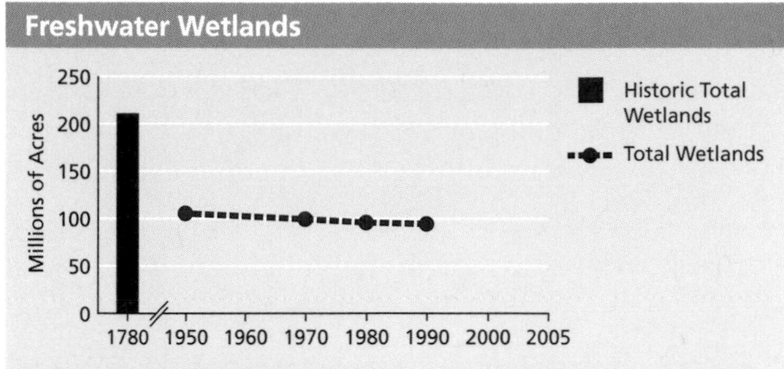

Data Source: U.S. Fish and Wildlife Service National Wetlands Inventory (NWI); Coverage: lower 48 states.

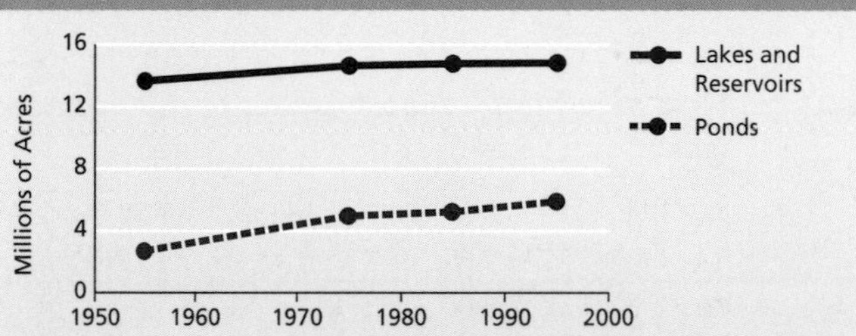

Data Source: U.S. Fish and Wildlife Service National Wetlands Inventory (NWI); Coverage: lower 48 states. Lake area does not include the Great Lakes, which cover about 60.2 million acres within the United States.

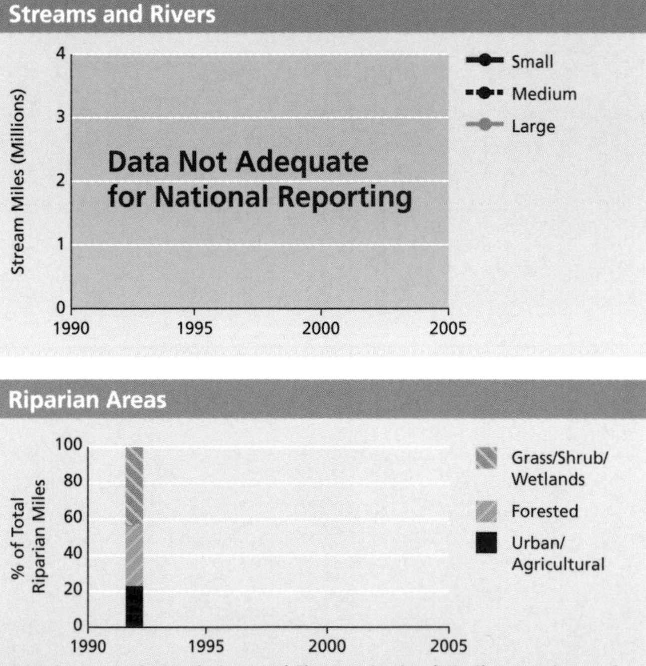

Data Source: Multi-Resolution Land Characterization (MRLC) Consortium and USGS National Hydrography Dataset, processed by U.S. EPA National Exposure Research Laboratory. Coverage: lower 48 states.

SYSTEM DIMENSIONS	CHEMICAL AND PHYSICAL	BIOLOGICAL COMPONENTS	HUMAN USES
Extent	Nutrients, Carbon, Oxygen	Plants and Animals	Food, Fiber, and Water
Pattern	Contaminants	Communities	Recreation and Other Services
	Physical	Ecological Productivity	

◐ Altered Freshwater Ecosystems

Altered Freshwater Ecosystems

Data Not Adequate for National Reporting on

- % of wetlands that are altered
- % of lake and pond shorelines that are altered
- % of stream-miles that are altered

Partial Indicator Data: Altered Riparian Areas

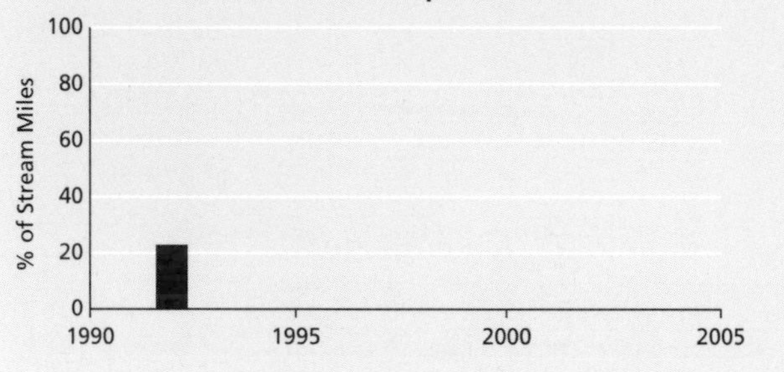

Data Source: Multi-Resolution Land Characterization (MRLC) Consortium and USGS National Hydrography Dataset, processed by U.S. EPA National Exposure Research Laboratory, Environmental Sciences Division. Coverage: lower 48 states.

What Is This Indicator, and Why Is It Important? This indicator of alteration reports the percentage of

- **Stream and river** miles that have been leveed, channelized, or impounded behind a dam
- **Ponds and lake** shoreline-miles that have agricultural or urban/suburban land cover within about 100 feet of the water's edge (reservoirs and constructed lakes are excluded)
- **Riparian zone** miles (the habitat at the edge of streams and rivers) that have agricultural or urban/suburban land cover within about 100 feet of the water's edge
- **Wetland** acres that have been excavated, impounded, diked, partially drained, or farmed

Physically altering a body of fresh water can affect the plants and animals that depend on it, as well as the goods and services people receive from it. Such areas are usually altered to achieve some benefit: flood control or easier navigation, erosion control to protect property, more land for farming or development, or supply of municipal, industrial, and irrigation water. However, these alterations can reduce fish and wildlife habitat, disrupt patterns and timing of water flows, serve as barriers to animal movement, and reduce or eliminate the natural filtering of sediment and pollutants.

Why Can't This Entire Indicator Be Reported at This Time? Data on the degree to which streams and rivers are channelized, leveed, or impounded behind dams are not available, nor are data on the extent of wetland alteration. In addition, available data on lake and pond shoreline alteration does not distinguish between natural and constructed bodies of water.

This indicator describes a few key types of alterations. As monitoring and reporting techniques improve, reporting on other alterations may be possible. Stream Habitat Quality (p. 149) and Changing Stream Flows (p. 142) also report on stream condition.

What Do the Data Show? About 23% of riparian areas have either farmlands or urban development in the narrow area (100-foot strip) immediately adjacent to the water's edge.

The technical note for this indicator is on page 247.

◐ Phosphorus in Lakes, Reservoirs, and Large Rivers

What Is This Indicator, and Why Is It Important? This indicator reports the average concentration of phosphorus in lakes, reservoirs, and large rivers. Total phosphorus concentrations are reported in four ranges: below 20 parts per billion (ppb), from 20 to 50 ppb, from 50 to 100 ppb, and 100 ppb or more.

Increased phosphorus concentrations are associated with increased algae growth in lakes, reservoirs, and large rivers. Algae are tiny aquatic plants that sustain the growth of most other aquatic life forms; when overabundant, however, they can contribute to reductions in dissolved oxygen, cause fish kills, and cause shifts in the number and type of fish and other aquatic animals. Algae blooms can also harm aesthetic and recreational values.

Lakes and reservoirs with phosphorus concentration of less than 20 ppb are generally free of negative effects; higher concentrations are accompanied by increasing effects. The U.S. Environmental Protection Agency's (EPA) recommended goal for preventing excess algae growth in streams that do not flow directly into lakes or other impoundments is 100 ppb. In 2000, EPA took steps to facilitate development of regional phosphorus criteria, but the regional criteria have not yet been adopted. There is no federal drinking water standard for phosphorus.

Phosphorus in Large Rivers

Source: U.S. Geological Survey National Water Quality Assessment (NAWQA) Program. Coverage: lower 48 states. Large rivers have flows greater than 1000 cubic feet per second (cfs).

Phosphorus in Lakes and Reservoirs

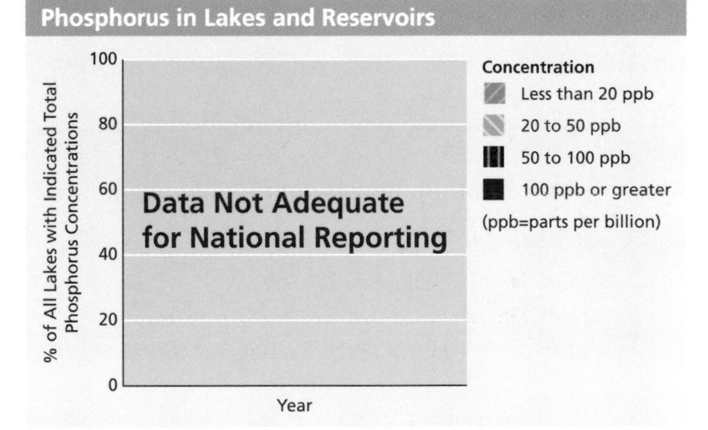

Why Can't This Entire Indicator Be Reported at This Time? There are no datasets that are known to provide representative phosphorus values for the nation's lakes and reservoirs. EPA's STORET data archive might serve as a source of data, but considerable research would be required to determine whether the samples reported there are representative of overall conditions.

This report also includes indicators for total phosphorus concentrations in farmland streams and urban and suburban streams (pp. 96 and 187).

What Do the Data Show? About half of all river sites tested had phosphorus concentration levels of 100 ppb or higher. About one-fourth of the tested sites had concentrations below 50 ppb. Since some areas have higher natural levels of phosphorus than others, interpreting this indicator will become much easier when trend information is available.

The technical note for this indicator is on page 248.

● Changing Stream Flows

Alteration of Key Flow Characteristics (compared with 1930-1949)

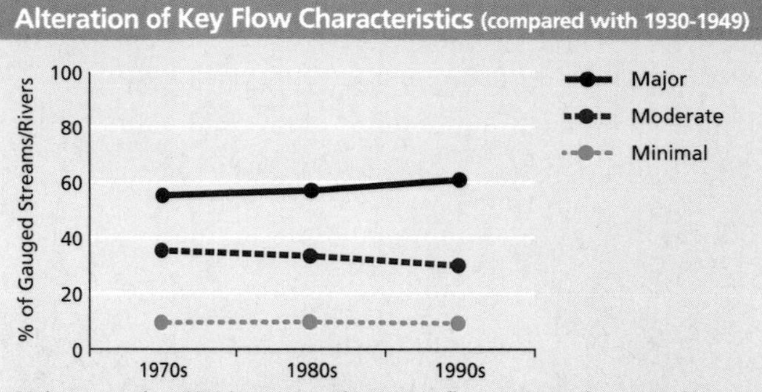

Major: more than 75% increase or decrease in flow, or more than a 60-day change in timing of low or high flow
Moderate: between 25% and 75% increase or decrease in flow, or a 30- to 60-day change in timing of low or high flow
Minimal/stable: less than 25% increase or decrease in flow, or less than 30-day change in timing of low or high flow

Data Source: U.S. Geological Survey; analysis by Colorado State University. Coverage: lower 48 states.

Major Changes in Low Flow (compared with 1930-1949)

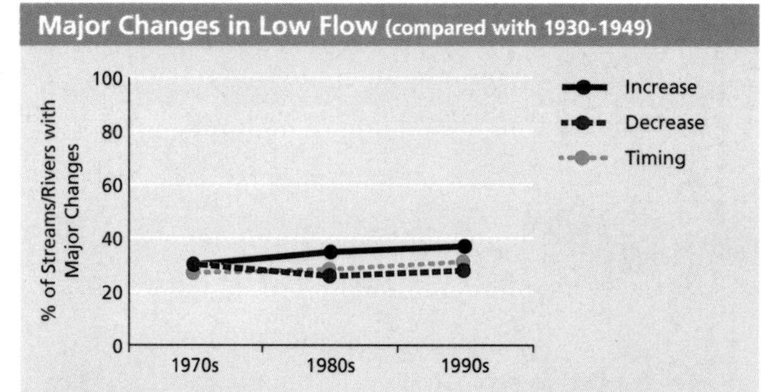

Low flow: Average annual 7-day low flow

Data Source: U.S. Geological Survey; analysis by Colorado State University. Coverage: lower 48 states.

Note: For the low-flow and high-flow graphs, totals may add to more than 100%, because both the timing and magnitude may change in a single stream or river.

Major Changes in High Flow (compared with 1930-1949)

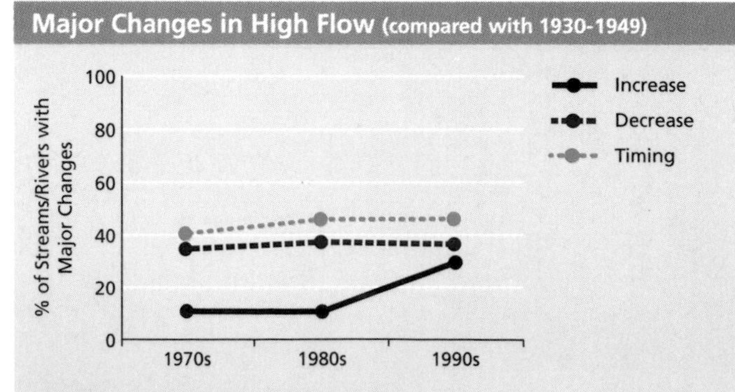

High flow: Average annual 1-day high flow

Data Source: U.S. Geological Survey; analysis by Colorado State University. Coverage: lower 48 states.

Note: For the low-flow and high-flow graphs, totals may add to more than 100%, because both the timing and magnitude may change in a single stream or river.

What Is This Measure, and Why Is It Important? This indicator describes changes in the amount and timing of river and stream flow by reporting the percentage of monitored streams or rivers with major, moderate, and minimal changes in low flow, high flow, and the timing of these two extreme events. The indicator also describes the nature of major flow changes. Flow characteristics were measured for three recent 10-year periods and compared against 1930–1949 as a reference period.

How a stream flows—the volume of its high and low flows, and when these extreme flows occur—is critical in determining what plants and animals live in the stream or river. For example, low flows define the smallest area the stream or rivers will occupy and thus the amount of fish habitat that will be available year-round; high flows shape the river channel and clear silt and debris; and some species require certain flows at specific periods, such as spawning season.

Changes in flow can be caused by dams; by pumping water for drinking, irrigation, or other uses; by groundwater pumping (which reduces flows into the stream); by changes in the type and amount of development and other land cover in the watershed; or by changes in long-term weather patterns, such as droughts or wet periods.

What Do the Data Show? The percentage of streams or rivers with major changes in the size of their highest or lowest flow, or in the timing of these flows, increased slightly from the 1970s to the 1990s. In addition, the number of streams or rivers whose high flows were well above those in the 1930–1949 reference period rose markedly from the 1980s to the 1990s.

The reference period used here included periods of relatively low rainfall, but it also predated much development activity (dam building, irrigation, etc.) that might affect flows. Therefore, it is more useful to focus this indicator on increases or decreases in the number of streams or rivers with major changes in flow, rather than on the actual number of streams or rivers with such changes. Finally, it is not possible to use these data to identify the cause of flow changes.

The technical note for this indicator is on page 249.

SYSTEM DIMENSIONS	CHEMICAL AND PHYSICAL	BIOLOGICAL COMPONENTS	HUMAN USES
Extent Pattern	Nutrients, Carbon, Oxygen Contaminants **Physical**	Plants and Animals Communities Ecological Productivity	Food, Fiber, and Water Recreation and Other Services

⊖ Water Clarity

What Is This Indicator, and Why Is It Important? This indicator would report the percentage of lake and reservoir area with low-, medium-, and high-clarity water (ponds are not included because of their shallow depth). A map would show regional patterns of change.

The simplest method for measuring clarity is to lower a standard painted disk (a "Secchi disk") until it cannot be seen; the clearer the lake or reservoir, the greater the "Secchi depth" (SD). Using this method, ranges for SD would be: low clarity (SD less than 3 feet), medium clarity (SD 3–10 feet), and high clarity (SD greater than 10 feet).

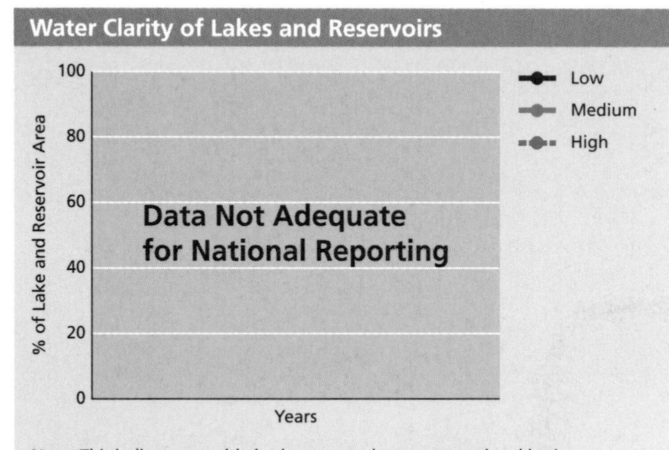

Note: This indicator would also be reported on an ecoregional basis (ecoregions are areas that are similar in climate, geography, and ecological conditions). A map would show each major ecoregion, colored to show changes in lake and reservoir water clarity.

Water clarity is important both to people and to ecological functioning—people like clearer water to swim in, to drink, and for esthetic reasons, and aquatic plants need light to grow and fish and other animals need light to feed and reproduce. Lakes and reservoirs can become cloudy when streams and runoff carry silt, clay, and organic materials into them. Runoff may also add phosphorus and other nutrients to lake or reservoir water; these nutrients fuel algae growth (see Phosphorus in Lakes, Reservoirs, and Large Rivers, p. 141), which also reduces water clarity.

Why Can't This Indicator Be Reported at This Time? Although considerable amounts of water clarity data are available from various sources, some areas are heavily sampled, while in other areas few or no lakes are tested. Thus, the available data do not provide representative coverage at a national level.

It is important to track water clarity through time, because lakes and reservoirs in different regions have different degrees of natural clarity. By tracking clarity over time, it will be possible to identify areas with declining or improving clarity and to distinguish these from naturally cloudy or clear areas.

The technical note for this indicator is on page 250.

◐ At-Risk Native Freshwater Species

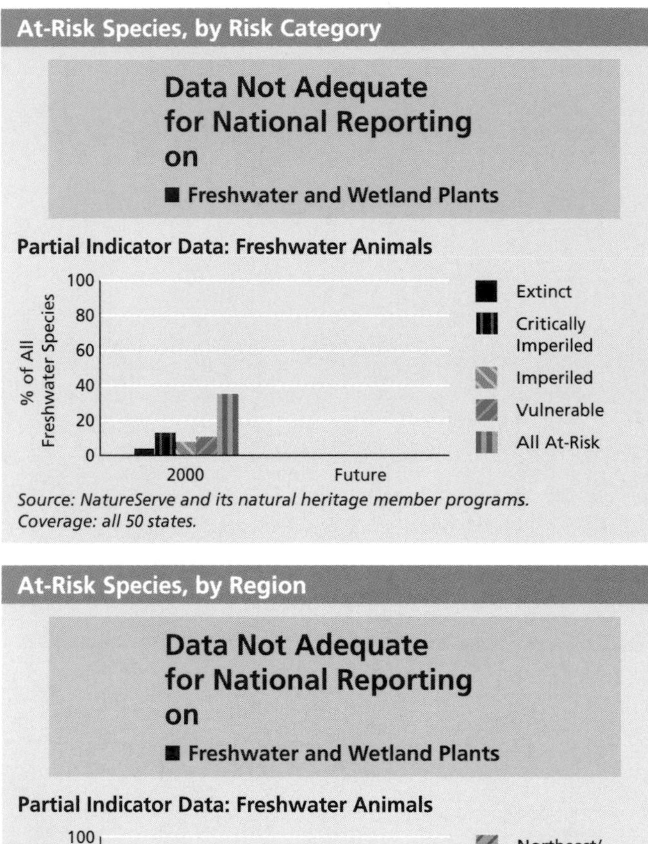

What Is This Indicator, and Why Is It Important? This indicator reports on the relative risk of extinction of native freshwater species. The risk categories are based on such factors as the number and condition of individuals and populations, the area occupied by the species, population trends, and known threats. Degrees of risk reported here range from very high ("critically imperiled" species often are found in five or fewer places or have experienced very steep declines) to moderate ("vulnerable" species often are found in fewer than 80 places or have recently experienced widespread declines). In all cases, a wide variety of factors contribute to the overall ratings. "Freshwater species" live in freshwater, wetland, or riparian habitats during at least part of their life cycle and depend on these habitats for survival.

Species are valued for a variety of reasons: they provide products, including food, fiber, and, more recently, genetic materials; they are key elements of ecosystems, which themselves provide valuable goods and services; and many people value them for their intrinsic worth or beauty.

Why Can't This Entire Indicator Be Reported? This indicator reports on fish; amphibians and reptiles; aquatic mammals; butterflies; mussels; snails; crayfishes; fairy, clam, and tadpole shrimp; dragonflies and damselflies; and mayflies, stoneflies, and caddisflies. Data on freshwater and wetland plants are not included because additional analyses are required to categorize correctly the habitats of all North American plants.

See also the national at-risk species indicator (p. 52) and the indicators for at-risk coastal, forest, and grassland and shrubland species (pp. 75, 124, and 168), as well as those for species in farmlands (p. 103) and urban and suburban areas (p. 191).

What Do the Data Show? About 13% of native freshwater species are critically imperiled, 8% are imperiled, and 4% are or may be extinct. When vulnerable species (11%) are counted, about a third of freshwater animal species are considered "at risk." Hawaii and the Southeast have a much larger percentage of at-risk freshwater species than any other region.

Interpreting these figures is complicated, however, because some species are naturally rare. Thus, the rankings are influenced by differences among regions and species groups in the number of naturally rare species, as well as by different types and levels of human activities that can cause species declines. Interpretation of these data will be greatly enhanced when information on population trends for these at-risk species becomes available.

The technical note for this indicator is on page 214.

◐ Non-native Species

What Is This Indicator, and Why Is It Important? This indicator reports the percentage of watersheds with different numbers of non-native species with established breeding populations. The number of such species is also shown for each watershed. "Non-native" includes species not native to North America and those that are native to this continent but are now found outside their historic range.

Non-native species are also called nonindigenous, exotic, or introduced; those that spread aggressively are termed invasive. They may act as predators or parasites of native species, cause diseases, compete for food or habitat, and alter essential habitat. They also may threaten human health and economic well-being—for example, the zebra mussel has damaged power plants, water treatment facilities, and other structures and significantly changed freshwater ecosystems. Watersheds with more non-natives are likely to experience greater ecological and economic disruption. In addition, non-native species may become established more easily in watersheds with other types of disturbance (such as degraded water quality, altered temperatures, and alterations to habitat or flows).

Some non-natives are introduced intentionally, for their desired characteristics. For example, brown trout are native to Europe, and rainbow trout to western North America; both are popular and widely stocked game fish throughout the nation.

Established Non-Native Species

Data Not Adequate for National Reporting on
- **Other Aquatic Animals or Plants**

Partial Indicator Data: Fish Only

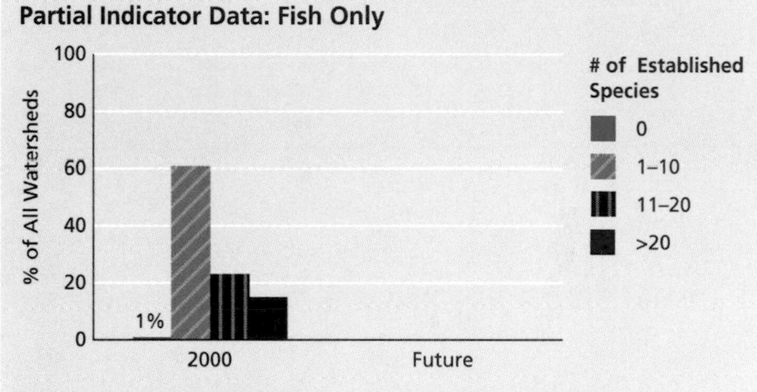

Data Source: U.S. Geological Survey. Coverage: lower 48 states only.

Established Non-native Fish Species, 2000

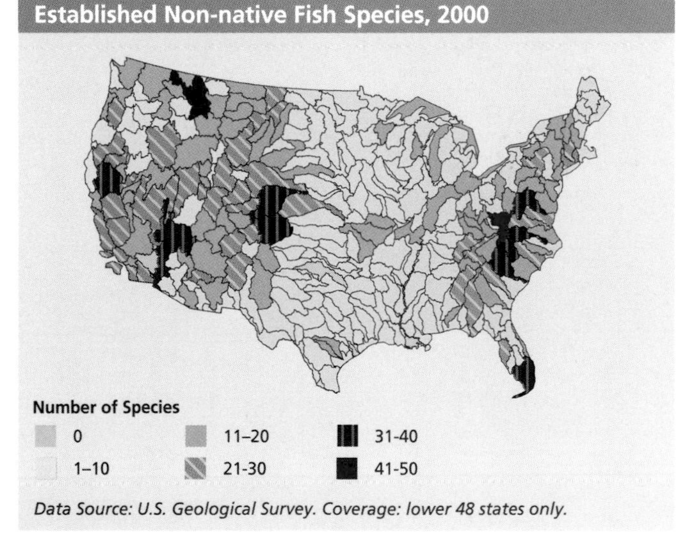

Data Source: U.S. Geological Survey. Coverage: lower 48 states only.

Why Can't This Entire Indicator Be Reported at This Time? Data on non-native fish are more widely available and of higher quality than data on other animal species like mollusks and amphibians, or on plants. When data become available, future reports will include these other species.

What Do the Data Show? Of 350 watersheds, only five have no established non-native fish. Sixty percent (213) have 1–10 non-native species, and two watersheds have 41–50 such species. Watersheds in the central United States—including those on the Gulf Coast—generally have the fewest non-natives.

Discussion Examples of native North American species found outside their historic range include bullfrogs and warmouth sunfish, both eastern natives now found in the West. Bullfrogs are associated with declines in native fish, bird, and amphibian populations in western lakes, and the warmouth has apparently contributed to the decline of some native frogs and salamanders.

The technical note for this indicator is on page 251.

SYSTEM DIMENSIONS	CHEMICAL AND PHYSICAL	**BIOLOGICAL COMPONENTS**	HUMAN USES
Extent Pattern	Nutrients, Carbon, Oxygen Contaminants Physical	**Plants and Animals** Communities Ecological Productivity	Food, Fiber, and Water Recreation and Other Services

◐ Animal Deaths and Deformities

Animal Mortality, by Size of Event

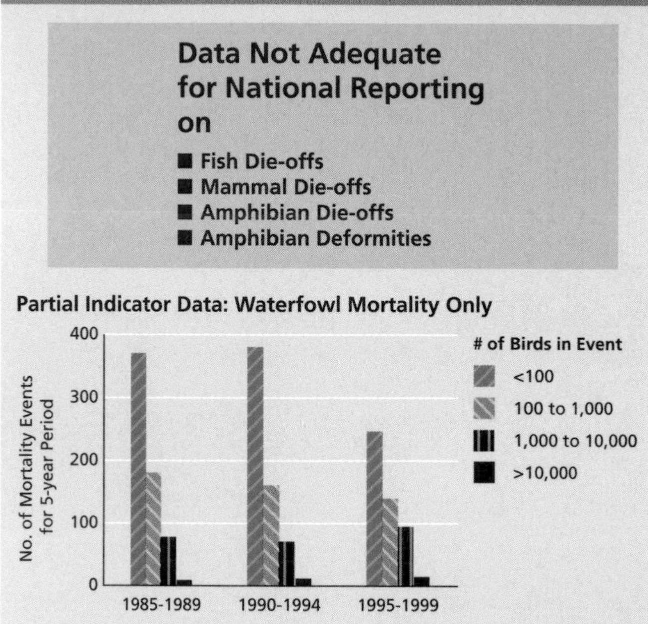

Data Source: U.S. Geological Survey. Coverage: all 50 states; Alaska and Hawaii are included in the Pacific region, and Puerto Rico, and the U.S. Virgin Islands are included in the Southeast region.

Animal Mortality, by Region

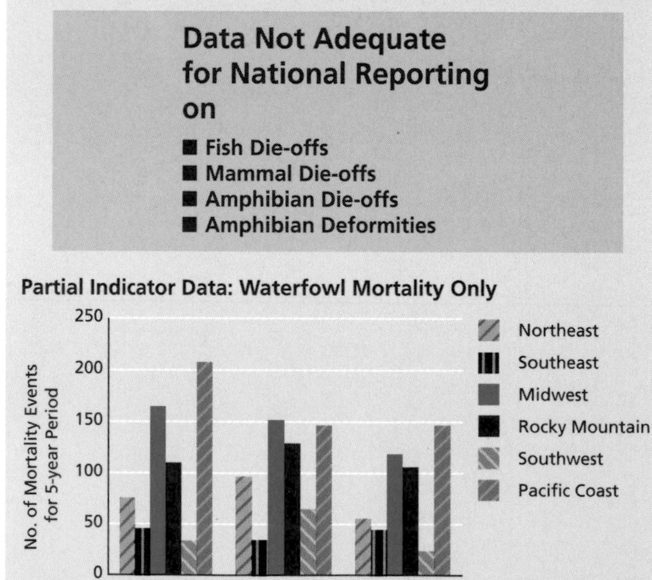

Data Source: U.S. Geological Survey. Coverage: all 50 states; Alaska and Hawaii are included in the Pacific region, and Puerto Rico, and the U.S. Virgin Islands are included in the Southeast region.

What Is This Indicator, and Why Is It Important? This indicator reports on unusual mortality events for waterfowl, fish, amphibians, and mammals, and on deformity events for amphibians. Only data on waterfowl mortality can be reported at this time.

Die-offs of fish, birds, and other freshwater animals generate considerable public concern. People may perceive a danger to their own health, or they may be concerned about disruptions to the ecosystem, loss of recreational opportunities and tourism income, and fish that cannot be eaten or sold. Die-offs can be caused by disease, too little oxygen or other imbalances in water chemistry, chemical pollution, extreme temperatures, or a combination of factors. Although the exact cause of an event is not always known, many scientists believe that die-offs indicate the presence of serious problems in an ecosystem. For information on mortalities in coastal waters, see p. 77.

Why Can't This Entire Indicator Be Reported at This Time? The U.S. Geological Survey (USGS), which provided the waterfowl data presented here, also gathers mortality information on mammals and amphibians. However, the data for these groups are less complete than for waterfowl. USGS also collects data on amphibian deformities, but there is no widespread monitoring program or systematic surveying for amphibian deformities. There is also no reporting mechanism for fish die-offs.

What Do the Data Show? From 1995 to 1999, about 500 incidents of unusual waterfowl mortality were reported in the United States. In half of these incidents, less than 100 birds died; in about 100 incidents, between 1,000 and 10,000 birds died, and 15 incidents involved more than 10,000 deaths. The total number of die-offs was about 20% lower in 1995–1999 than in 1985–1989 and 1990–1994. In general, there are more die-offs in the Pacific and Midwest and fewer in the Southwest and Southeast.

The technical note for this indicator is on page 252.

⊖ Status of Freshwater Animal Communities: Fish and Bottom-Dwelling Animals

What Is This Indicator, and Why Is It Important? This indicator reports on "biological integrity"—the degree to which the suite of fish and bottom-dwelling animals in a lake or stream resembles what one might find in a relatively undisturbed lake or stream in the same region. Tests assess the number of different species, the number and condition of individuals, and food chain interactions for fish and bottom-dwelling (or benthic) animals, which include insects, worms, mollusks, and crustaceans. High scores indicate close resemblance to "natural" conditions, and low scores indicate significant deviation from them.

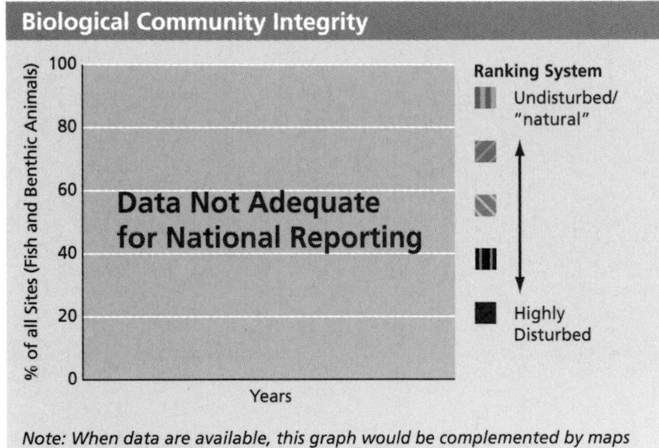

Note: When data are available, this graph would be complemented by maps showing the percentage of watersheds with fish and bottom-dwelling animal communities resembling those in undisturbed conditions.

Undisturbed lakes and streams in a particular region have a relatively predictable set of fish and bottom-dwelling animals, which occur in predictable proportions. Alterations to the stream or lake can change the composition and condition of these biological communities from this undisturbed or "reference" condition. Alterations that affect biological integrity include decreased water quality, introduction of non-native species, changes in the amount or timing of water flows, and modification of the lake or stream bed or shoreline. Some lakes and streams are so modified that, for example, both the number of species and the number of individuals are very low when compared with undisturbed areas, and many of those that remain are diseased or otherwise damaged. Ecosystems that are "healthy," or show high integrity, are more likely to withstand natural and man-made stresses.

Why Can't This Indicator Be Reported at This Time? The tests of biological integrity now in use have been developed primarily for streams and wadeable rivers; methods for lakes and larger rivers are not as well developed. In addition, these tests must be tailored to each region of the country to ensure that each stream or lake is compared with an appropriate reference. Only a handful of states regularly conduct quantitative tests of the condition of fish or bottom-dwelling animal communities. Thirty states are developing such tests, and five states already use such tests in regulating water quality.

The technical note for this indicator is on page 253.

◐ At-Risk Freshwater Plant Communities

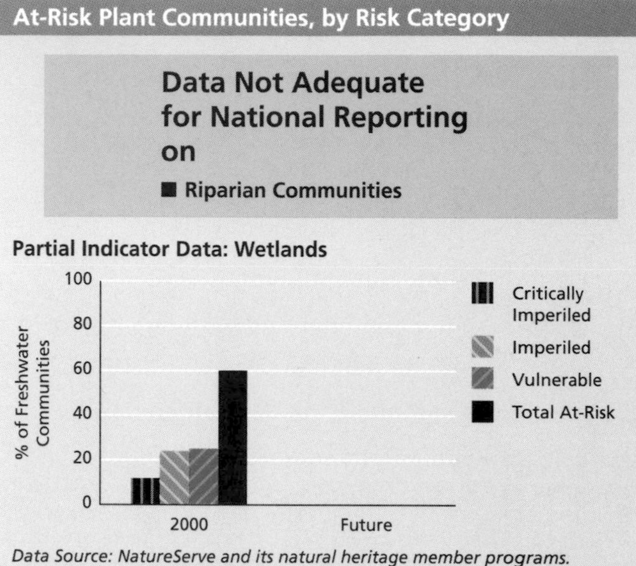

Data Source: NatureServe and its natural heritage member programs. Coverage: excludes Alaska.

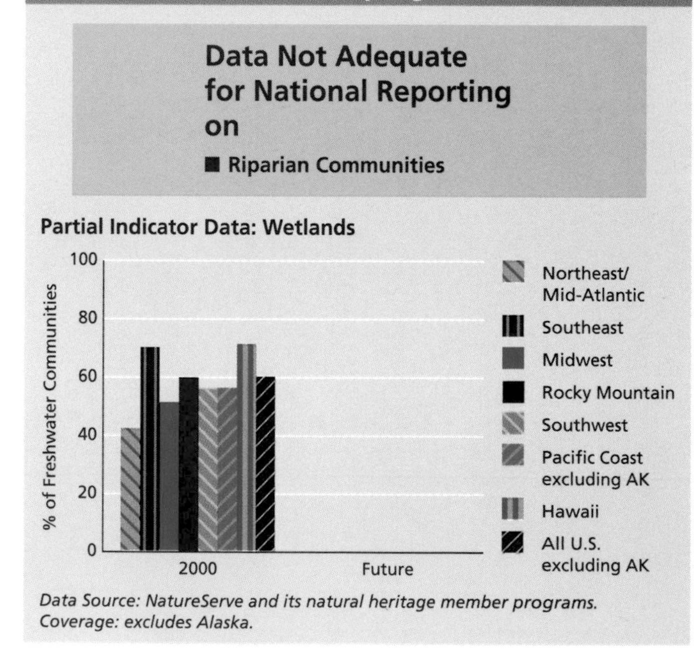

Data Source: NatureServe and its natural heritage member programs. Coverage: excludes Alaska.

What Is This Indicator, and Why Is It Important? This indicator reports on the percentage of wetland and riparian plant communities that are at different degrees of risk of elimination. These status ranks are based on such factors as the remaining number and condition of occurrences of the community, the remaining acreage, and the severity of threats to the community type. Degrees of risk reported here range from very high ("critically imperiled" communities often are found in five or fewer places or have experienced very steep declines) to moderate ("vulnerable" communities often are found in 80 or fewer places or have experienced widespread declines). Communities ranked as "secure" or "apparently secure" are not listed. In all cases, a wide variety of factors contribute to overall ratings.

Different plant communities (groups of plant species that tend to occur in similar environmental conditions) support distinct species combinations and may provide unique ecosystem values. One community might provide habitat for several rare plant and animal species; another might sequester an especially large amount of carbon.

Why Can't This Entire Indicator Be Reported at This Time? Riparian areas also have characteristic plant communities, but these are less distinct than for wetlands, and there are technical challenges to creating a classification system for riparian areas. Work is under way to develop such a system, which will facilitate future reporting.

What Do the Data Show? About 60% of the 1560 wetland communities ranked here are considered to be at-risk: about 12% are critically imperiled, about 24% are imperiled, and 25% are vulnerable. Hawaii and the Southeast have a larger percentage of at-risk wetland communities, but in all regions except the Northeast, more than 50% of wetland communities are at risk.

Interpreting these figures is complicated, however, because some of these wetland community types have never been widely distributed, while others once covered much larger areas and have been reduced in area by conversion of wetlands to other uses. Because the data do not distinguish between naturally rare community types and those that are declining, this indicator will be much more informative when trend information becomes available. At present, the at-risk plant communities reported here generally occupy small areas and thus probably represent less than 60% of total wetland acreage.

The technical note for this indicator is on page 253.

⑦ Stream Habitat Quality

Stream Habitat Quality

Indicator Development Needed

What Is This Indicator, and Why Is It Important? This indicator would describe stream habitat quality by comparing the habitat in any given stream segment against the habitat that would be found in a relatively undisturbed stream in the same region. The index used for comparison would incorporate the presence of riffles and pools, the size of streambed sediments and the degree to which larger gravel and cobbles are buried in silt, the presence of branches, tree trunks, and other large woody pieces, and the stability of the bank. A companion indicator would report on stream habitat quality in farmland streams (p. 105).

Streams with higher condition ratings—that is, they closely resemble undisturbed streams—have a more natural and diverse array of underwater and bank habitats and are therefore capable of supporting diverse native species. These streams are also more likely to have relatively undisturbed flow patterns (see p. 142) and to have vegetation along their banks. Both these features help maintain the conditions necessary to support a healthy biological community over the long term.

Stream-dwelling animals and plants require specific habitat conditions in order to survive and reproduce. Because each species has its own particular habitat requirements, a variety of habitats along a stream are needed to maintain the stream's natural complement of plants and animals.

Why Can't This Indicator Be Reported at This Time? Scientists generally agree on the key stream attributes that should be measured to evaluate stream habitat quality (riffles and pools, streambed sediments, and so on), and there is considerable work under way by the U.S. Environmental Protection Agency, the U.S. Geological Survey, and state agencies to gather data and develop ranking methods. However, there is still no generally accepted method for combining data on individual attributes into a single index. In addition, habitat values for any particular stream must be evaluated in relation to the plants and animals in that region, so any stream habitat index would have to be tailored for different regions.

The technical note for this indicator is on page 237.

SYSTEM DIMENSIONS	CHEMICAL AND PHYSICAL	BIOLOGICAL COMPONENTS	**HUMAN USES**
Extent Pattern	Nutrients, Carbon, Oxygen Contaminants Physical	Plants and Animals Communities Ecological Productivity	**Food, Fiber, and Water** Recreation and Other Services

Water Withdrawals

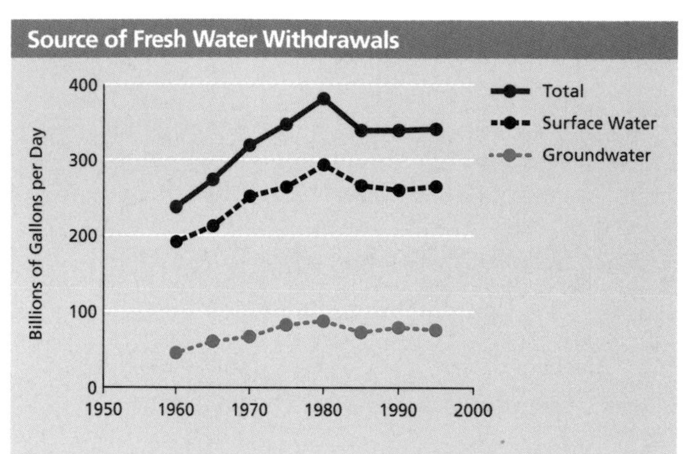

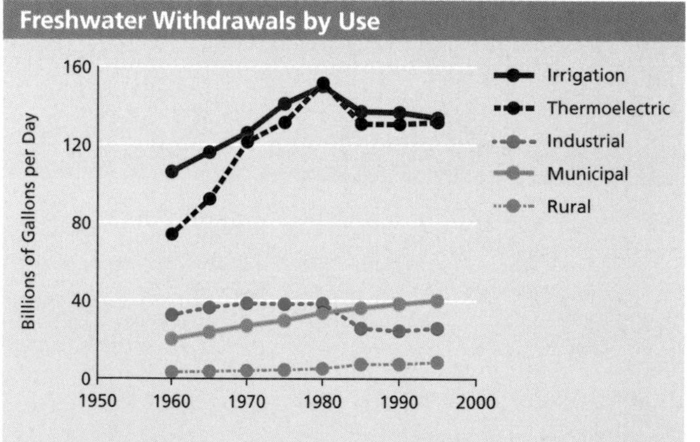

What Is This Indicator, and Why Is It Important? This indicator reports the total amount of surface water and groundwater withdrawn for use in the municipal, rural, industrial, thermoelectric, and irrigation sectors (see the technical note for a description of these categories).

Accurate information about the amount of water being used and what it is being used for will help planners and managers make better decisions about the nation's water resources. Information on water withdrawals can help them assess the effectiveness of alternative water management policies, regulations, and conservation activities and project future demand.

What Do the Data Show? Groundwater and surface water withdrawals increased from 1960 to 1980, and these increases are attributed to increasing demand from all major sectors. Total water withdrawals declined about 10% between 1980 and 1985, then grew slightly from 1985 to 1995. Reduced demand for irrigation, thermoelectric power generation, and self-supplied industrial use was responsible for the decline in total withdrawals between 1980 and 1985; demand in these three sectors was nearly flat from 1985 to 1995. Demand for municipal and rural use has grown steadily over the past few decades, with municipal demand increasing more rapidly.

Discussion For most categories of use, very little water is actually consumed—that is, most of the water withdrawn is returned to the environment for subsequent use by others, although its quality may be lower than when it was initially withdrawn, reducing its suitability for some uses. So, for example, most of the water withdrawn to cool an electric power plant is returned to the river for use downstream for irrigation, municipal water supply, and so on.

The technical note for this indicator is on page 254.

SYSTEM DIMENSIONS	CHEMICAL AND PHYSICAL	BIOLOGICAL COMPONENTS	HUMAN USES
Extent	Nutrients, Carbon, Oxygen	Plants and Animals	**Food, Fiber, and Water**
Pattern	Contaminants	Communities	Recreation and Other Services
	Physical	Ecological Productivity	

⊖ Groundwater Levels

What Is This Indicator, and Why Is It Important? This indicator would report the percentage of the area of the nation's major regional aquifers in which water levels are increasing, decreasing, or stable. The indicator would report what fraction of the aquifer area declined, increased, or remained stable in comparison to a previous period, and it would be reported every 5 years.

Groundwater provides about 40% of the nation's municipal water supply and is the source of much of the water used for irrigation. For most people in rural America, groundwater from their own wells is their only source of water. Groundwater is a major contributor to flow in many streams and rivers, and it has a strong influence on river and wetland habitats for plants and animals.

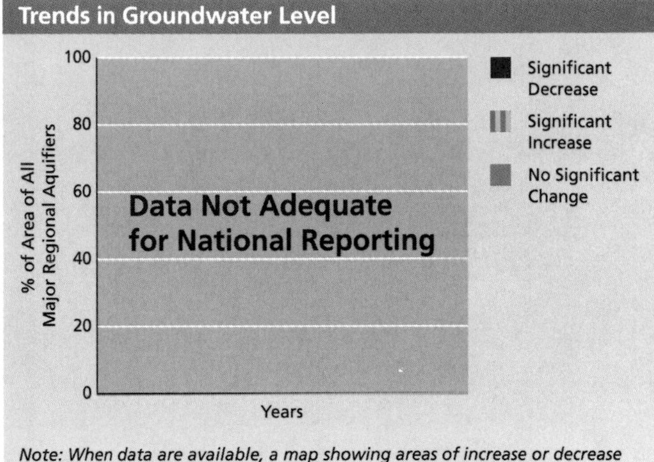

Note: When data are available, a map showing areas of increase or decrease would complement this graph.

Why Can't This Indicator Be Reported at This Time? Data on groundwater levels are collected by federal, regional, state, and local agencies. All states have some coverage, but there are areas of the country for which very little information is available. The data that do exist have not been aggregated to provide systematic measurements of water levels in a significant portion of the nation's major aquifers.

The first step in producing national coverage would be to locate and assess the quality and consistency of existing data. It would then be necessary to aggregate those data and determine where there is sufficient geographic coverage of the major aquifers and adequate characterization of conditions in those aquifers. In areas where data coverage is inadequate, additional measurements would be necessary.

Discussion Changes in water levels reflect changes in the amount of groundwater pumped from or returned to major aquifers; changes may also reflect climate variability or climate change. The measure reports on changes in the quantity of groundwater: it does not address the quality of that water or its suitability for use.

The technical note for this indicator is on page 255.

SYSTEM DIMENSIONS	CHEMICAL AND PHYSICAL	BIOLOGICAL COMPONENTS	HUMAN USES
Extent Pattern	Nutrients, Carbon, Oxygen Contaminants Physical	Plants and Animals Communities Ecological Productivity	**Food, Fiber, and Water** Recreation and Other Services

● Waterborne Human Disease Outbreaks

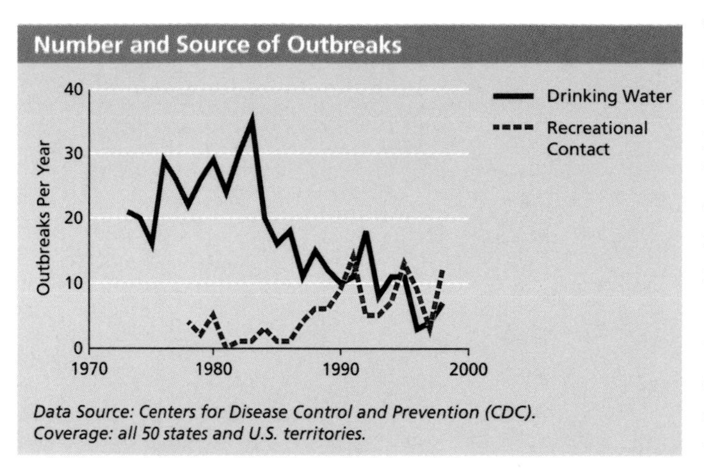

Data Source: Centers for Disease Control and Prevention (CDC). Coverage: all 50 states and U.S. territories.

What Is This Indicator, and Why Is It Important? This indicator reports the number of disease outbreaks—which generally means at least two people getting sick—attributed to drinking water that is untreated or where treatment has failed to remove disease-causing organisms, or to swimming or other recreational contact at lakes, streams, and rivers (see p. 84 for an indicator of coastal recreational water quality).

Ensuring that water is fit to drink and swim in without fear of disease is a basic societal objective. The number of disease outbreaks that can be attributed to contaminated water is a direct measure of the fitness of the nation's waters for these two key uses.

What Do the Data Show? The number of disease outbreaks attributable to contaminated drinking water declined significantly overall from 1973 to 1998. Since 1986, the average number of outbreaks per year was lower than the average during the 1973–1985 period, although there was notable year-to-year variation. There is also notable variation in the number of outbreaks associated with recreational contact, which have increased significantly since 1978. Since 1990, the number of outbreaks associated with drinking water and the number associated with recreational contact have followed a similar pattern.

Discussion This indicator reports outbreaks, not the number of people who become ill. Thus, depending on the location of contamination problems, the size and type of water delivery system, and other factors not related to environmental quality, the trend in the number of people affected may be different from the trend in the number of outbreaks. Doctors and state and local public health officials report data on outbreaks to the Centers for Disease Control and Prevention. Whether an outbreak is identified and reported depends on many factors, so these reports are best considered an indication—rather than a perfect record—of the true incidence of waterborne disease outbreaks.

The technical note for this indicator is on page 255.

⊖ Participation in Freshwater Recreational Activities

What Is This Indicator, and Why Is It Important? This indicator shows the number of days that people took part in a variety of freshwater activities. A "recreation day" for this measure is any day during which a person was engaged in the activity, whether for only a few minutes or for many hours.

Americans enjoy recreation in and around water, from birdwatching and fishing to sailing and swimming. Information on trends in participation documents the demand for recreation opportunities and can be useful in planning for recreational facilities.

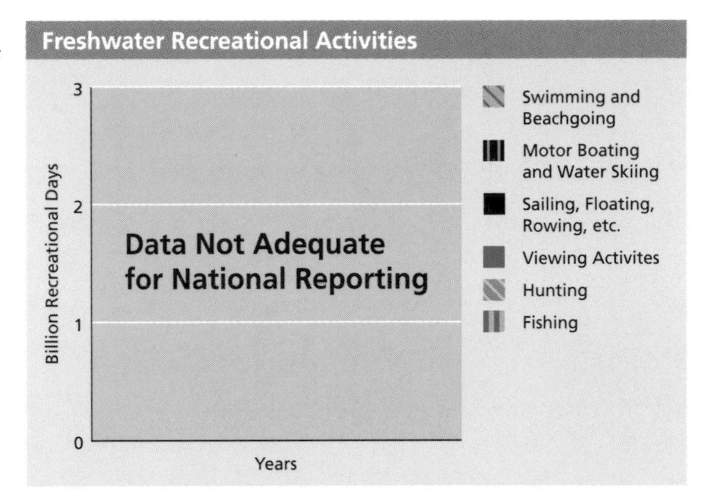

Why Can't This Entire Indicator Be Reported at This Time? Data on national recreation participation have been collected since 1960, but earlier surveys are not compatible with the most recent, and most thorough, survey. Data from 1995 are available for the activities shown in the figure (see the core national recreation indicator, page 60), but except for freshwater fishing, these data do not distinguish what portion of the activities took place in fresh water rather than salt water. In 1995, Americans spent a total of 886 million days fishing in fresh water. The second National Survey on Recreation and the Environment is under way, but it is not clear if it will provide all of the data needed for this indicator.

There is no technical note for this indicator; the technical note (p. 217) for the core national recreation indicator lists the specific activities in each category on the graph above.

Grasslands and Shrublands

	What Indicators are used to describe Grasslands and Shrubands?		Can we report trends? Are there other useful reference points?
SYSTEM DIMENSIONS			
● All Necessary Data Available	Area of Grasslands and Shrublands	How much land is covered by grasslands and shrublands?	Current data only
◐ Partial Data Available	Land Use	How are grasslands and shrublands used? How many acres are used for livestock grazing; oil, gas, and mineral development; rural residences; intensive recreation; "protected areas"; and the Conservation Reserve Program?	Trends
⊖ Data Not Adequate for National Reporting	Area and Size of Grassland and Shrubland Patches	What fraction of grasslands and shrublands is found in patches of various sizes?	No data reported
CHEMICAL AND PHYSICAL CONDITIONS			
⊖ Data Not Adequate for National Reporting	Nitrate in Groundwater	How much nitrate is there in groundwater in grasslands and shrublands?	No data reported
⊖ Data Not Adequate for National Reporting	Carbon Storage	How much carbon is stored in grasslands and shrublands?	No data reported
● All Necessary Data Available	Number and Duration of Dry Periods in Streams and Rivers	How many streams have zero flow for at least one day a year? For these streams, are the dry periods getting longer or shorter?	Trends
⊖ Data Not Adequate for National Reporting	Depth to Shallow Groundwater	What is the depth to shallow groundwater in various parts of the country?	No data reported
BIOLOGICAL COMPONENTS			
◐ Partial Data Available	At-Risk Native Species	How many grassland and shrubland native species are at different levels of risk of extinction?	Current data only
⊖ Data Not Adequate for National Reporting	Non-native Plant Cover	What percentage of grassland and shrubland plant cover is not native to the region?	No data reported
● All Necessary Data Available	Population Trends in Invasive and Noninvasive Birds	Are invasive bird populations increasing more than other bird populations?	Trends
⊖ Data Not Adequate for National Reporting	Fire Frequency Index	Are grassland and shrubland fires occurring more or less frequently than in presettlement times?	No data reported
? Indicator Development Needed	Riparian Condition	What is the condition of stream banks (riparian areas) in grassland and shrubland areas?	No data reported
HUMAN USES			
● All Necessary Data Available	Production of Cattle	How many cattle are fed on grasslands and shrublands?	Trends
⊖ Data Not Adequate for National Reporting	Recreation on Grasslands and Shrublands	How much recreational activity takes place on the nation's grasslands and shrublands?	No data reported

● All Necessary Data Available ◐ Partial Data Available ⊖ Data Not Adequate for National Reporting ? Indicator Development Needed

Chapter 9: Indicators of the Condition and Use of Grasslands and Shrublands

Grasslands and shrublands cover vast tracts of the United States, especially in the West. The sagebrush steppes of the northern Rockies, the prairies of the Midwest and the Great Plains, and the deserts of the Southwest and the intermountain West are all part of this system, as are the Alaskan tundra and shrublands, and the scrublands of Florida. Pastures and haylands, which are also part of this system, are often treated as part of the farmlands landscape. From bare-rock desert to lush coastal meadows, the diversity of grasslands and shrublands is staggering, but their value—ecological, economic, and social—is often overlooked.

Lands dominated by grasses and shrubs are also widely referred to as "rangelands." The definition of rangeland has evolved slowly over the past half-century, from one tied closely to livestock production, to definitions that focus on the natural vegetation found on the land—grasses, grasslike plants, other small broadleaf plants, or shrubs—and on the lack of intensive agricultural management, regardless of how the land is used. Therefore, even though the terms "rangelands" and "grasslands and shrublands" are nearly synonymous, the latter may be less open to misinterpretation by those who continue to associate the term "rangelands" with livestock grazing.

What can we say about the condition and use of grasslands and shrublands?

Fourteen indicators describe the condition and use of grasslands and shrublands in the United States. Partial or complete data are available for six of these indicators, four of which have a long enough data record to enable reporting on trends. Of the eight indicators for which data are not reported, seven have inadequate data for national reporting, while one indicator requires additional development before it will be possible to determine whether data are available.

After the following brief summaries of the findings and data availability for each indicator, the remainder of this chapter consists of the indicators themselves. Each indicator page offers a graphic representation of the available data, defines the indicator and explains why it is important, and describes either the available data or the gaps in those data.

System Dimensions

Three indicators describe the dimensions of the grassland and shrubland system. The first reports the acreage of the major types of land cover: grasslands, shrublands, pasture, and tundra. The second tracks such major land uses as livestock raising, intensive recreation, or rural residences. The third indicator reports the fraction of grassland area and shrubland area that is in patches of different sizes.

- **How much land is covered by grasslands and shrublands?** There are 683 million acres of grasslands and shrublands in the lower 48 states, or about 36% of the total land area. In addition, there are 178 million acres of pastures, some of which—especially if they are not cultivated—have many of the qualities of more "natural" grasslands. There are about 200 million acres of shrubland and tundra in Alaska. It is clear that substantial areas of grassland and shrubland have been converted to other uses since European settlement: in fact, from 1982 to 1997 alone, 11 million acres of nonfederal grasslands and shrublands were converted to other uses.
- **How are grasslands and shrublands used?** Data are not available to report on the acreage used for livestock grazing; oil, gas, and mineral development; rural residences; "protected areas"; and intensive recreation. About 29 million acres, or about 4% of all grasslands/shrublands in the lower

48 states (excluding pastures), are under Conservation Reserve Program (CRP) contracts involving planting to grassland/shrubland cover types.

- **What fraction of grasslands and shrublands is found in patches of various sizes?** Grasslands and shrublands are commonly found intermingled with each other and with forests and woodlands. Changes in the mix of patch sizes and types can affect the value of habitat for different species, may encourage the spread of non-native species, and can change fire frequency and intensity. Data are not adequate to report nationally on this indicator.

Chemical and Physical Conditions

Four indicators describe the chemical and physical condition of grasslands and shrublands. Two are quite similar to two forest indicators: nitrate in water and carbon storage. We track nitrate in groundwater because elevated concentrations of this nutrient can be a sign that inputs from human sources have increased or that plants are under stress. We track carbon storage because carbon is the major building block of grasslands and shrublands and because increased carbon storage can play a role in offsetting emissions of carbon dioxide from burning fossil fuels.

Water—or the lack of it—is important in many grassland and shrubland regions. Two indicators focus on the availability of water in these low-rainfall regions: we track how often streams go dry and, when they do, for how long; and we report depth to groundwater.

- **How much nitrate is there in groundwater in grasslands and shrublands?** Increasing levels of nitrate (a common form of the essential plant nutrient, nitrogen) in groundwater is a signal that plants in the ecosystem are under stress or that the amount of nitrogen entering the system is increasing. Such increases can come from fertilizer use and disposal of animal waste and from rain and snowfall (acid rain), or from changes in vegetation. Data are not adequate for national reporting on this indicator.
- **How much carbon is stored in grasslands and shrublands?** Soil organic matter (decaying plants and animals) consists primarily of carbon. Organic matter helps soils hold water and can be a source of plant nutrients and a deterrent to erosion. Increased carbon storage by ecosystems can offset emissions of carbon dioxide, of concern because of climate change. Data are not adequate for national reporting on this indicator.
- **How many streams have zero flow for at least one day a year? For these streams, are the dry periods getting longer or shorter?** Flowing water in streams and rivers is important for human needs such as drinking water and irrigation; it also sustains plants and animals, both in the stream and nearby. When this flow stops—even for short periods—both human uses and ecological functioning can be disrupted. In the 1970s, 1980s, and 1990s, the percentage of streams experiencing periods of zero flow was noticeably lower than in 1950s and 1960s. Fewer streams and rivers had longer than average zero-flow periods in the 1970s, 1980s, and 1990s, compared with the 1950s and 1960s.
- **What is the depth to shallow groundwater in various parts of the country?** Shallow aquifers provide water for streamflow and maintain water levels in lakes, are used for a variety of human uses, and can be used directly by some plants. Increased groundwater pumping and reduced percolation of water through soils due to development or soil compaction can lead to lower groundwater availability. Data are not adequate for national reporting on this indicator.

Biological Components

Five indicators describe the biological condition of grasslands and shrublands. As with several of the other ecosystems, we track the fraction of native grassland and shrubland species that are at different levels of risk of extinction. Non-native species often crowd out more desirable natives, so the second indicator tracks the percentage of grassland /shrubland area covered by non-native plants. A third indicator compares population trends in invasive bird species to those of non-invasive birds. A fourth focuses on fire frequency, while the final indicator, still under development, will focus on the condition of riparian areas (stream edges), which serve as cover and feeding habitat for much of the wildlife in these systems.

- **How many grassland and shrubland species are at different levels of risk of extinction?** About 9% of 1700 native animal species that depend on grasslands and shrublands are considered "imperiled" or "critically imperiled," and about 0.5% may already be extinct. When "vulnerable" species are counted, a total of about 17% of grassland and shrubland species are considered "at risk." Interpretation of these data will be greatly enhanced when information on population trends for these species becomes available.
- **What percentage of grassland and shrubland plant cover is not native to the region?** Some non-native species can spread aggressively, reducing habitat for native species. In other cases, non-native plants have been used to control erosion and fire and as livestock feed. Data are not adequate for national reporting on this indicator.
- **Are invasive bird populations increasing more than non-invasive populations?** For most of the past 35 years, about the same proportion of invasive and native, non-invasive bird species were increasing. However, from 1996 to 2000, significantly more populations of invasive species increased—a disparity that should be interpreted as a sign of changing ecosystem conditions only if it persists into the future, because many factors can cause short-term fluctuations in bird populations.
- **Are grassland and shrubland fires occurring more or less frequently than in presettlement times?** The frequency with which fires burn is an important factor in determining the kind of vegetation in many grasslands and shrublands. Data are not adequate for national reporting on this indicator.
- **What is the condition of stream banks (riparian areas) in grassland and shrubland areas?** Riparian areas provide habitat for many grassland and shrubland species. This indicator requires further development.

Human Use

We track two human uses of grasslands and shrublands: production of cattle and recreational use, such as hunting and hiking.

- **How many cattle are fed on grasslands and shrublands?** The number of cattle on grasslands and shrublands declined from about 100 million to 93 million from 1994 to 2001.
- **How much recreational activity takes place on the nation's grassland and shrublands?** A variety of recreational activities, including hunting, fishing, and driving off-road vehicles, takes place on the nation's grasslands and shrublands. Data are not adequate for national reporting on this indicator.

What do we mean by "grasslands and shrublands"?

The name of this system is quite descriptive: lands in which the dominant vegetation is grasses and other nonwoody vegetation, or where shrubs and scattered trees are the norm. Grasslands and shrublands are the parts of the terrestrial landscape that are not generally recognized as forests, cropland, or urban and suburban areas. Examples of grasslands and shrublands include

- Tall, mid-, and shortgrass prairies of the Midwest and Great Plains
- Sagebrush steppes of the northern Rockies
- Palouse prairies of Oregon and Washington
- Florida scrublands
- Coastal grasslands of the Atlantic and Gulf coasts
- Chaparral and savanna in California
- Deserts of the Southwest and intermountain West
- Mountain shrublands
- Shrubland and tundra in Alaska
- Pastures, as long as they are not cultivated

Map 9.1. Ecoregional Scheme Used for the Dry Streams Indicator

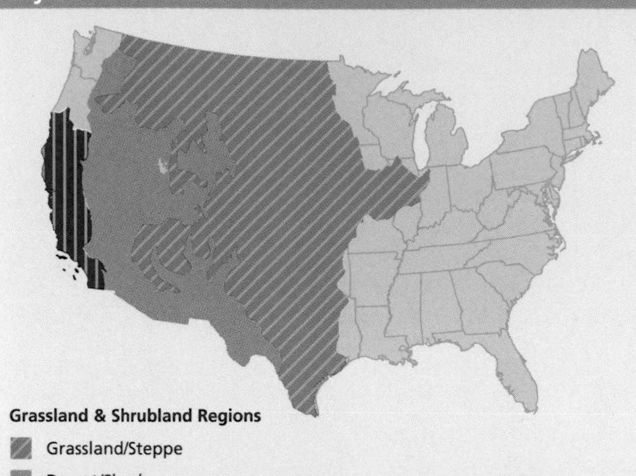

The ecoregions shown are aggregations of the ecoregional divisions from Bailey's system. Obviously, these ecoregions include many areas that are not grasslands or shrublands. However, only the grassland and shrubland areas within these ecoregions were analyzed for the indicator. It is also obvious that many grassland and shrubland areas, particularly in the East, are not shown on this map and were not analyzed for this indicator. Most U.S. grasslands and shrublands, however, are included in the three regions.

Map 9.2. Ecoregional Scheme Used for At-Risk Native Species Indicator

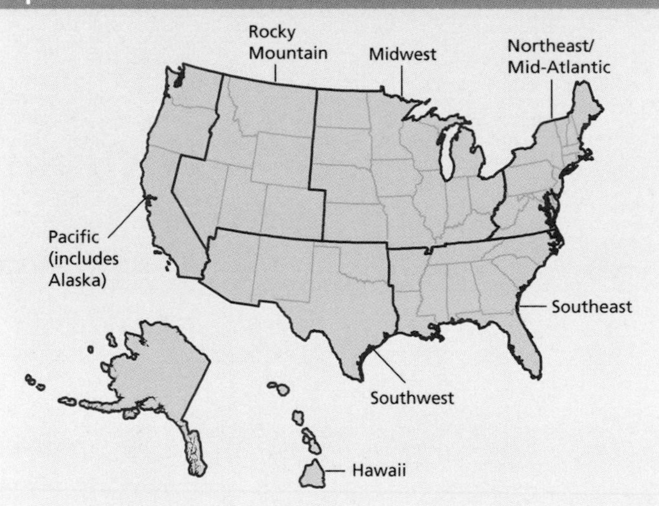

There are overlaps between lands considered here as "grasslands and shrublands" and lands considered elsewhere as either forest or farmlands. For example, lands in the Southwest where pinyon–juniper and chaparral grow are considered shrublands in this chapter, but they are also, based on USDA Forest Service definitions, classified as "forests." And as noted above, pastures are considered both farmlands (since they are clearly part of farming operations and the farmland landscape) and grasslands and shrublands (since, by definition, they are generally covered with grass, with scattered trees or shrubs).

A Note about Regions

In this section, two indicators are reported on a regional basis. The Number and Duration of Dry Periods in Grassland/Shrubland Streams (p. 166) uses a three-region approach based on an ecoregion map developed by the USDA Forest Service and known as Bailey's system, after its author (Map 9.1). The at-risk native species indicator (p. 168) also uses a regional scheme, this one developed by The Heinz Center for use with all at-risk species indicators (Map 9.2).

SYSTEM DIMENSIONS	CHEMICAL AND PHYSICAL	BIOLOGICAL COMPONENTS	HUMAN USES
Extent	Nutrients, Carbon, Oxygen	Plants and Animals	Food, Fiber, and Water
Pattern	Contaminants	Communities	Recreation and Other Services
	Physical	Ecological Productivity	

● Area of Grasslands and Shrublands

What Is This Indicator, and Why Is It Important? This indicator reports the acreage of U.S. grasslands and shrublands (although data are not available for Hawaii). Because grasslands and shrublands are one of the largest ecosystem types in the United States, it is especially important to document changes in their extent.

"Grasslands and shrublands" are any lands that are dominated by grass or shrubs. This includes not only the grasslands and shrublands of the American West, but also coastal meadows, grasslands and shrublands in Florida, mountain meadows, hot and cold deserts, tundra, and similar areas in all states. It also includes pasture- and haylands, which share important characteristics with less-managed grasslands. However, since these areas are also important in describing the area of farmland, they are also included in the extent figures for farmlands (p. 91); see also the national extent indicator (p. 40).

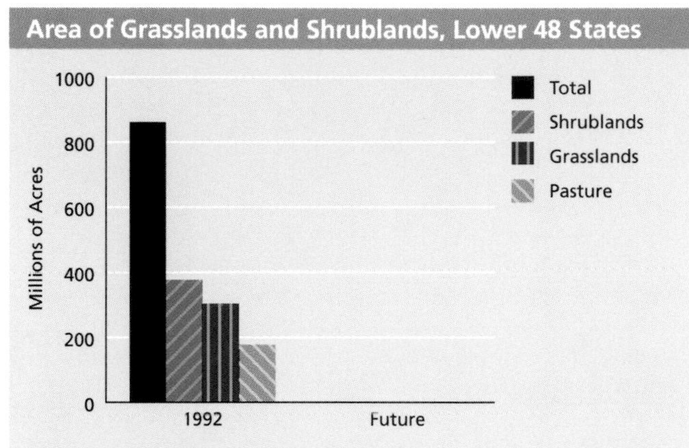

Data Source: Multi-Resolution Land Characteristics (MRLC) Consortium, USGS EROS Data Center. Coverage: lower 48 states.

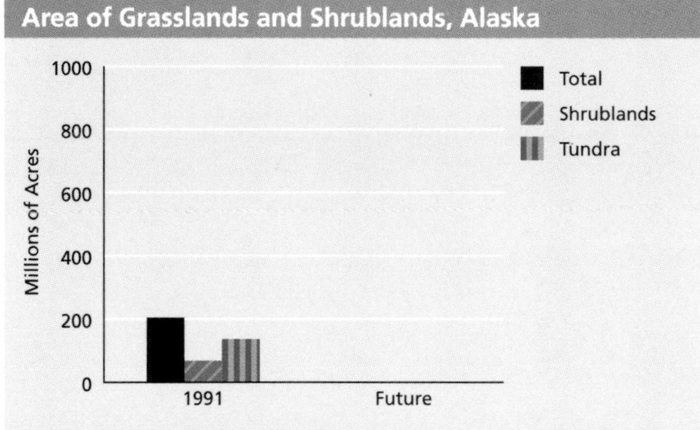

Data Source: Flemming (1996), USGS EROS Data Center. Coverage: Alaska.

What Do the Data Show? In 1992, there were 861 million acres of grasslands and shrublands in the lower 48 states, and 205 million acres in Alaska, for a total of just over 1 billion acres. In the lower 48 states, there were 377 million acres of shrubland, 306 million acres of grassland, and 178 million acres of pasture- and haylands. In Alaska, tundra occupied about 135 million acres and other shrublands about 70 million acres.

Discussion No consistent, nationwide data are available on the change in acreage of grasslands and shrublands. Researchers have estimated that there were between 900 million and 1 billion acres of grasslands and shrublands in the lower 48 states before European settlement, so between 40 million and 140 million acres had been converted to other uses by 1992. However, many pastures are managed in such a way that little of their original grassland character remains. Thus, the area of relatively unmanaged, "natural," grasslands and shrublands has declined more—perhaps substantially more—than the overall figures would indicate. In addition, U.S. Department of Agriculture data indicate that from 1982 to1997, nonfederal grasslands and shrublands declined by about 11 million acres, although the rate of conversion to other land uses slowed substantially after 1992.

The technical note for this indicator is on page 256.

SYSTEM DIMENSIONS	CHEMICAL AND PHYSICAL	BIOLOGICAL COMPONENTS	HUMAN USES
Extent Pattern	Nutrients, Carbon, Oxygen Contaminants Physical	Plants and Animals Communities Ecological Productivity	Food, Fiber, and Water Recreation and Other Services

◐ Land Use

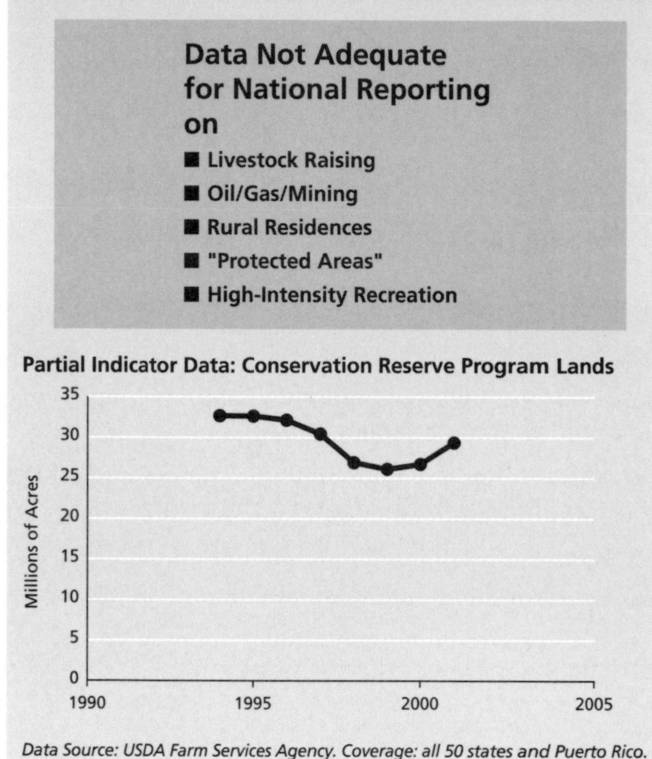

What Is This Indicator, and Why Is It Important? This indicator will describe how much grasslands and shrublands is devoted to six major land uses: livestock raising, rural residences, oil and gas development and mining, Conservation Reserve Program (CRP) lands (see below), "protected areas" (see below), and high-intensity recreation. Because grasslands and shrublands may be used for other purposes as well, the total acreage eventually reported here would not equal the total reported in the extent indicator (p. 161).

Within grasslands and shrublands, differing land uses create very different landscapes. Each of the land uses identified here is associated with specific goods and services and with certain impacts on grasslands and shrublands.

Two land use categories merit brief explanation. The Conservation Reserve Program provides for 10-year lease payments to farmers to remove sensitive lands from production; this indicator includes only acreage on which grass, shrubs, or similar cover (i.e., not trees) are established. "Protected areas" include lands that are primarily managed to maintain biodiversity and natural processes; these are sometimes referred to as "conservation lands."

Why Can't This Entire Indicator Be Reported at this Time? There is no consistent reporting of the amount of land in the categories shown here, with the exception of CRP lands. For example, while data are available on the number of livestock raised (see p. 173), data on the acreage used for this purpose are not available for either public or private lands.

In addition, reporting on this indicator would require the development of consistent definitions for the land use categories used here. For example, what level of recreational use qualifies an area as "high-intensity"? What housing density, over how large an area, qualifies an area as "rural residences"? Which federal, state, and private lands are to be considered "protected areas"?

What Do the Data Show? Currently, data are available only for CRP acreage since 1994. In 2001, there were approximately 29 million acres of lands under active CRP contracts that include planting of grassland or shrubland cover types. This is about 3 million acres more than 1999 acreage and 3 million less than 1994 and 1995 levels.

The technical note for this indicator is on page 257.

SYSTEM DIMENSIONS	CHEMICAL AND PHYSICAL	BIOLOGICAL COMPONENTS	HUMAN USES
Extent	Nutrients, Carbon, Oxygen	Plants and Animals	Food, Fiber, and Water
Pattern	Contaminants	Communities	Recreation and Other Services
	Physical	Ecological Productivity	

⊖ Area and Size of Grassland and Shrubland Patches

What Is This Indicator, and Why Is It Important? This indicator will describe the fraction of grassland area and shrubland area that is in patches of different sizes. Patches of grassland or shrubland are identified separately, and the total area occupied by patches of a certain size will be reported as a percentage of the total area of either grasslands or shrublands.

Patches of grasslands and shrublands are often naturally intermingled with each other and with forest or woodland. Each part of the country has a characteristic mix of small and large patches, and these intermingled patches provide the diversity of habitat types needed by the animals native to a region. (These patches are not static; they may shift over time, so that any single location may switch, for example, from grassland to shrubland, or from shrubland to forest, while maintaining the region's characteristic mix of land cover.) Activities such as fire suppression, grazing, agriculture, and residential, commercial, and industrial development can change this typical pattern, resulting in more or less of an area's grasslands or shrublands being found in large or small patches.

These alterations can create conditions that favor wildfires and affect wildlife populations. For example, fire suppression allows ponderosa pine to invade grasslands. The grassland plants are shaded out, and the grassland animals in the area are restricted to the smaller acreage of grasslands that remains. Non-native cheatgrass can expand into sagebrush (shrubland) following fire, thereby altering future susceptibility to fire and fire frequency patterns and reducing habitat for shrubland species (see the fire frequency indicator, p. 171)

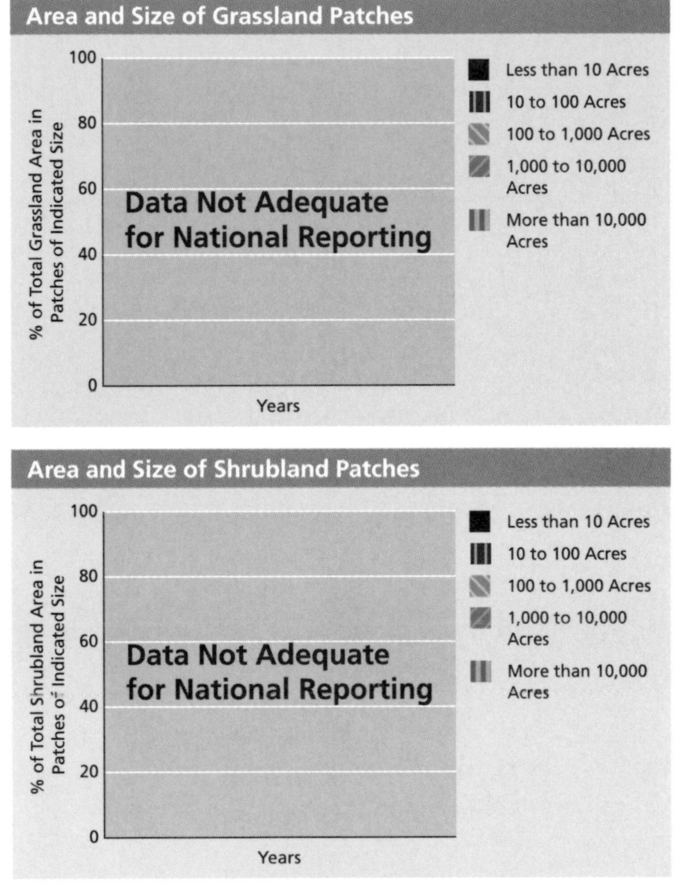

Why Can't This Indicator Be Reported at This Time? The same satellite data used to report on the total acreage of grass and shrublands (see Area of Grasslands and Shrublands, p. 161) can be used to determine the size of patches and thus the total area found in patches of different sizes. However, these data have not been used for this purpose, in part because the methods required for such analyses are not fully developed.

The technical note for this indicator is on page 258.

⊖ Nitrate in Grassland and Shrubland Groundwater

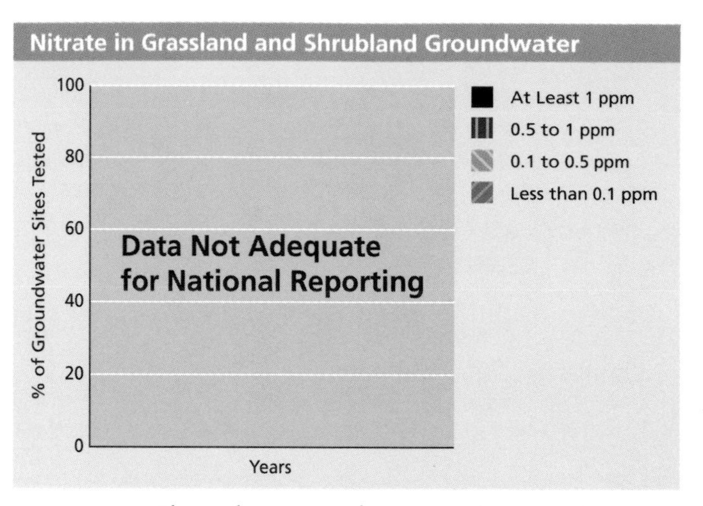

What Is This Indicator, and Why Is It Important? This indicator reports on the concentration of nitrate in groundwater in grassland and shrubland areas. Specifically, the indicator reports the percentage of groundwater sites with average nitrate concentrations in one of four ranges, in areas that are primarily grassland or shrubland.

Nitrate is a naturally occurring form of nitrogen and an important plant nutrient; it is often the most abundant of the forms of nitrogen that are usable by plants. Elevated nitrate in drinking water is a health threat to young children and is of particular concern for people using household groundwater wells; municipal water supply systems typically take steps to remove nitrate.

Elevated amounts of nitrate in the groundwater are a sign that inputs from human sources have increased or that that plants in the system are under stress. Nitrogen is a critical plant nutrient, and most nitrogen is used and reused by plants within an ecosystem. Thus, in less-disturbed grassland or shrubland ecosystems, there is very little "leakage" into either surface runoff or groundwater, and concentrations are very low. Elevated amounts might come from fertilizer use or disposal of animal waste, from rain and snowfall (acid rain), or from changes in vegetation associated with fire suppression or overgrazing.

Why Can't This Indicator Be Reported at This Time? Data on nitrate concentrations in groundwater are available in fragmentary form, collected by many different agencies and institutions using different methods, but they have not been aggregated to enable national reporting. The U.S. Geological Survey's National Water Quality Assessment program, which provides consistent water quality data, is expected to provide sufficient data in the future to allow reporting at a national level.

See also the national nitrogen indicator (p. 46) and the farmlands, forests, and urban and suburban nitrate indicators (pp. 95, 122, and 186).

The technical note for this indicator is on page 258.

SYSTEM DIMENSIONS	CHEMICAL AND PHYSICAL	BIOLOGICAL COMPONENTS	HUMAN USES
Extent	**Nutrients, Carbon, Oxygen**	Plants and Animals	Food, Fiber, and Water
Pattern	Contaminants	Communities	Recreation and Other Services
	Physical	Ecological Productivity	

⊖ Carbon Storage

What Is This Indicator, and Why Is It Important? This indicator will report the total amount of carbon stored in soil and plants in grasslands and shrublands.

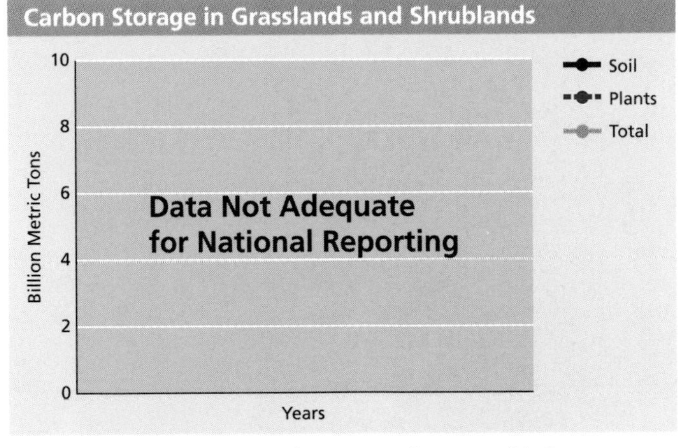

Carbon storage has become an important issue in international negotiations on the management of greenhouse gas emissions, because increased carbon storage can be useful in offsetting emissions of carbon from fossil fuel burning and other sources. The amount of carbon stored in grasslands and shrublands generally changes very slowly. It may be affected by changes in fire frequency, changes in grazing intensity, by the introduction of non-native species, or by conversion of these lands to other uses (like agriculture). In grasslands (including alpine and arctic tundra), more than two-thirds of all carbon is stored in the soil. This contrasts with forests, where significant amounts of carbon are stored in trees (see Forest Carbon Storage, p. 123). Some grassland and shrubland soils normally have low levels of stored carbon; however, at least globally, they are thought to store about half as much carbon as is stored by forests and as much as is stored in croplands.

Carbon in soil—in the form of organic matter, or partially decayed plant and animal matter—helps the soil hold water and supply nutrients to plants; it also protects against erosion and helps support a healthy and diverse set of microscopic plants and animals. Soil carbon is indicative of soil fertility, and some grassland soils are among the most fertile on earth. Further, soil organic matter also stores nitrogen for hundreds and even thousands of years. This helps limit the effect of increasing atmospheric nitrogen deposition, by ensuring that nitrogen does not leach into groundwater (see Nitrate in Groundwater, p. 164).

See also Farmlands Soil Organic Matter (p. 99).

Why Can't This Indicator Be Reported at This Time? There are baseline estimates of soil carbon, but there is no mechanism for regular monitoring of and reporting on carbon storage.

The technical note for this indicator is on page 259.

Number and Duration of Dry Periods in Grassland/ Shrubland Streams and Rivers

Streams That Have Zero-Flow Periods

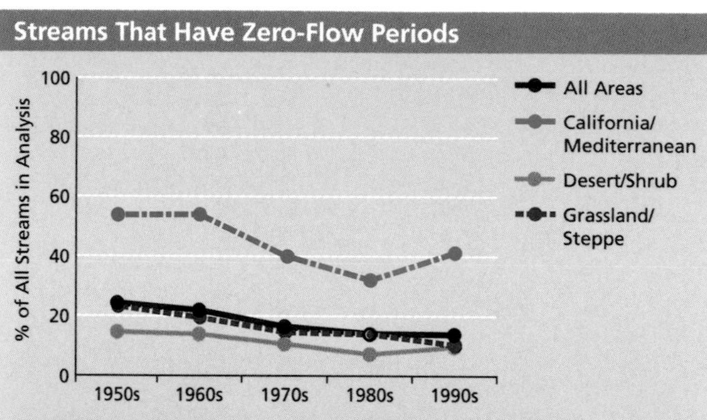

Data Source: U.S. Geological Survey; analysis by Colorado State University. Coverage: grassland/shrubland regions (see map on p. 160) in lower 48 states.

Duration of Zero-Flow Periods (Compared to 50-Year Average)

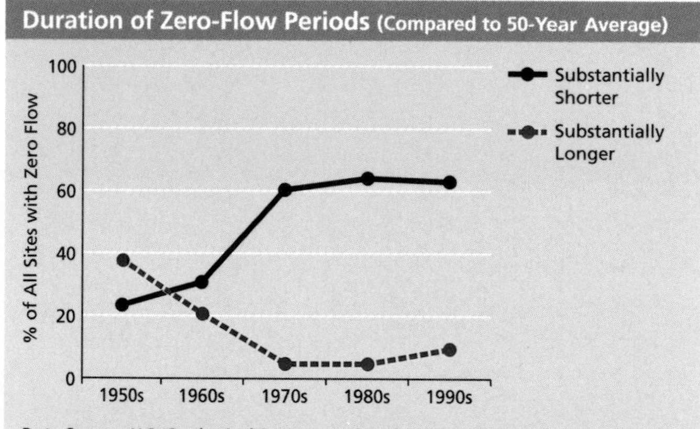

Data Source: U.S. Geological Survey; analysis by Colorado State University. Coverage: grassland/shrubland regions (see map on p. 160) in lower 48 states.

Note: zero-flow periods are "substantially longer" than the long-term average if they are at least twice as long as the average and "substantially shorter" if they are 50% or less of the average.

What Is This Indicator, and Why Is It Important? The indicator tracks the frequency and duration of zero-flow conditions for streams and rivers in grassland/shrubland regions. It reports the percentage of streams and rivers that have at least one no-flow day per year, and the percentage where the duration of zero-flow periods for a given period is substantially longer or shorter than the long-term average.

Stream flow is the lifeblood of uncountable plant and animal species, as well as a major source of water for agricultural, municipal, and other uses. Changes in stream flow can affect plants and animals accustomed to particular levels of flow. No-flow periods may lead to a loss of fish and aquatic animals (although some will survive short periods of zero flow in pools). Depending on the length of the no-flow period, streamside vegetation and the wildlife habitat it provides will gradually be lost. In other cases, the absence of a no-flow period (as in regulated flow below a dam) may also lead to shifts in the animals and plants living in and around streams and rivers.

Some no-flow periods occur naturally. Others occur because of increased water use for domestic, irrigation, or other purposes, or because of changes in land use (e.g., grazing or development) or vegetation that modify the flow of surface water and the recharge of groundwater (e.g., expansion of deep-rooted vegetation such as pinion-juniper woodlands can draw down surface aquifers). No-flow periods may also be due to changing weather or climate, such as the longer periods of drought in recent decades (e.g., mid-1970s), while return of year-round flows may coincide with wet periods (e.g., mid-1980s).

What Do The Data Show? The percentage of streams with no-flow periods has decreased in all grassland/shrubland regions of the West. The 1950s and 1960s showed similar percentages of no-flow, while the 1970s, 1980s, and 1990s recorded noticeably lower percentages. During the relatively wet 1980s, both the California/Mountain and the Desert/Shrub ecoregions had a noticeably lower percentage of streams and rivers with no-flow periods, although the California/Mountain region consistently has the highest percentage of no-flow streams.

The number of streams and rivers with longer than average zero-flow periods decreased in the 1970s, 1980s, and 1990s, compared to the 1950s and 1960s.

The technical note for this indicator is on page 259.

SYSTEM DIMENSIONS	CHEMICAL AND PHYSICAL	BIOLOGICAL COMPONENTS	HUMAN USES
Extent	Nutrients, Carbon, Oxygen	Plants and Animals	Food, Fiber, and Water
Pattern	Contaminants	Communities	Recreation and Other Services
	Physical	Ecological Productivity	

⊖ Depth to Shallow Groundwater

What Is This Indicator, and Why Is It Important? This indicator will describe the depth to shallow groundwater in grassland and shrubland areas. Specifically, it will report the percentage of grassland and shrubland areas where the depth to groundwater falls within several ranges (less than 5 feet, 5 to 10 feet, 10 to 20 feet and more than 20 feet). (The freshwater groundwater level indicator, p. 151, deals with deeper regional aquifers.)

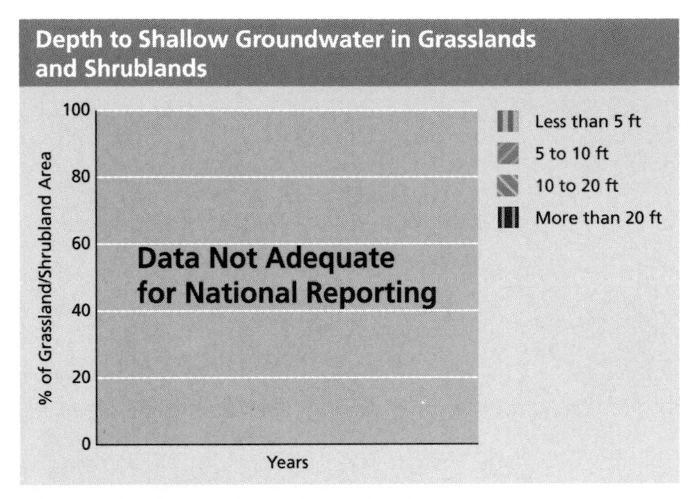

When groundwater levels drop, wetland and streamside (or riparian) plant communities decline, springs and streams dry up, and lake levels drop.

Shallow groundwater aquifers are generally the primary water source for springs, seeps, wetlands, potholes, and riparian areas, all of which provide habitat for plants and animals. Groundwater levels can increase, or be recharged, directly from streams and rivers, or from the percolation through soil of rainwater or melted snow. This recharge is reduced when the ground is compacted or when it is covered completely (by development, for example), and less water can seep into the soil. Groundwater pumping can cause aquifer levels to drop, as can expansion of deep-rooted vegetation, such as pinyon–juniper and western juniper woodlands. Less commonly, higher water tables have provided additional flows to streams, wetlands, and springs.

Why Can't This Indicator Be Reported at This Time? Although depth to deep groundwater or regional aquifers is regularly measured in monitoring and withdrawal wells across the country, there are limited data on shallow aquifers. A few states have mapped shallow aquifer levels, but these data have not been integrated.

Integration of data on shallow groundwater from different studies, complemented by expanded monitoring, is needed to support reporting for this indicator. Because shallow groundwater depth is particularly important for the maintenance of riparian and wetland communities, measuring shallow groundwater depth along rivers and streams should be a higher priority than measuring it in other areas.

The technical note for this indicator is on page 260.

◐ At-Risk Native Grassland and Shrubland Species

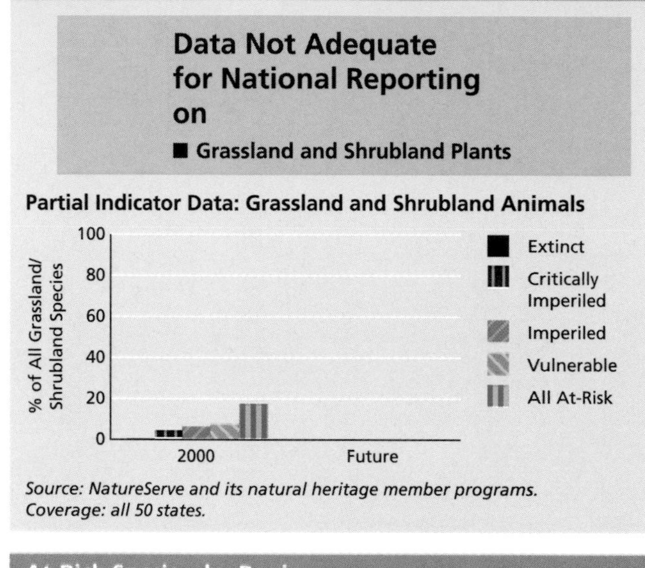

Source: NatureServe and its natural heritage member programs. Coverage: all 50 states.

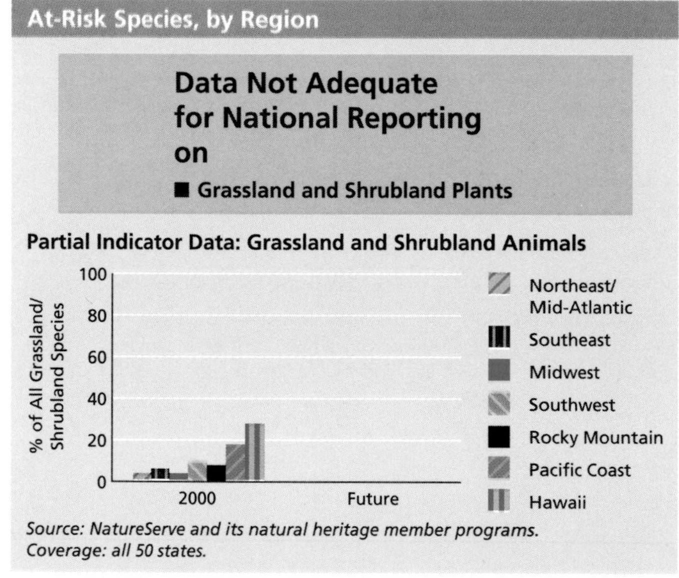

Source: NatureServe and its natural heritage member programs. Coverage: all 50 states.

What Is This Indicator, and Why Is It Important? This indicator reports on the status of native grassland and shrubland species with respect to their *relative risk of extinction.* These status ranks are based on multiple factors: the number and condition of individuals and populations, the area occupied by the species, population trends, and known threats. Degrees of risk reported here range from very high ("critically imperiled" species often are found in five or fewer places or have experienced very steep declines) to moderate ("vulnerable" species often are found in fewer than 80 places or have recently experienced widespread declines). In all cases, a wide variety of factors contribute to overall ratings. "Grassland and shrubland species" live in these habitats during at least part of their life cycle and depend on them for survival.

Species are valued for a variety of reasons: they provide valuable products, including food, fiber, and, more recently, genetic materials; they are key elements of ecosystems, which themselves provide valuable goods and services; and many people value them for their intrinsic worth.

Why Can't This Entire Indicator Be Reported? This indicator reports on mammals, birds, reptiles, amphibians, grasshoppers, and butterflies. Data on other groups have not been included either because too little is known to assign to risk categories or, as with most plants, because determinations as to which are associated with forests, or grasslands, or other habitats has not been completed.

What Do the Data Show? About 3.5% of native grassland/shrubland animal species are critically imperiled, 6% are imperiled, and 0.5% are or may be extinct. When vulnerable species (7%) are counted, about 17% of grassland/shrubland animal species are considered "at risk." Hawaii has a much larger percentage of at-risk grassland and shrubland species than any other region.

Interpreting these figures is complicated, however, because some species are naturally rare. Thus, the rankings are influenced by differences in the number of naturally rare species among regions and species groups as well as different types and levels of human activities that can cause species declines. Interpretation of these data will be greatly enhanced when it is possible to present information on population trends for these at-risk species.

See also the national at-risk species indicator (p. 52), plus those for coastal, forest, and freshwater species (pp. 75, 124, and 144), and for species in farmland (p. 103) and urban/suburban areas (p. 191).

The technical note for this indicator is on page 214.

SYSTEM DIMENSIONS	CHEMICAL AND PHYSICAL	**BIOLOGICAL COMPONENTS**	HUMAN USES
Extent	Nutrients, Carbon, Oxygen	**Plants and Animals**	Food, Fiber, and Water
Pattern	Contaminants	Communities	Recreation and Other Services
	Physical	Ecological Productivity	

⊖ Non-native Plant Cover

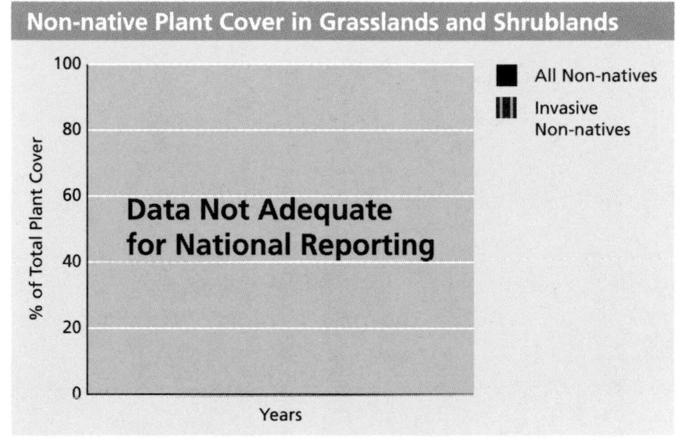

What Is This Indicator, and Why Is It Important? This indicator will report the percentage of plant cover in grasslands and shrublands that is made up of non-native species. The indicator will report on both invasive non-native species (those that spread aggressively) and all non-native species.

Plants that are not native to an area may be highly invasive, crowding out native plants, making areas more susceptible to catastrophic fire, and radically changing the way an ecosystem functions. However, some non-natives can help stabilize eroding soils, serve as part of a grazing system, and act as a barrier to fire. Non-native species such as crested wheatgrass are intentionally seeded for these purposes, although overuse may result in reducing natural ecosystem function.

Some of the most troublesome non-native plants—such as cheatgrass—are much more likely than native plants to increase fire frequency. Exacerbating the problem, cheatgrass easily colonizes recently burned land, further increasing an area's flammability. Some invasives are known as "noxious" plants (examples include leafy spurge, spotted knapweed, and Canada thistle)—they cause only problems and are of generally agreed to provide no benefit in grassland/shrubland management. See also the invasive bird indicator, p. 170.

Nearly all grassland and shrubland areas in the western United States have been adversely affected by invasive species like the yellow star thistle, European wild oats, tamarisk, African lovegrass, purple loosestrife, and Russian olive. As non-native plants cover more of the landscape, they make it increasingly difficult to manage native grassland/shrubland resources and to conserve natural ecosystems and associated ecosystem services.

Why Can't This Indicator Be Reported at This Time? Although many state and federal agencies, nongovernmental organizations, and universities collect data on non-native plants, these data have not yet been brought together to provide consistent information over large areas. Many states do collect data on "noxious" plants on grasslands and shrublands, but this is only a subset of the data needed for this indicator.

The technical note for this indicator is on page 261.

● Population Trends in Invasive and Non-invasive Grassland/Shrubland Birds

Population Trends of Invasive and Native, Non-invasive Birds

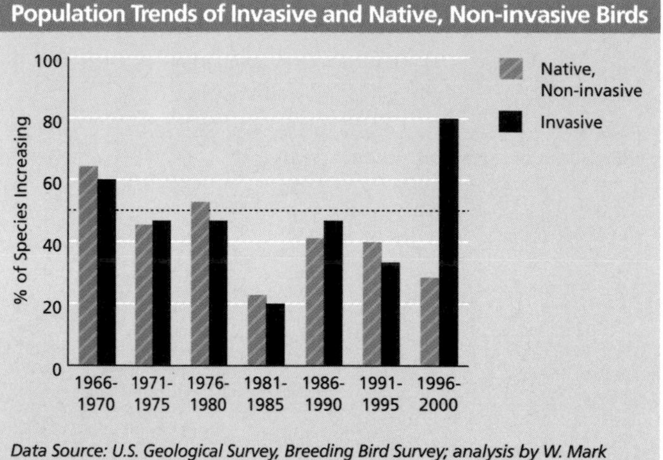

Data Source: U.S. Geological Survey, Breeding Bird Survey; analysis by W. Mark Roberts. Coverage: selected grassland and shrubland areas (see the technical note).

What Is This Indicator, and Why Is It Important? This indicator describes population trends for selected grassland/shrubland bird species by comparing trends for selected "invasive" species with those that are not invasive. Invasive species spread aggressively and can disrupt established native bird populations. While many invasive species are non-natives, they can also be native species that, because of a change in conditions, are able to spread aggressively. This indicator reports the percentage of selected invasive and non-invasive native bird species that had increasing populations in grassland/shrubland ecosystems during 5-year intervals.

Birds respond quickly to environmental change. Because they are highly mobile (they can fly to a new location), birds will simply leave grassland and shrubland areas that no longer meet their environmental needs. The invasive species reported here are believed to be indicative of agricultural conversion, landscape fragmentation due to suburban and rural development, and the spread of exotic vegetation (see the non-native plant cover indicator, p. 169). Native, non-invasive species depend on relatively intact, high-quality native grasslands and shrublands.

What Do the Data Show? For most of the past 35 years, populations of invasive and native, non-invasive bird species were increasing in about the same proportion. However, from 1996 to 2000, significantly more populations of invasive species increased—a disparity that should be interpreted as a sign of changing ecosystem conditions only if it persists: many factors can cause short-term fluctuations in bird populations.

Discussion Bird populations fluctuate normally, even if only to a small degree. When about half the species are increasing and half decreasing over a given period, there will be no consistent increase or decline for the group of species. Populations of invasive birds that consistently increase compared to native birds would be interpreted as a sign that conditions favoring invasives—agricultural conversion, landscape fragmentation due to suburban and rural development, and the spread of exotic vegetation—are increasing. A higher percentage of native, non-invasive birds with increasing populations could indicate that conditions have not changed, or that these birds are adapting to changed conditions.

The technical note for this indicator is on page 262.

⊖ Fire Frequency

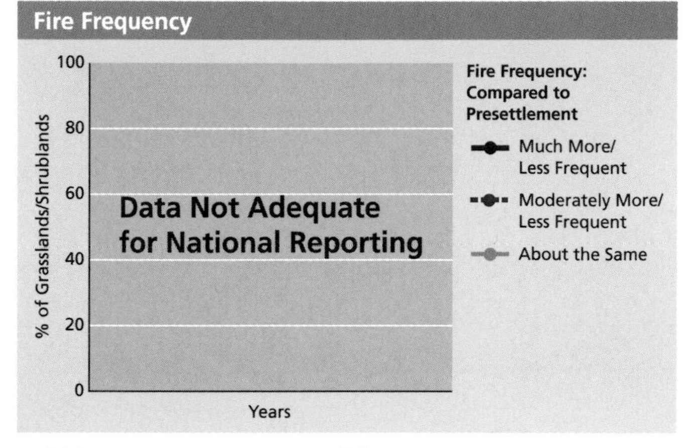

What Is This Indicator, and Why Is It Important? This indicator will describe how often grassland and shrublands are burned by wildfire. Specifically, it will report the fraction of grassland and shrubland areas that burn much more or less often, moderately more or less often, or about as often as before European settlement. So, for example, an area that, historically, burned every 5 years on average might be considered moderately altered if it burns every 10 years and significantly altered if it burns only every 25 years. An area that, historically, burned every 80 years might be considered moderately altered if it burns every 40 years and significantly altered if it burns every 20 years. (Presettlement conditions are used here as a reference against which to compare current conditions, not as an implied management goal.)

Periodic fire helps determine the makeup of grasslands and shrublands, by allowing certain "fire-adapted" species to thrive, while removing other, less fire-tolerant, plants. Since the last Ice Age (about 10,000 years ago), most grasslands and shrublands in the lower 48 states have burned regularly, with fires started by lightning or by American Indians for agricultural and other reasons. Different areas burned at different intervals, ranging from 2 years in eastern grasslands to about every 80 years in intermountain shrub areas.

Active fire suppression or suppression due to the reduction in available fuel resulting from heavy grazing can increase tree and shrub density, decrease the extent of certain "soil-forming" grasses, and enhance the spread of species formerly controlled by fire. For example, a decrease in fire frequency in some sites in the Great Basin is resulting in conversion from mountain big sagebrush and Idaho fescue to western juniper and pinyon–juniper. In addition, some non-native species, such as cheatgrass, increase the frequency and intensity of fires. Since native plants and animals did not evolve under these conditions, these new fire regimes can give non-native species an additional advantage.

See also the fire frequency in forests indicator (p. 128).

Why Can't This Indicator Be Reported at This Time? This indicator requires information on both current and historic fire frequency. Satellite data can be used to measure current fire frequencies. Field-based measurements of historic fire frequency, upon which this indicator depends, are difficult and may not be possible to obtain for many grasslands and shrublands. Fire frequency data have been measured (from tree ring scars and similar evidence) at only a few sites.

The technical note for this indicator is on page 243.

⑦ Riparian Condition

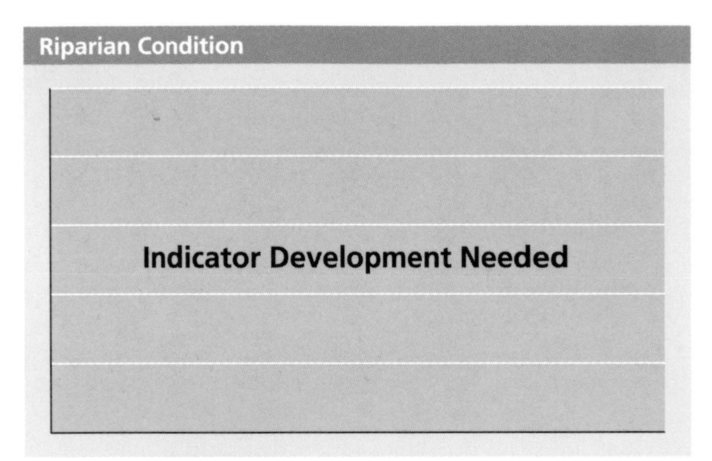

What Is This Indicator, and Why Is It Important? This indicator will describe the condition of riparian (streamside) areas. The condition of these areas will be rated using an index that combines key factors such as water flows, streambed physical condition, riparian vegetation composition and structure, and use by various species.

Riparian areas, the usually vegetated areas along streams and rivers, provide habitat for a variety of wildlife. They serve as cover and feeding habitat for a high percentage of the animal species in grasslands and shrublands and provide important services, such as trapping sediment, modifying flood flows, and increasing groundwater recharge. Changes in riparian condition can enhance or degrade these functions. See related farmland, freshwater, and urban/suburban indicators (pp. 105, 149, and 185)

The condition of riparian areas often reflects influences from outside the immediate area, so they serve to indicate changes throughout a watershed. For example, shifts in vegetation or increased suburban development in a watershed can change the amount and timing of stream flows (see stream flow indicators, pp. 142 and 166), which affects both the streambed and the riparian zone. Other potential influences include the regulation of water flow by dams, bank stabilization, diversions of water for irrigation and other uses, changes in land use in the watershed (such as increases in agriculture or grazing), and changes in vegetation (including the establishment of non-native species) or fire frequency in the watershed.

Why Can't This Indicator Be Reported at This Time? There is no adequate and generally accepted single measure of "riparian condition," although researchers have developed several approaches to the design of such an index. As noted above, such a measure should take into account multiple factors, including hydrology (e.g., relationship to natural flow patterns), geomorphology (e.g., stream sediment transport), and biology (e.g., canopy cover) to provide an overall index of condition.

The technical note for this indicator is on page 263.

SYSTEM DIMENSIONS	CHEMICAL AND PHYSICAL	BIOLOGICAL COMPONENTS	HUMAN USES
Extent Pattern	Nutrients, Carbon, Oxygen Contaminants Physical	Plants and Animals Communities Ecological Productivity	**Food, Fiber, and Water** Recreation and Other Services

Production of Cattle

What Is This Indicator, and Why Is It Important? This indicator reports the number of cattle grazing on grasslands and shrublands (including pastures), rather than at feedlots, during July of each year.

Cattle are reported as an indicator of overall use of these lands for raising livestock. Cattle production is an important economic use of grasslands and shrublands and also remains an important part of the community identity of many parts of the country. Over 90% of beef cattle graze on grasslands and shrublands, at least during the summer months.

What Do the Data Show? The number of cattle on grasslands and shrublands declined from about 100 million in 1994 to 93 million in 2001.

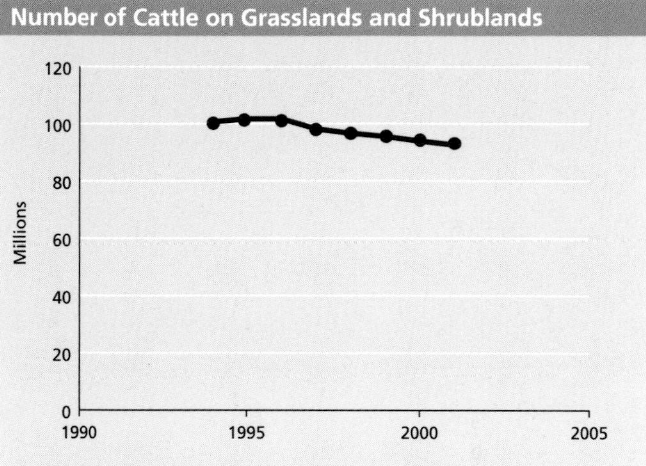

Data Source: USDA National Agricultural Statistics Service. Coverage: all 50 states.

Discussion The changes shown here may be part of a repeating pattern of roughly 10-year cycles that have occurred regularly since the 1880s. Alternatively, there are other data (also not shown here) suggesting that an historical peak in cattle production occurred in the mid-1970s, and that the national herd size has declined by about one-quarter since then. (These earlier data are not comparable to the numbers presented here, because they are based on January inventories, which are believed to underrepresent the number of cattle on grasslands and shrublands.)

July inventories of cattle grazing on grasslands and shrublands (including pastures) are believed to be the most representative of overall grazing use. While nearly all cattle spend some time in feedlots before slaughter, those that graze on grasslands or shrublands (including pastures) are likely to be in these areas during July. (Some cattle spend part of the summer in forests rather than on grasslands and shrublands, but their numbers are unknown. Finally, the distribution of cattle on public and private grasslands and shrublands may change over time for a variety of economic and policy reasons. In the future, tracking this split may be desirable.

The technical note for this indicator is on page 264.

⊖ Recreation on Grasslands and Shrublands

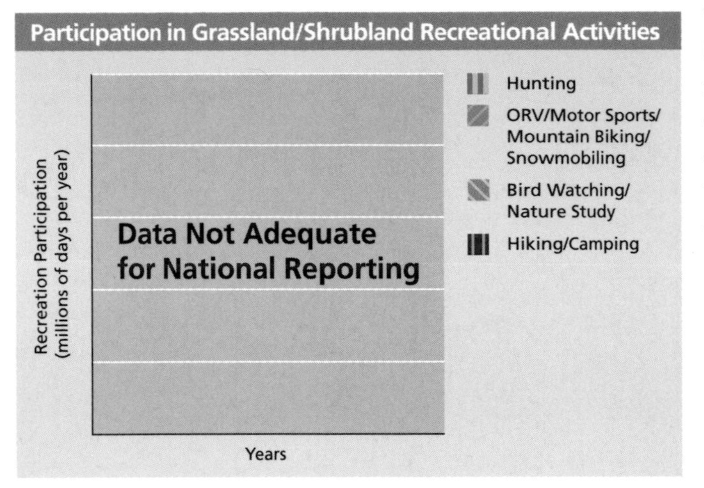

What Is This Indicator, and Why Is It Important? This indicator will report the number of days per year that people engage in a variety of recreational activities on the nation's grasslands and shrublands. Activities will include hunting; off-road vehicle (ORV) driving, motorsports, mountain biking, and snowmobiling; bird watching and nature study; and hiking and camping. (Other categories necessary to describe grassland/shrubland recreation more fully may be added when data become available.)

A great deal of recreation takes place on grasslands and shrublands. These lands provide a benefit to society through recreation in much the same way that they support the production of cattle (p. 173).

Why Can't This Indicator Be Reported at This Time? There are no national data sets that document the type and amount of recreation on grasslands and shrublands. The National Survey of Fishing, Hunting, and Wildlife-Associated Recreation (http://fa.r9.fws.gov/surveys/surveys.html) and the National Survey on Recreation and the Environment (http://www.srs.fs.fed.us/trends/nsre.html) both provide reliable data on these activities, but neither survey identifies whether these activities take place on grasslands or shrublands, in forests, on farmlands, or elsewhere.

Adequate reporting would require modification of existing surveys to elicit information either on the location of recreational activities or on the amount of recreation in grassland/shrubland areas.

See also the indicators of recreational activity in farmlands (p. 109), forests (p. 132), and fresh waters (p. 153), as well as the core national recreation indicator (p. 60).

There is no technical note for this indicator.

Urban and Suburban Areas

What Indicators Are Used To Describe Urban and Suburban Areas?			Can we report trends? Are there other useful reference points?
SYSTEM DIMENSIONS			
●	Area of Urban and Suburban Lands	How much land do "urban and suburban areas" occupy? How much of this land is developed, and how much is forest, grasslands and shrublands, wetlands, and croplands?	Current data only, regional comparison
?	Suburban/Rural Land Use Change	How are patterns of development changing at the boundary between suburban and rural areas?	No data reported
●	Patches of Forest, Grassland and Shrubland, and Wetlands	How large are urban/suburban forests, grasslands and shrublands, and wetlands, which provide green space and wildlife habitat?	Current data only, regional comparison
⊖	Total Impervious Area	How much urban/suburban land is covered with buildings, concrete, asphalt, and other "hard," or impervious, surfaces?	No data reported
?	Stream Bank Vegetation	What fraction of urban/suburban stream banks are vegetated?	No data reported
CHEMICAL AND PHYSICAL CONDITIONS			
●	Nitrate in Urban and Suburban Streams	How much nitrate is found in urban/suburban streams?	Current data only, federal standard
●	Phosphorus in Urban and Suburban Streams	How much phosphorus is found in urban/suburban streams?	Current data only, federal goal
●	Air Quality (High Ozone Levels)	How common are air pollution levels that exceed federal guidelines in urban/suburban areas?	Trends, federal standard
◐	Chemical Contamination	What levels of artificial compounds and heavy metals are found in water and soil?	Current data only, federal standards and guidelines
?	Urban Heat Island	How much hotter are urban/suburban areas than less-developed areas nearby?	No data reported
BIOLOGICAL COMPONENTS			
⊖	Species Status	How many of the plants and animals that once inhabited areas that are now urban/suburban are locally at risk or absent?	No data reported
⊖	Disruptive Species	Are there more or fewer "disruptive species" such as white-tailed deer and Scotch broom in urban/suburban areas?	No data reported
⊖	Status of Animal Communities in Urban and Suburban Streams	What is the condition of fish and bottom-dwelling animals in urban/suburban streams?	No data reported
HUMAN USES			
⊖	Publicly Accessible Open Space per Resident	How much public open space is there per urban/suburban resident?	No data reported
?	Natural Ecosystem Services	What other important ecosystem services are provided by urban/suburban areas?	No data reported

● All Necessary Data Available ◐ Partial Data Available ⊖ Data Not Adequate for National Reporting ? Indicator Development Needed

Chapter 10: Indicators of the Condition and Use of Urban and Suburban Areas

Thinking of America's cities and suburbs as an ecosystem does not come automatically to many people. "Ecosystems" are usually defined by plants, animals, naturally occurring attributes like landscape type, and the interaction among these elements. People, by contrast, create cities and suburbs, and it is the built environment—houses, office buildings, factories, roads, sidewalks, piers, parking lots—that defines them. Although they occupy less than 2% of the land area of the lower 48 states, cities and suburbs are home to more than 75% of all Americans. Characterized by a great many people living in a very small area, much of it covered by a variety of hard surfaces, cities and suburbs nevertheless provide a range of goods and services not unlike those provided by the other, more "natural," ecosystems. In fact, 20% of urban and suburban land is forest, farmland, wetlands, or grassland and shrubland; streams run through cities and suburbs, and many cities lie on the coast. It is in this unconventional ecosystem that people interact most often with nature.

What can we say about the condition and use of urban and suburban areas?

Fifteen indicators describe the condition and use of urban and suburban areas. Partial or complete data are available for six of these indicators. Only one (air quality) has a long enough data record to judge trends, while four can be compared to a regulatory standard or guideline. For five indicators, data are not available for reporting on a national basis, and four indicators require additional development before it will be possible to assess the availability of data.

After the following brief summaries of the findings and data availability for each indicator, the remainder of this chapter consists of the indicators themselves. Each indicator page offers a graphic representation of the available data, defines the indicator and explains why it is important, and describes either the available data or the gaps in those data.

System Dimensions

Five key indicators describe the dimensions of the urban/suburban system. The first and most basic is how much land these areas occupy, and how much is developed or remains as forest, grassland or shrubland, or other undeveloped land. A second indicator, still requiring development, would track conversion of land from rural to suburban. Three other indicators provide further detail on the character of urban and suburban lands. One tracks the size of the patches of forest, grasslands and shrublands, and other natural areas that provide green space and wildlife habitat; a second tallies the fraction of urban and suburban lands covered by asphalt, buildings, and other impervious surfaces that prevent the penetration of rainfall and on which plants cannot grow. A final measure will track the percentage of urban streams that are lined with vegetation, which can have a significant effect on water quality and which also serves as wildlife habitat.

- **How much land do "urban and suburban areas" occupy? How much of this land is developed, and how much is forest, grasslands and shrublands, wetlands, and croplands?** In 1992, urban and suburban areas, as defined by this report, accounted for about 32 million acres in the lower 48 states, or about 1.7% of total land area. About 22% of urban and suburban land in the South, Northeast, and West was undeveloped; in the Midwest, this figure was 17%.
- **How are patterns of development changing at the boundary between suburban and rural areas?** When suburban development expands into rural areas, the pattern of development—how dense or spread out it is; how transportation, water, sewer, and other infrastructure are integrated, and so

on—can affect both wildlife and people living in and around newly developed areas. This indicator requires further development.

- **How large are patches of "natural" lands (forests, grasslands and shrublands, and wetlands), which provide green space and wildlife habitat?** Natural lands are important for urban and suburban recreation and quality of life, and they are also important as wildlife habitat. The value of these patches for both people and wildlife can be affected by their size. About half of all natural lands in urban and suburban areas are in patches smaller than 10 acres. A progressively smaller percentage are found in larger patches—nationally, less than 5% of urban/suburban natural lands are found in patches of 1,000 acres or more.
- **How much of urban and suburban areas is covered with buildings, concrete, asphalt, and other "hard," or impervious, surfaces?** Places that have a higher percentage of impervious surfaces often have more, and dirtier, runoff, than places with less "hard" surface. Data are not adequate for national reporting on this indicator.
- **What fraction of urban and suburban stream banks are vegetated?** Vegetation along streams can reduce the effects of runoff and serve as wildlife habitat. This indicator requires further development.

Chemical and Physical Conditions

As is the case for several of the other systems, many of the indicators of chemical and physical condition of urban and suburban areas focus on streams. Because streams receive runoff from the land surface, they are powerful indicators of conditions on that land surface. Two indicators track concentrations of nitrate and phosphorus, nutrients that can, in excess, cause problems. A third indicator tracks contaminants such as pesticides, PCBs, and heavy metals in stream water and soils, while a fourth tracks urban air quality, particularly concentrations of ozone, a key component of smog. Finally, one indicator, requiring further development, tracks the differences in temperature between cities and their surrounding regions.

- **How much nitrate is there in urban and suburban streams?** Nitrate is an important plant nutrient, but it also contributes to water quality problems. Nitrate in drinking water is a health threat for young children, and it must be removed at significant cost by municipalities that rely on river water. Sources of nitrate include sewage treatment plants, animal wastes, and fertilizers. About 60% of urban and suburban stream sites tested have concentrations of nitrate below 1 part per million (ppm); all samples were below the federal drinking water standard of 10 ppm. No trend data are available, but nitrate levels are lower in urban/suburban streams than in streams in agricultural areas, but higher than in forest streams.
- **How much phosphorus is there in urban and suburban streams?** About two-thirds of urban and suburban stream sites had phosphorus levels of at least 0.1 part per million, the upper limit recommended by the Environmental Protection Agency to prevent nuisance algae growth in fresh waters. Phosphorus concentrations in urban and suburban streams are similar to those in farmland streams, and much higher than those in forest streams. Sources of phosphorus in urban streams include sewage treatment plants, animal wastes, some detergents, and fertilizers.
- **How common are air pollution (ozone) levels that exceed federal guidelines in urban and suburban areas?** In 1999 about 55% of monitoring stations in urban and suburban areas recorded high ozone levels on at least 4 days, a number that generally triggers violations of federal air quality regulations. Throughout the 1990s, about 50% of monitoring stations recorded high ozone levels on at least 4 days each year. During the same period, the number of monitors recording high levels on 25 or more days per year declined, to about 5% in 1999.
- **What levels of contaminants (primarily artificial compounds and heavy metals) are found in stream water and soil?** All urban and suburban stream sites had at least one chemical that exceeded guidelines for protection of aquatic life, and 5% of sites had contaminants that exceeded human health standards or guidelines. About 85% of stream sites in urban and suburban areas had an

average of at least five detectable contaminants throughout the year. Data are not adequate to report on contamination in soils in urban and suburban areas.

- **How much hotter are urban and suburban areas than less developed areas nearby?** Asphalt, concrete, and other constructed materials in developed areas absorb solar energy, often leading to higher temperatures than in undeveloped areas. This can raise summertime cooling costs and cause human health problems where air conditioning is not available. This indicator requires further development.

Biological Components

The biological condition of urban and suburban regions is captured by three indicators. The first reports the percentage of presettlement species that are now rare or missing from urban areas. The second tracks the number of disruptive species, like white-tailed deer and Tartarian honeysuckle. The third indicator, also used in the freshwater system, compares the fish and bottom-dwelling communities in urban and suburban streams to those in relatively undisturbed streams.

- **How many of the plants and animals that once inhabited areas that are now urban and suburban are locally at risk or absent?** Conversion of land from rural to urban or suburban is often accompanied by declines in the populations of native plants and animals or loss of species. These declines and losses may be influenced by the kind and amount of development, and by how sensitive different species are to disruption. Data are not adequate for national reporting on this indicator.
- **Are there more or fewer "disruptive species," like white-tailed deer and Scotch broom, in urban and suburban areas?** Disruptive species are those that—whether they are native or non-native—cause problems for people, property, or wildlife. For example, deer are so numerous in some suburban areas that damage to gardens, car accidents, and increased incidence of Lyme disease have become serious political, health, and safety issues. Data are not adequate for national reporting on this indicator.
- **What is the condition of fish and bottom-dwelling animals in urban/suburban streams?** Modifying a stream, through pollution, changes to the streambed or bank, flow modification, or other means can change the number and diversity of fish and bottom dwelling animals. Data are not adequate for national reporting on this indicator.

Human Use

Natural areas in cities and suburbs provide many benefits to people. The most basic measure of the potential for such benefits is the amount of public open space per resident. A second indicator would tally natural ecosystem services, such as purifying air and water, lowering energy consumption, and reducing stormwater runoff.

- **How much public open space is there per urban/suburban resident?** Open space is valued by many urban dwellers for recreation and general "quality of life" reasons. Data are not adequate for national reporting on this indicator.
- **What other important natural ecosystem services are provided by urban and suburban areas?** Undeveloped lands provide a variety of services of value to people, including purification of stormwater by forested areas and wetlands and cooling and noise reduction by shade trees. This indicator requires further development.

What do we mean by the "urban and suburban ecosystem"?

Urban and suburban areas are those places where most of the land is devoted to buildings, houses, roads, concrete, grassy lawns, and other elements of human use and construction.

This system spans a range of density, from high-rise-dominated downtowns to the suburban fringe, where residential tract development gradually thins to a rural landscape. This transition is neither smooth nor uniform. At this outer edge, new developments—some quite dense—may appear in otherwise rural areas and leave intervening areas undeveloped, at least for a time.

This report uses a newly developed approach to define urban and suburban lands. It relies upon the physical characteristics of the land, rather than population density, as is commonly done, and employs two basic criteria. First, a substantial portion of the land must be covered with buildings, roads, concrete, and the like, and second, these areas must be sufficiently large (about 270 acres or more) to be considered "urban / suburban." This method excludes scattered or isolated areas such as small settlements, large parking lots, or single residences, but includes large "natural" areas, such as city parks, which are surrounded by otherwise-urban lands. Details of the methods used for identifying these areas are provided in the Area of Urban and Suburban Lands technical note (p. 264).

A Note about Undeveloped Lands in Urban and Suburban Areas

Three indicators describe undeveloped urban and suburban lands (Area of Urban and Suburban Lands; Patches of Forest, Grassland and Shrubland, and Wetlands; and Publicly Accessible Open Space per Resident). Since these indicators focus on different aspects of undeveloped land, they count different types of such lands. In the Area of Urban and Suburban Lands indicator (p. 181), we classify forests, croplands (including pastures), grasslands and shrublands, and wetlands as "undeveloped lands." In the Patches of Forest, Grassland and Shrubland, and Wetlands indicator (p. 183), we focus on "natural" lands, a subset of undeveloped lands that includes forests, grasslands and shrublands, and wetlands, but not croplands or pastures. Finally, in the per capita open space indicator (p. 194), we report on all lands that are publicly accessible, as long as they are not paved. This includes publicly accessible forests, grasslands and shrublands, and wetlands (natural lands), but also areas such as parks with manicured lawns, ballfields, beaches, and the like. Farms, which provide significant amounts of open space in many areas, are generally privately owned and not typically accessible to the general public.

Map 10.1. Census Bureau Regions

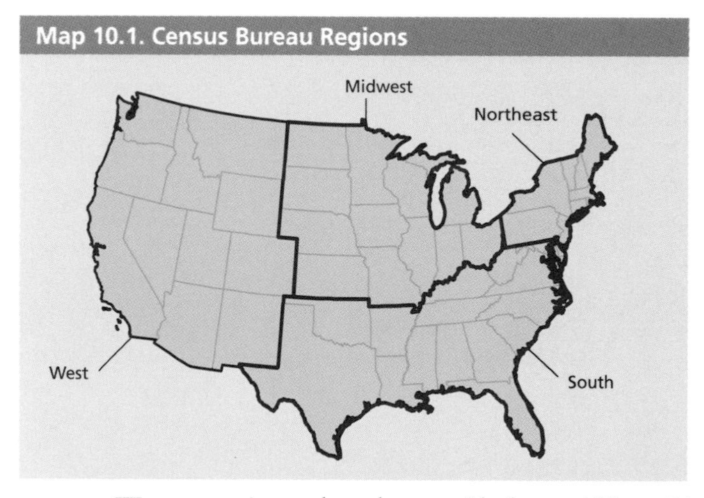

A Note about Regions

Two indicators—Area Of Urban and Suburban Lands (p. 181) and Patches of Forest, Grassland and Shrubland, and Wetlands (p. 183)—are reported on the basis of multistate regions adopted by the Census Bureau (Map10.1). While the data presented for these indicators do not include Alaska and Hawaii, the Census Bureau includes these states in the Western region; when data on Alaska and Hawaii become available, future reports will include them in this region. The air quality indicator is presented in map form. Several other indicators would, if data were available, also be presented in regional or map form.

Finally, many of the indicators included in this report would probably require local financial resources and expertise, which may not be available for all areas meeting the "urban and suburban" definition we used. Therefore, several indicators—Urban Heat Island (p. 190) and Species Status (p. 191), among others—are presented on the basis of the "percent of all metropolitan areas," which is intended to imply that the reporting will be focused on major cities and their surrounding areas. Implementing these indicators might best be achieved by identifying a suite of metropolitan areas (perhaps defined by size or population) to serve as the basis for national and regional reporting.

SYSTEM DIMENSIONS	CHEMICAL AND PHYSICAL	BIOLOGICAL COMPONENTS	HUMAN USES
Extent	Nutrients, Carbon, Oxygen	Plants and Animals	Food, Fiber, and Water
Pattern	Contaminants	Communities	Recreation and Other Services
	Physical	Ecological Productivity	

● Area of Urban and Suburban Lands

What Is This Indicator, and Why Is It Important? This indicator reports the extent of urban and suburban lands, both in acres and as a percentage of all land area in a region; it also reports on the extent and composition of undeveloped lands, such as wetlands, croplands, forest, or grassland and shrubland, contained within urban and suburban areas.

About 75% of all Americans live on land that is urban or suburban in character, which is less than 2% of the lower 48 states. Increases in urban/suburban area are generally permanent and may affect the use and character of surrounding lands (see the land use change indicator, p. 182). Describing the amount and composition of undeveloped lands provides a coarse view of how intensely developed urban and suburban lands are, which is related to the amount and type of open space available to a region's residents (see p. 194), the extent of impervious surfaces (see p. 184), and the services provided by the "natural" systems in urban and suburban areas (see p. 195).

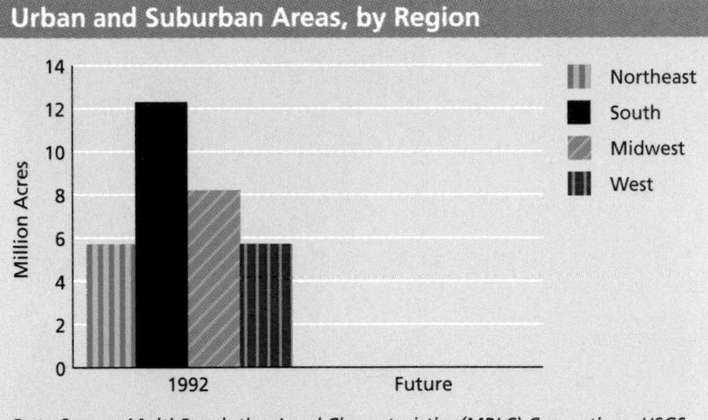

Data Source: Multi Resolution Land Characteristics (MRLC) Consortium, USGS EROS Data Center. Coverage: lower 48 states.

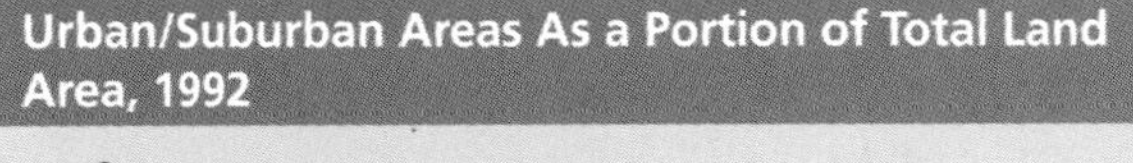

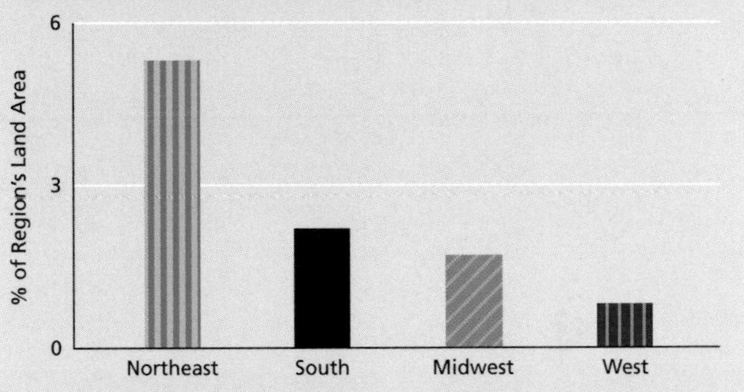

Data Source: Multi-Resolution Land Characteristics (MRLC) Consortium, USGS EROS Data Center. Coverage: lower 48 states.

What Do the Data Show? In 1992, urban and suburban areas occupied 32 million acres in the lower 48 states, or 1.7% of total land area. Most of the land designated urban or suburban is in the South and Midwest, but cities and suburbs account for less than 2% of the land area in those regions. In comparison, urban and suburban lands in the Northeast made up over 5% of the landscape.

The South, Northeast, and West had nearly identical percentages of undeveloped land within their urban and suburban areas (about 22%), while the Midwest had less (17%). In the Northeast and South, forests dominate these undeveloped areas; in the Midwest, farmlands dominate, and in the West grasslands and shrublands dominate.

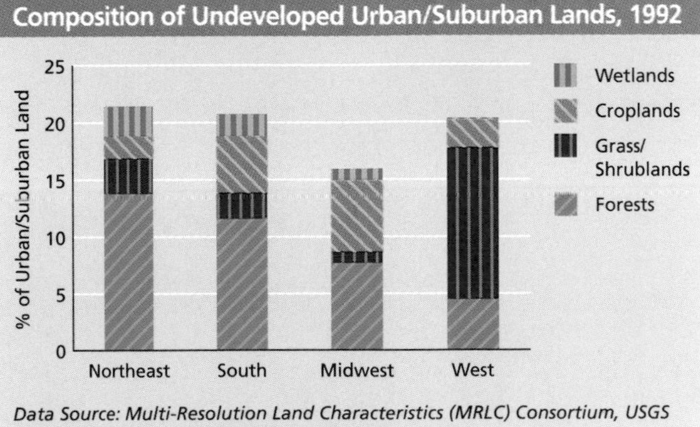

Data Source: Multi-Resolution Land Characteristics (MRLC) Consortium, USGS EROS Data Center. Coverage: lower 48 states.

Discussion The definition of urban and suburban areas used here is fairly restrictive. It focuses on highly urbanized areas and their surrounding suburbs, plus developed outlying areas above a minimum size. It covers residential areas, commercial and industrial areas, parks and golf courses, and the like. It is not delineated on the basis of jurisdictional boundaries, but rather on actual land cover as identified using satellite data, and can be applied repeatedly over time. Other programs (see technical note), such as those that tally all developed lands, whether or not they are sufficiently aggregated to be considered "suburban," identify more developed lands than are reported here.

The technical note for this indicator is on page 264.

SYSTEM DIMENSIONS	CHEMICAL AND PHYSICAL	BIOLOGICAL COMPONENTS	HUMAN USES
Extent	Nutrients, Carbon, Oxygen	Plants and Animals	Food, Fiber, and Water
Pattern	Contaminants	Communities	Recreation and Other Services
	Physical	Ecological Productivity	

⑦ Suburban/Rural Land Use Change

Suburban/Rural Land Use Change

Indicator Development Needed

What Is This Indicator, and Why Is It Important? This indicator will describe the pattern and intensity, or density, of development, both at the outer edge of suburban development around cities, and in rural areas that, despite the lack of a large town center, are growing rapidly toward suburban densities.

Citizens and policymakers alike have expressed strong interest in the nature and pace of suburban development. Patterns of development can directly affect wildlife and the people living in and around newly developed areas. Concerns often focus on the conversion of natural or agricultural land to low-density housing or commercial development, often accompanied by loss of open space; demands for more roads and sewers; increased crowding in public schools; and longer travel times to jobs and stores. Landowners, however, often resist efforts to control or channel development, and some jurisdictions favor continued growth as a means of ensuring steady or increasing tax revenues.

Why Can't This Indicator Be Reported at this Time? Reporting on this indicator will require agreement among land use professionals on the most appropriate measure of changes in suburban and rural land use, and on monitoring of these changes using consistent methods.

Public debate often focuses on "sprawl" or "smart growth," but there is no consensus on how best to measure—and thus to track—these phenomena. Issues include change in overall density, the appropriate mix of commercial and low- and high-density housing, and the degree to which new development is located near existing development or in more remote undeveloped areas.

One type of candidate indicator focuses on the degree to which patches of forests, grasslands, and wetlands are reduced in size and isolated from each other, affecting the amount of wildlife habitat and open space values they provide (see the "natural area" patches indicator, p. 183, the open space indicator, p. 194, and the ecosystem services indicator, p. 195). Another approach focuses on such issues as the amount of time residents spend traveling to stores, jobs, and schools, perhaps measured in vehicle-hours.

There is no technical note for this indicator.

SYSTEM DIMENSIONS	CHEMICAL AND PHYSICAL	BIOLOGICAL COMPONENTS	HUMAN USES
Extent	Nutrients, Carbon, Oxygen	Plants and Animals	Food, Fiber, and Water
Pattern	Contaminants	Communities	Recreation and Other Services
	Physical	Ecological Productivity	

Patches of Forest, Grassland and Shrubland, and Wetlands

What Is This Indicator, and Why Is It Important? This indicator reports how much of the "natural" area within urban and suburban lands is in patches of varying size, from less than 10 acres to greater than 10,000 acres. Natural areas include forests, grasslands and shrublands (including most pasturelands—especially in the west), and wetlands.

Smaller patches of natural habitat generally provide lower-quality habitat for plants and animals (although this is not necessarily true for wetlands) and provide less solitude and fewer recreational opportunities for people. Smaller patches of habitat favor common, human-tolerant species like squirrels, white-tailed deer, starlings, and sparrows, over less common species that require larger areas, such as some birds (pileated woodpeckers, broad-winged hawks, and many warblers), mammals (bears, mountain lions, wolves, coyotes, mink, otters, and weasels), and amphibians.

"Natural" Area in Patches of Different Sizes (nationally)

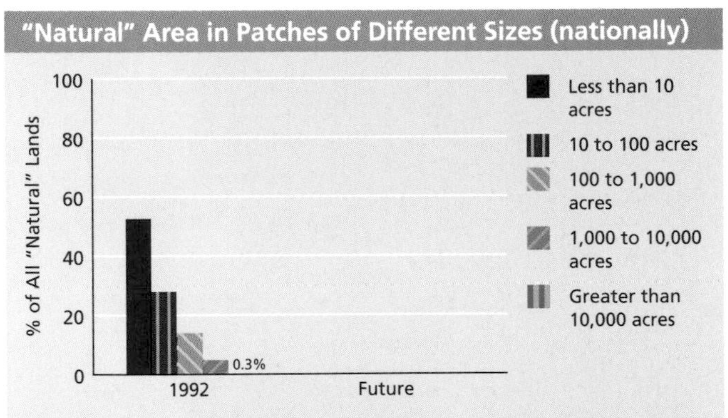

Data Source: Multi-Resolution Land Characteristics (MRLC) Consortium, USGS EROS Data Center. Coverage: lower 48 states.

"Natural" Area in Patches of Different Sizes, 1992 (by region)

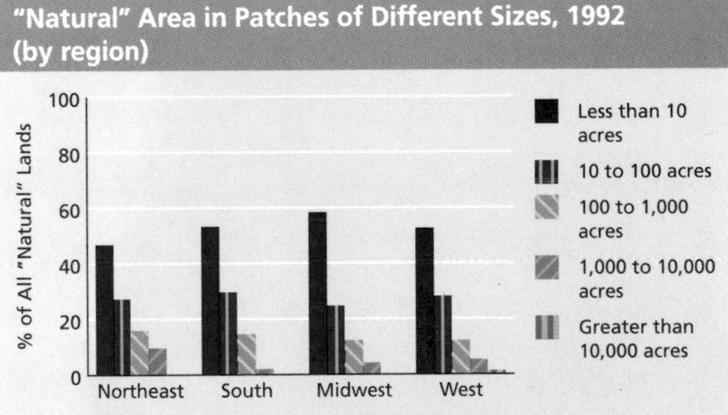

Data Source: Multi-Resolution Land Characteristics (MRLC) Consortium, USGS EROS Data Center. Coverage: lower 48 states.

What Do the Data Show? About half of all natural lands in urban and suburban areas are in patches smaller than 10 acres. A progressively smaller percentage of natural areas are found in larger patches, so that, nationally, less than 5% of the total is found in patches of 1,000 acres or more. The Northeast has a higher percentage of large "natural" patches (100 to less than 1,000 acres and 1,000 to less than 10,000 acres) than the other regions, while very large patches (greater than 10,000 acres) are found only in the West; these patches account for 0.3% of all natural lands in urban and suburban areas.

Discussion In addition to size (shown here), the quality of habitat and recreational value of natural areas is influenced by other factors, such as the shape of patches and how isolated they are from other natural areas.

The technical note for this indicator is on page 266.

SYSTEM DIMENSIONS	CHEMICAL AND PHYSICAL	BIOLOGICAL COMPONENTS	HUMAN USES
Extent	Nutrients, Carbon, Oxygen	Plants and Animals	Food, Fiber, and Water
Pattern	Contaminants	Communities	Recreation and Other Services
	Physical	Ecological Productivity	

⊖ Total Impervious Area

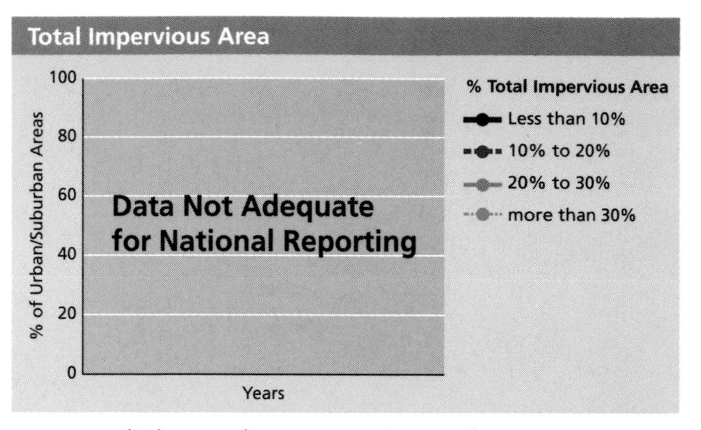

What Is This Indicator, and Why Is It Important? This indicator classifies urban and suburban areas according to the percentage of impervious surface—roads, parking lots, driveways, sidewalks, rooftops, and the like—they contain. The indicator uses several thresholds: less than 10% impervious surface in the region, at least 10%, at least 20%, and at least 30%.

The amount of impervious surface is a direct measure of the degree of urbanization, and it strongly affects both water quality in urban and suburban areas and replenishment of groundwater. Areas with more impervious, or nonporous, surfaces generate more runoff, which not only can contaminate and warm stream waters but also can degrade stream channels and banks. These changes have major impacts on the fish and wildlife that inhabit streams. In general, the impact on streams increases as the percentage of impervious surface in a watershed increases.

Why Can't This Indicator Be Reported at This Time? Total impervious area is difficult to measure. Measurements must be made on a fine scale to account for small areas such as sidewalks and driveways, but the finest-scale satellite information generally available cannot distinguish features of this size. Many local planning and environmental management programs collect this information, but the data have not been compiled regionally or nationally, nor are there standard methods for estimating the amount of impervious surface.

The technical note for this indicator is on page 266.

SYSTEM DIMENSIONS	CHEMICAL AND PHYSICAL	BIOLOGICAL COMPONENTS	HUMAN USES
Extent	Nutrients, Carbon, Oxygen	Plants and Animals	Food, Fiber, and Water
Pattern	Contaminants	Communities	Recreation and Other Services
	Physical	Ecological Productivity	

⑦ Stream Bank Vegetation

What Is This Indicator, and Why Is It Important? This indicator will describe the percentage of miles of stream (stream-miles) in urban and suburban areas that are lined with trees, shrubs, and other plants.

The amount of vegetation along a stream bank strongly affects both water quality and the kinds of fish and other animals that live in and along the stream. Plants lining a stream bank shade the stream, making it cooler in summer, and they serve as habitat for animals. Plants drop leaves and branches into the stream, providing food and habitat for stream animals; they also trap sediments and pollutants washing in from adjacent areas, preventing them from reaching the water and helping to maintain good water quality.

Stream Bank Vegetation

Indicator Development Needed

Why Can't This Indicator Be Reported at This Time? Reporting on this indicator would require agreement on a standard approach for defining and measuring streamside vegetation, including the minimum width of plant cover for a stream to be considered "vegetated," and agreement on whether and to what degree "unnatural" plant cover such as lawns should be counted. Once such definitions are established, data sources could be evaluated. These include satellite-based measures, which currently cannot distinguish very small landscape features, and locally generated information, which can be quite detailed, although it may be incompatible from location to location and very expensive to obtain.

There is no technical note for this indicator.

Nitrate in Urban and Suburban Streams

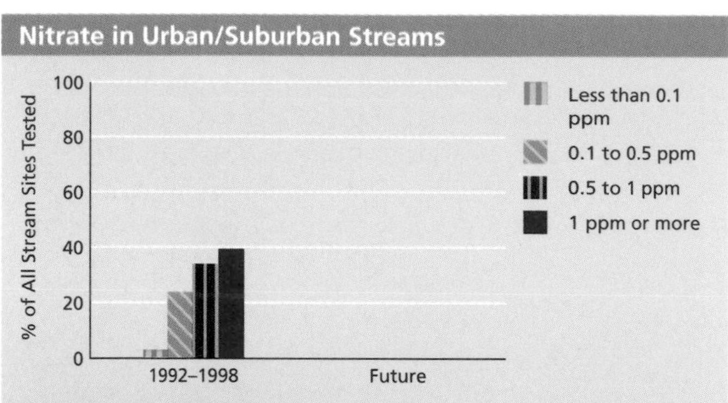

Data Source: USGS National Water Quality Assessment. Coverage: lower 48 states. Each sampling area was sampled intensively for approximately 2 years during 1992–1998.

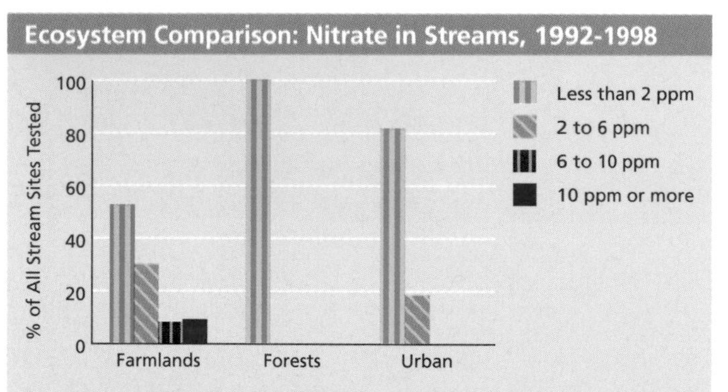

Data Source: USGS National Water Quality Assessment. Coverage: lower 48 states. Each sampling area was sampled intensively for approximately 2 years during 1992–1998.

What Is This Indicator, and Why Is It Important? This indicator reports the concentration of nitrate in streams in representative urban areas. Specifically, the indicator reports the percentage of streams with average nitrate concentrations in one of four ranges, for streams draining watersheds that are primarily urban.

Nitrate is a naturally occurring form of nitrogen and an important plant nutrient; it is often the most abundant of the forms of nitrogen that are readily usable by plants, including algae. Increased nitrate in streams that ultimately empty into coastal waters can lead to algal blooms in those waters; these blooms decrease recreational and aesthetic values and help deplete oxygen needed by fish and other animals (see the national nitrogen indicator and the hypoxia indicator, pp. 46 and 71). Nitrate in drinking water is also a health threat for young children, and it must be removed at significant cost by municipalities that rely on river water.

Sources of nitrogen in urban streams include effluent from sewage treatment plants, animal wastes, and fertilizers used on lawns, gardens, golf courses, and agricultural fields.

What Do the Data Show? About 60% of the stream sites in areas dominated by urban and suburban land use had concentrations of nitrate below 1 part per million, about 25% had concentrations below 0.5 part per million (ppm), and about 3% had concentrations that were less than 0.1 ppm. The federal drinking water standard for the protection of human health is 10 ppm of nitrate, which is exceeded in streams only in agricultural areas.

Concentrations in streams in areas dominated by urban land use are lower than those from agricultural areas but higher than those from forests (see pp. 95 and 122). There is also a core national indicator for nitrogen (p. 46).

The technical note for this indicator is on page 267.

● Phosphorus in Urban and Suburban Streams

What Is This Indicator, and Why Is It Important? This indicator reports the concentration of phosphorus in representative streams in urban areas. Specifically, the indicator reports the percentage of streams with average annual concentrations in one of four ranges, for streams draining watersheds that are primarily urban.

Phosphorus is an essential nutrient for all life forms and occurs naturally in soils and aquatic systems; phosphate is the most biologically active form of phosphorus. At high concentrations in freshwater systems, however, phosphorus can lead to algal blooms, which can decrease recreational and aesthetic values and help deplete oxygen needed by fish and other animals.

Sources of phosphorus in urban streams include effluent from sewage treatment plants, animal wastes, some detergents, and fertilizers used on lawns, gardens, golf courses, and agricultural fields.

What Do the Data Show? About two-thirds of stream sites in urban areas had concentrations of phosphorus that were at least 0.1 part per million (ppm), and about 10% of urban streams sites had concentrations of at least 0.5 ppm.

Streams in urban areas have similar average phosphorus concentrations to streams draining farmland watersheds.

Total Phosphorus in Urban/Suburban Streams

% of All Stream Sites Tested
100
80
60
40
20
0
1992–1998
Future
Less than 0.1 ppm
0.1 to 0.3 ppm
0.3 to 0.5 ppm
0.5 ppm or more

Data Source: USGS National Water Quality Assessment. Coverage: lower 48 states. Each sampling area was sampled intensively for approximately 2 years during 1992–1998.

Ecosystem Comparison: Total Phosphorus in Streams, 1992-1998

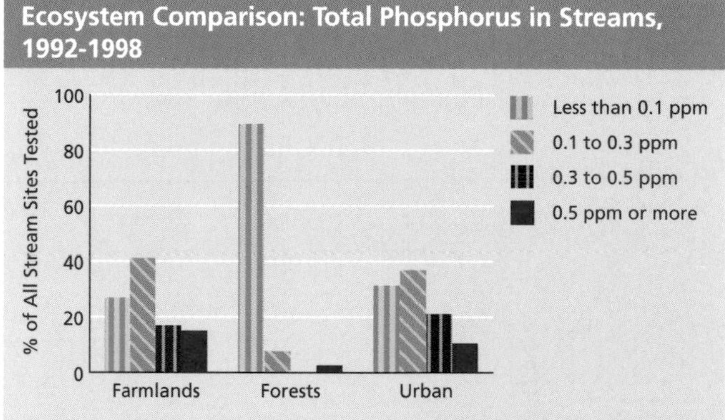

Data Source: USGS National Water Quality Assessment. Coverage: lower 48 states. Each sampling area was sampled intensively for approximately 2 years during 1992–1998.

Discussion The U.S. Environmental Protection Agency (EPA) has recommended 0.1 ppm as a goal for preventing excess algae growth in streams. In 2000, EPA took steps to facilitate development of regional criteria, but these regional criteria have not yet been adopted. There is no federal drinking water standard for phosphorus.

The technical note for this indicator is on page 267.

Air Quality (High Ozone Levels)

Exceedances of 8-hour Ozone Threshold (0.08 ppm)

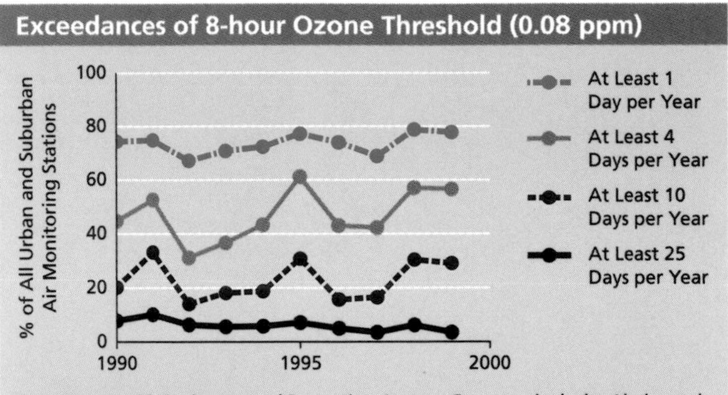

Data Source: U.S. Environmental Protection Agency. Coverage includes Alaska and Hawaii. Data from before 1990 exist, but they are not directly comparable. Only monitoring stations located in urban/suburban areas were included in the analysis.

Air Monitoring Stations Exceeding 8-hour Ozone Threshold (0.08 ppm), 1999

220 Stations with 1–3 Exceedances (45% of Stations)

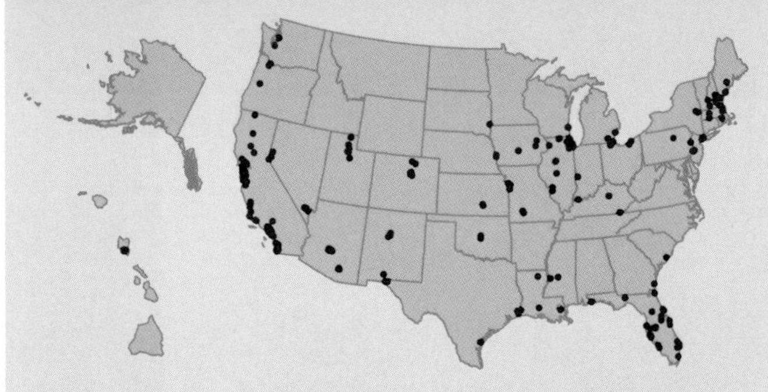

266 Stations With at Least 4 Exceedances (55% of Stations)

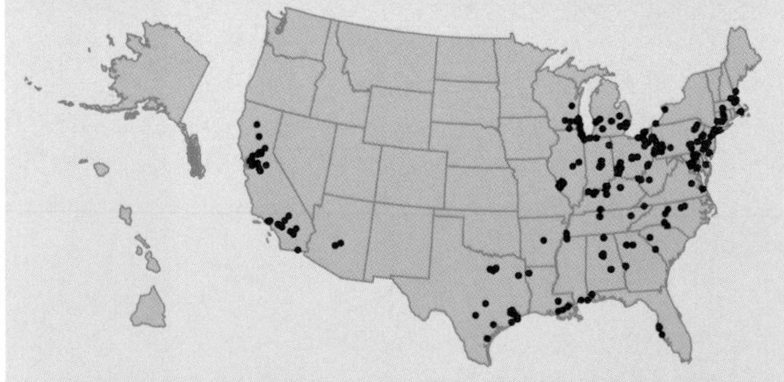

Data Source: U.S. Environmental Protection Agency. Coverage includes Alaska and Hawaii. Data from before 1990 exist, but they are not directly comparable. Only monitoring stations located in urban/suburban areas were included in the analysis.

What Is This Indicator, and Why Is It Important? This indicator reports the percentage of air pollution monitoring stations in urban and suburban areas with "high" ozone concentrations at least 1, 4, 10, and 25 days a year. Ground-level ozone is considered high when the 8-hour average concentration exceeds 0.08 parts per million (ppm). Violations of federal air quality standards are generally triggered by ozone concentrations exceeding this level for 4 or more days. For this reason, the maps show monitoring stations with less than 4 days and 4 or more days of high concentrations in 1999.

Ground-level ozone is one of the most pervasive air quality problems in the United States. Children and adults who are active outdoors, and people with respiratory diseases, are most likely to be harmed. Ozone can inflame the lungs, make people more susceptible to respiratory infection, and aggravate respiratory diseases such as asthma; repeated exposure may lead to permanent lung damage. High concentrations can harm trees, other plants, wildlife, and pets, and can damage painted surfaces, plastics, and rubber materials. In contrast, ozone in the upper atmosphere absorbs harmful ultraviolet radiation.

What Do the Data Show? In 1999, about 55% of monitoring stations in urban and suburban areas recorded high ozone levels on 4 or more days. The percentage of urban and suburban monitors with high ozone levels on 4 or more days per year fluctuated around 50% during the 1990s. The fluctuations are in large part due to year-to-year variability in weather conditions. The percentage of monitors recording high levels 25 or more times per year declined over the same period, to about 5% in 1999. Many of these monitors were in southern California, Houston, and Atlanta.

Discussion Ground-level ozone forms when pollutants from vehicles, paints and solvents, unburned fuel, and industrial sources "bake" in hot, sunny, stagnant weather. Ground-level ozone is one of six common air pollutants considered harmful to human health and the environment (the others are lead, carbon monoxide, nitrogen dioxide, sulfur dioxide, and particulate matter). While high concentrations of other pollutants do occur in some urban and suburban areas, ozone is responsible for more than 95% of all days with violations of any air quality standard.

The technical note for this indicator is on page 267.

SYSTEM DIMENSIONS	CHEMICAL AND PHYSICAL	BIOLOGICAL COMPONENTS	HUMAN USES
Extent	Nutrients, Carbon, Oxygen	Plants and Animals	Food, Fiber, and Water
Pattern	**Contaminants**	Communities	Recreation and Other Services
	Physical	Ecological Productivity	

◐ Chemical Contamination

What Is This Indicator, and Why Is It Important? This indicator reports on contaminants found in urban and suburban streams and soils. The graph on the top reports the average number of contaminants detected throughout the year at urban and suburban stream sites, and the middle graph reports the percentage of stream sites with concentrations of chemical contaminants that exceeded standards or guidelines for the protection of human health or aquatic life. Compounds reported here include many pesticides, selected pesticide breakdown products, ammonia, and nitrate (because nitrate and ammonia occur naturally, they are not included in the graphs showing contaminant occurrence).

In sufficient quantities, contaminants such as pesticides can harm people as well as fish and other wildlife. The number of contaminants detected is important, but the presence of pesticides does not necessarily mean that the levels are high enough to cause problems. Comparison with standards and guidelines (benchmarks) provides a useful reference to help judge the significance of contamination.

However, appropriate benchmarks do not exist for many compounds: for example, there are no drinking water benchmarks for 33 of the 76 pesticides analyzed here and no aquatic life benchmarks for 48 of the 76. Current benchmarks do not account for mixtures of chemicals or seasonal pulses of high concentrations. In addition, potential effects on the reproductive, nervous, and immune systems, as well as on particularly sensitive individuals, are not yet well understood.

Contaminant Occurrence in Urban/Suburban Streams

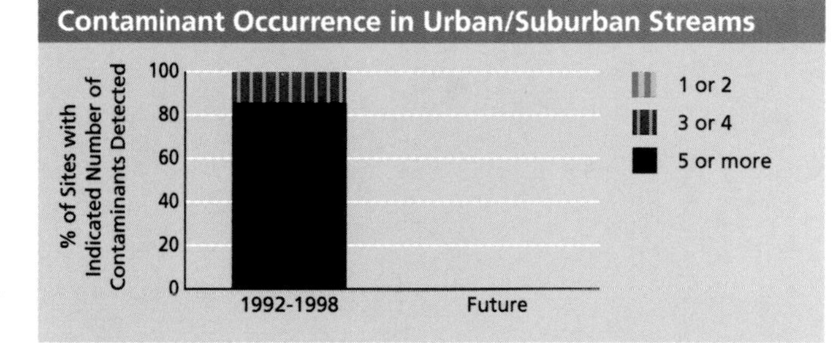

Contaminants above Standards or Guidelines in Urban/Suburban Streams

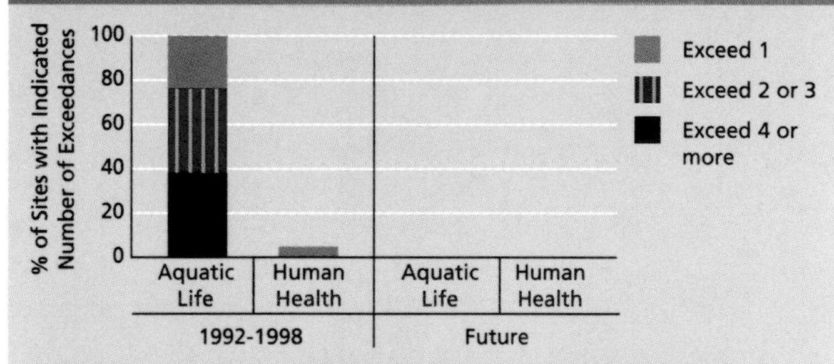

Data Source: U.S. Geological Survey, National Water Quality Assessment Program (NAWQA). Coverage: lower 48 states. Note: Each sampling area was sampled intensively for approximately 2 years during 1992-1998.

Contaminants in Urban/Suburban Soils

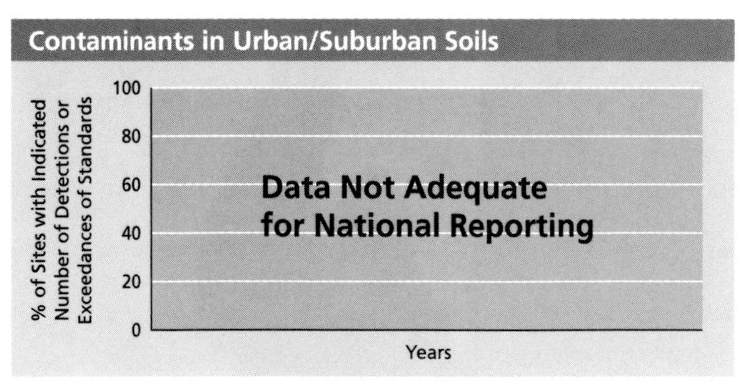

Why Can't This Entire Indicator Be Reported At This Time? Data are not currently available to report in a consistent manner on chemical contamination in urban and suburban soils.

What Do the Data Show? About 85% of stream sites in urban and suburban areas had an average of at least five detectable contaminants throughout the year. All sites had at least one chemical that exceeded guidelines for protection of aquatic life, and about 5% of sites had a contaminant that exceeded human health standards or guidelines.

Discussion The data shown here do not represent assessments of the risks posed to people or ecosystems in any specific location, since they do not incorporate factors such as whether the water tested is actually used as a drinking water source or whether aquatic animals are biologically active at the time of year when the contaminants are found.

Guidelines for the protection of aquatic life are often numerically lower than similar benchmarks for human health. Aquatic animals spend much or all of their life in water and may be more sensitive to specific contaminants.

See also the national, coastal, and farmland contaminants indicators (pp. 48, 72, and 97).

The technical note for this indicator is on page 268.

Urban Heat Island

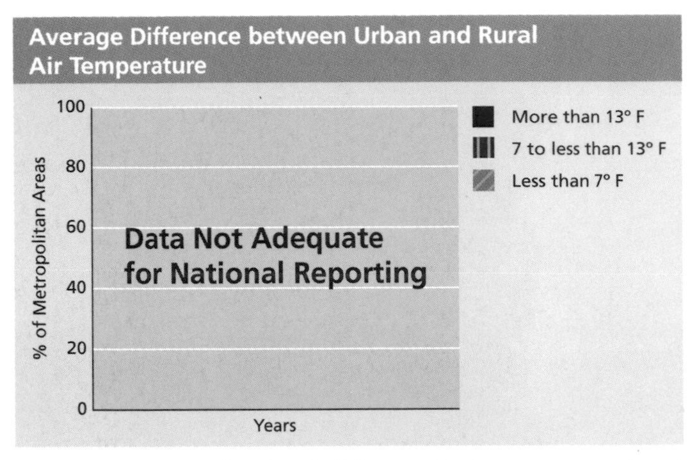

What Is This Indicator, and Why Is It Important? This indicator will describe the difference between urban and rural air temperatures for major U.S. metropolitan areas. Temperatures within urban areas will be compared to those in less-developed surrounding areas.

Extremely hot weather is responsible for greater loss of human life in the United States than hurricanes, lightning, tornadoes, floods, and earthquakes combined. Building density and type, amount of road surface, and energy use, as well as local topography and regional weather patterns, all work together to modify a city's climate. The urban heat island effect is often noticed most at night when buildings and other constructed surfaces radiate the heat they have accumulated during the day. Beyond posing a threat to human health (through heat stroke, for example) and raising air conditioning costs, the heat island effect can cause physiological stress in other animals, change the mix of plants and animals that live in the area, and even lead to changes in the distribution of pathogens. Elevated temperatures also accelerate the formation of ground-level ozone (see Air Quality, p. 188) and other air pollutants that adversely affect human health.

Why Can't This Indicator Be Reported at This Time? There is no single metric that has been adopted by the scientific community as an indicator of the heat island effect. One possible presentation would report the percentage of all U.S. metropolitan areas where the average annual difference between urban and rural air temperatures is relatively small (less than 7°F), moderate (7° to less than 13°F), or large (more than 13°F). National Weather Service data could be used to determine current and historic heat island effects in many locations. However, there is no program in place either to retrieve and analyze historic information or to identify appropriate pairs of urban and rural sites necessary to make calculations of the heat island effect.

The technical note for this indicator is on page 268.

⊖ Species Status

What Is This Indicator, and Why Is It Important? This indicator will report on the degree to which "original" plants and animals are either absent entirely or are at risk of being lost from metropolitan areas. Original species are those that, before European settlement, inhabited the lands now occupied by metropolitan areas. Specifically, the indicator will report on the fraction of metropolitan areas where 25% or more, 50% or more, and 75% or more of original species are at risk of being displaced or are absent.

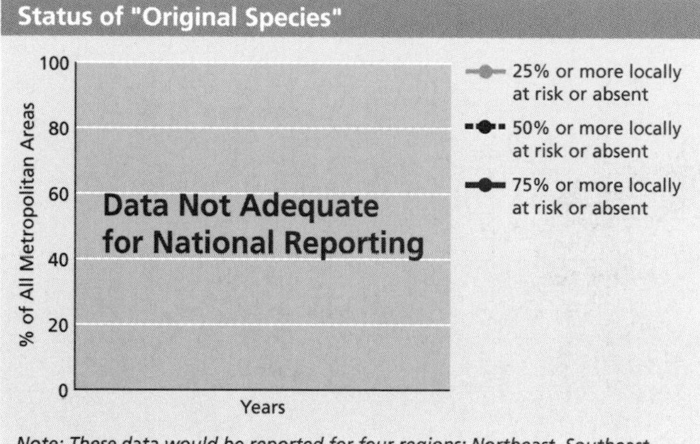

Note: These data would be reported for four regions: Northeast, Southeast, Midwest, and West.

Species differ in their ability to cope with urban/suburban development, and cities and suburbs differ in their capacity to accommodate those species common in the area before European settlement. How thoroughly an area is developed, and whether there are areas and corridors maintained for wildlife, has an influence on whether some species—those less tolerant of people and development—become reduced in population or displaced from the local area. Some of these species may have healthy populations outside cities and their suburbs, but in other cases the loss of habitat in urban and suburban areas can contribute to the overall decline of a species' population. Efforts to improve or restore habitat within urban/suburban areas can increase the likelihood that original species will re-inhabit these areas.

Why Can't This Indicator Be Reported at This Time? The historical data necessary to establish lists of original species are incomplete, and current information on their status, especially within cities and their suburbs, is not systematically collected and reported. When available, the amount, quality, and format of such data are extremely variable.

Discussion Note that it is difficult to distinguish between at-risk and absent in this context, so both are included. In addition, presettlement is used as a benchmark simply as a way to track changes, not because the full suite of original species would necessarily be desirable in any given metropolitan area.

This indicator would not be calculated for all urban and suburban areas, as defined in this report (see Area of Urban and Suburban Lands, p. 181), as it is likely that information, expertise, and financial resources will be available only for larger metropolitan areas. Thus, it might be appropriate to base reporting for this indicator on data from a suite of cities (and their suburbs) whose population exceeds 100,000 or that cover at least 50 square miles, for example.

The technical note for this indicator is on page 269.

⊖ Disruptive Species

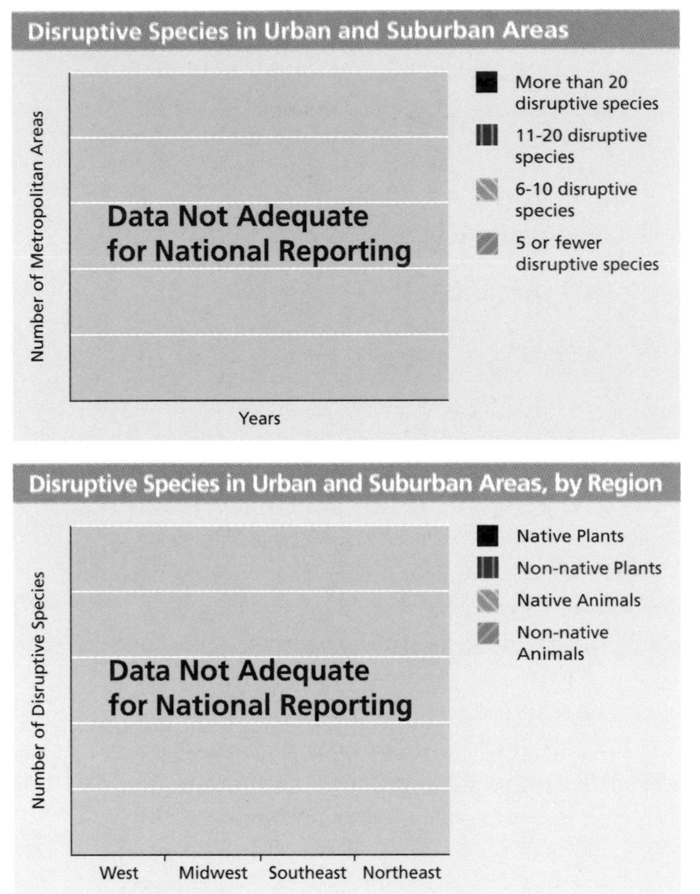

What Is This Indicator, and Why Is It Important? This indicator would report the number and type of "disruptive" species found in metropolitan areas. Disruptive species are those that have negative effects on natural areas and native species or cause damage to people and property. Specifically, the indicator will report the number of larger metropolitan areas with 5 or fewer, from 6 to 10, from 11 to 20, and more than 20 disruptive plant and animal species. It would also report the number of disruptive native and non-native plant and animal species on a regional basis, for the most current year.

Some species of plants and animals are so abundant in urban and suburban areas that they disrupt other species and cause problems for people. In the Northeast, for example, white-tailed deer are major suburban pests. They damage native vegetation in natural areas, destroy crops and gardens, and are involved in countless automobile accidents. In and around Portland, Oregon, Scotch broom, native to the British Isles, is spreading rapidly, often growing in dense, nearly impenetrable clusters that make maintenance of roads, ditches, canals, and power and telephone lines difficult and costly. Minneapolis, among other cities in the Midwest, is taking action against disruptive woody plants like buckthorn, Tartarian honeysuckle, and mulberry, which are taking over the city's woods and wetlands.

Why Can't This Indicator Be Reported at This Time? Regional lists of disruptive species do not exist. Creating them requires definition of thresholds that distinguish truly disruptive species from those that cause fewer problems, as well as consistent policies for including species based on their potential to cause damage, as shown by experiences in other locations.

In addition, monitoring and reporting programs need to be put in place to track the occurrence of disruptive species. Many knowledgeable individuals and institutions could participate, but no entity currently has the mandate to coordinate such an activity.

Discussion Disruptive species may be native, or they may have been introduced from other regions or other countries. The altered landscape in urban and suburban areas encourages the growth of these species, which tolerate and even thrive around built-up areas. At the same time, populations of more sensitive species shrink, reducing competition and further encouraging the spread of disruptive species.

There is no technical note for this indicator.

SYSTEM DIMENSIONS	CHEMICAL AND PHYSICAL	**BIOLOGICAL COMPONENTS**	HUMAN USES
Extent Pattern	Nutrients, Carbon, Oxygen Contaminants Physical	Plants and Animals **Communities** Ecological Productivity	Food, Fiber, and Water Recreation and Other Services

⊖ Status of Animal Communities in Urban and Suburban Streams

What Is This Indicator, and Why Is It Important? This indicator reports on "biological integrity" in streams in urban and suburban areas. Biological integrity is a measure of the degree to which the suite of fish and bottom-dwelling (or benthic) animals (including insects, worms, mollusks, and crustaceans) resembles what one might find in a relatively undisturbed stream in the same region. Tests assess the number of different species, number and condition of individuals, and food chain interactions. High scores indicate close resemblance to "reference" or undisturbed conditions, and low scores indicate significant deviation from them. (See also Status of Freshwater Animal Communities, p. 147.)

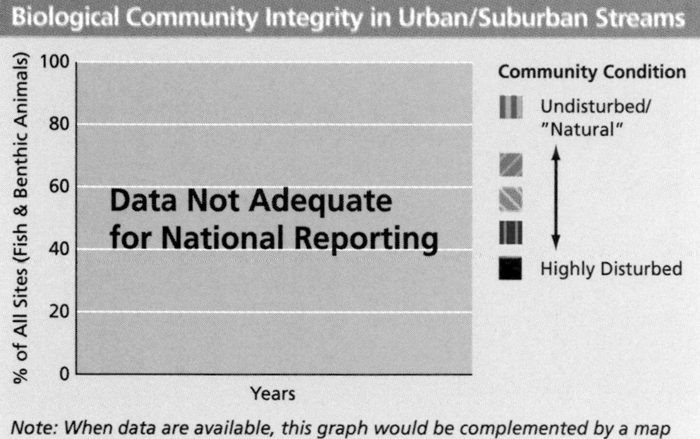

Note: When data are available, this graph would be complemented by a map showing the percentage of urban/suburban watersheds with fish and bottom-dwelling animal communities resembling those in undisturbed conditions.

Undisturbed streams in a particular region have a relatively predictable set of fish and bottom-dwelling animals, in predictable proportions. The composition and condition of these biological communities may be altered, often as a result of development in the stream's watershed. Sources of degradation include contaminated runoff from streets, driveways, lawns, golf courses, and the like, increased stream temperature caused by runoff that is warmed as it flows over paved surfaces, and channelizing or other modifications of the streambed. Some streams are so modified that, for example, both the number of species and the number of individuals are very low when compared to undisturbed areas, and many of those that remain are diseased or otherwise damaged. Ecosystems that are "healthy," or show high integrity, are more likely to withstand natural and man-made stresses.

Why Can't This Indicator Be Reported at This Time? The tests of biological integrity now in use must be tailored to ensure that each stream is compared with an appropriate reference from within the same region, but outside of the urban/suburban area. Only a handful of states regularly conduct quantitative tests of condition of fish or bottom-dwelling animal communities, and these are not specific to urban streams.

The technical note for this indicator is on page 269.

SYSTEM DIMENSIONS	CHEMICAL AND PHYSICAL	BIOLOGICAL COMPONENTS	**HUMAN USES**
Extent Pattern	Nutrients, Carbon, Oxygen Contaminants Physical	Plants and Animals Communities Ecological Productivity	Food, Fiber, and Water **Recreation and Other Services**

⊖ Publicly Accessible Open Space per Resident

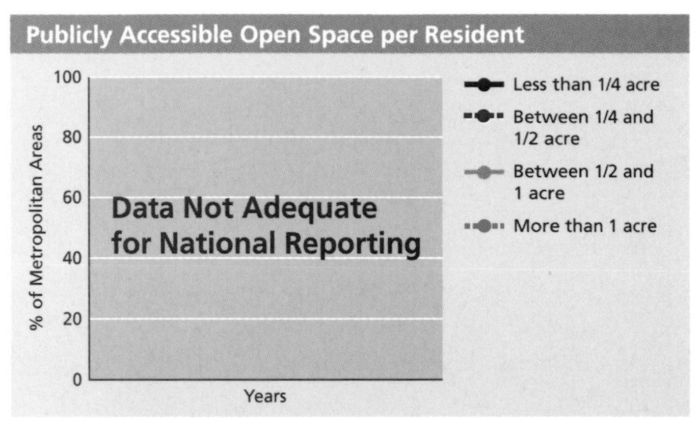

What Is This Indicator, and Why Is It Important? This indicator would report the amount of open space—land that is dominated by "natural" surfaces, like grass or woods, along with lakes, rivers, beaches, and wetlands—that is accessible to the general public in large metropolitan areas. Specifically, the indicator would report the percentage of metropolitan areas with different amounts of open space per resident.

Americans enjoy outdoor recreation, and urban and suburban residents place a high value on access to public spaces where they can picnic, play ball, swim, hike, fish, walk their dogs, enjoy nature, and engage in any of a myriad of other outdoor activities. The amount of such open space per resident often determines how intensely such places will be used and how crowded they will be.

Why Can't This Indicator Be Reported at This Time? There are no consistent or comprehensive surveys of the amount of publicly accessible open space in cities and suburban areas. A combination of satellite remote sensing and local tax and land records would be required for reporting on this indicator.

Discussion This indicator focuses on public areas that are natural or relatively undeveloped. In practice, this means that areas dominated by grass, woods, dirt, or other unpaved surfaces would be counted, while predominately paved areas would not (paved walkways in a park that is primarily grass would not "disqualify" the area). In addition, areas counted in this indicator are those that are accessible to the general public, even if fees (such as for a county-run golf course) are charged. Thus, a public golf course and even some cemeteries would qualify, but a farm or a country club would not. Note that a change in population without a change in open space would change the value of this indicator. Area of Urban and Suburban Lands (p. 181) provides a context for this indicator, because it reports the overall percentage of natural lands in the urban/suburban landscape; however, it does not distinguish between publicly accessible and inaccessible lands.

The technical note for this indicator is on page 269.

SYSTEM DIMENSIONS	CHEMICAL AND PHYSICAL	BIOLOGICAL COMPONENTS	HUMAN USES
Extent	Nutrients, Carbon, Oxygen	Plants and Animals	Food, Fiber, and Water
Pattern	Contaminants	Communities	**Recreation and Other Services**
	Physical	Ecological Productivity	

⑦ Natural Ecosystem Services

What Is This Indicator, and Why Is It Important? Urban and suburban areas are defined by what people have built, but the remaining "natural" components—trees, meadows, streams, wetlands, and the like—provide valuable services to the residents of these developed areas. Ecosystem services are the benefits, both tangible and intangible, that these natural elements provide. For example, forested areas reduce stormwater runoff, when compared to paved areas, and trees cool streets and buildings, reducing energy consumption; trees also reduce urban noise levels. Natural areas, including forests, grasslands and shrublands, beaches, lakes, streams, and wetlands, also provide recreational opportunities, increase property values and community amenities, and are aesthetically pleasing.

Natural Ecosystem Services

Indicator Development Needed

In urban and suburban areas, the loss of ecosystem services is often not recognized until a functioning ecosystem has already been altered, and millions or even billions of dollars are needed for technological fixes. Suburban development in the Catskill Mountains, the primary source of water for New York City, has jeopardized water quality, potentially requiring a filtration system costing billions of dollars to construct and millions of dollars a year to operate, in order to provide the same water quality as was provided before development. In another example, the nonprofit organization American Forests found that trees in the Denver/Front Range area provide the equivalent of a $44 million stormwater management system.

Why Can't This Indicator Be Reported at This Time? Scientists are uncertain about how to measure ecosystem services in urban, and other, ecosystems. They may rely, for example, on tree canopy as a surrogate measure because of its influences on air quality, water flow, property values, microclimates, and aesthetics. Scientists are working to quantify the relationship, which is likely to be stronger in some areas than in others, between the amount of tree canopy and the levels of services provided. Other components of the urban/suburban ecosystem, like wetlands, streams, and grasslands, also provide important services, and these should be incorporated as well.

Discussion Considerable scientific effort needs to be invested in understanding the relationship between various ecosystem components and the kinds of services they provide. Recently, the National Science Foundation's Long Term Ecological Research Network established research sites in Baltimore, Maryland, and Phoenix, Arizona, to study the ecology of cities (see http://lternet.edu/). These sites seek to understand the nature and functioning of urban and suburban ecosystems, and how people influence and are influenced by them. Through such detailed studies and the accompanying long-term observation of changes in urban and suburban areas, it will be possible to quantify ecosystem services and understand how urbanization alters these services.

There is no technical note for this indicator.

Part III:
Appendix and Technical Notes

Appendix:

Data Availability and Gaps

As explained in Chapter 2, we attempted to locate data of sufficient quality and coverage to support national reporting for each of the 103 indicators included in this report. As might be expected, we found a wide variation in the availability of data. Throughout the report, we highlight those indicators for which sufficient data are not available and, in doing so, identify needs for additional monitoring. We also highlight a number of measures for which additional research is needed to define more fully the specific value to be reported. This appendix summarizes the "state of the data" for the indicators in the report.

Guidelines for Including Data

Once an indicator was chosen and relevant sources of data were identified, the first screen for inclusion was *scientific credibility.* Again, data were not used simply because they were the "best available" but, based on the professional judgment of the members of each work group, they had to meet the highest standards of the appropriate discipline. The judgments of the work groups were then extensively peer-reviewed.

The second criterion for including a data source was that it *provide information on a substantial majority of the resource or issue in question.* The practical result is that we relied on data sources that covered a majority of states or a significant fraction of coastline. For some indicators, complete coverage is available (such as is provided by remote sensing data). For others, regionally or nationally representative samples are used (such as is provided by monitoring programs employing statistical sampling techniques).

The first draft/prototype of this report, released in 1999, included many examples of data for small areas of the country, as illustrations of the types of results we had hoped to include. Feedback from readers led us to conclude that while such examples are interesting, they obscure the fact that data are not available for a significant fraction of the desired indicators.

Third, to be included in this report, data sources must be from ongoing programs—that is, there must be a reasonable chance that the *measurements will be repeated at regular intervals in the future.* Although all monitoring and reporting programs are subject to changes in funding and priorities, established programs are clearly different from one-time studies conducted by individual researchers or groups. One-time efforts are extremely valuable because they often break new ground scientifically, and they may provide baselines against which data collected later may be compared, but they do not necessarily form a solid foundation for periodic reporting.

Ideally, data sources used in this report have time trends, but the lack of trends was not a criterion for eliminating data. Where possible, we have attempted to use data from 1950 to the present, with longer historical perspectives included as needed to provide an ecological context for current reporting. These longer-term perspectives include reporting on conditions before European settlement, in the early 20th century, or for other relevant time periods. Many data sources, particularly those based on remote sensing, cover shorter time periods but will illuminate longer-term trends as time goes on and measurements are repeated.

Note as well that there were a number of cases where we were not able to determine which of several possible indicators for an important ecosystem attribute was best, and thus we could not judge whether adequate data are available. We have highlighted these cases, and we hope to work closely with the relevant scientific communities to narrow the range of possible indicators.

Table A.1. Indicators According to Data Availability

● Indicators with ALL Required Data

Core National Indicators
- Movement of Nitrogen
- Plant Growth Index
- Production of Food and Fiber and Water Withdrawals

Coasts and Oceans
- Sea Surface Temperature
- Commercial Fish and Shellfish Landings

Farmlands
- Total Cropland
- The Farmland Landscape
- Nitrate in Farmland Streams and Groundwater
- Phosphorus in Farmland Streams
- Pesticides in Farmland Streams and Groundwater
- Soil Erosion
- Major Crop Yields
- Agricultural Inputs and Outputs
- Monetary Value of Agricultural Production

Forests
- Forest Area and Ownership
- Forest Types
- Forest Management Categories
- Nitrate in Forest Streams
- Forest Disturbance: Fire, Insects, and Disease
- Timber Harvest
- Timber Growth and Harvest

Fresh Waters
- Changing Stream Flows
- Water Withdrawals
- Waterborne Human Disease Outbreaks

Grasslands and Shrublands
- Area of Grasslands and Shrublands
- Number and Duration of Dry Periods in Streams and Rivers
- Population Trends in Invasive Birds
- Production of Cattle

Urban and Suburban Areas
- Area of Urban / Suburban Lands
- Patches of Forest, Grasslands and Shrublands, and Wetlands
- Nitrate in Urban and Suburban Streams
- Phosphorus in Urban and Suburban Streams
- Air Quality

⊖ Indicators with INADEQUATE DATA FOR NATIONAL REPORTING

Core National Indicators
- None

Coasts and Oceans
- Areas with Depleted Oxygen
- Coastal Erosion
- At- Risk Marine Species
- Selected Contaminants in Fish and Shellfish
- Recreational Water Quality

Farmlands
- Fragmentation of Farmland Landscapes by Development
- Shape of "Natural" Patches in the Farmland Landscape
- Soil Organic Matter
- Soil Salinity
- Soil Biological Condition
- Recreation on Farmlands

Forests
- Non-native Plants
- Fire Frequency
- Forest Community Types with Significantly Reduced Area
- Recreation in Forests

Fresh Waters
- Water Clarity
- Status of Freshwater Animal Communities
- Groundwater Levels
- Freshwater Recreation Activities

Grasslands and Shrublands
- Area and Size of Grassland and Shrubland Patches
- Nitrate in Groundwater
- Carbon Storage
- Depth to Shallow Groundwater
- Non-native Plant Cover
- Fire Frequency
- Recreation on Grasslands and Shrublands

Urban and Suburban Areas
- Total Impervious Area
- Species Status
- Disruptive Species
- Status of Animal Communities in Urban and Suburban Streams
- Publicly Accessible Open Space per Resident

◐ Indicators with PARTIAL DATA

Core National Indicators
- Ecosystem Extent
- At-Risk Native Species
- Chemical Contaminants
- Outdoor Recreation

Coasts and Oceans
- Coastal Living Habitats
- Shoreline Types
- Contamination in Bottom Sediments
- Unusual Marine Mortalities
- Condition of Bottom-Dwelling Animals
- Chlorophyll Concentration
- Status of Commercially Important Fish Stocks

Farmlands
- None

Forests
- Forest Pattern and Fragmentation
- Carbon Storage
- At-Risk Native Species
- Forest Age

Fresh Waters
- Extent of Freshwater Ecosystems
- Altered Freshwater Ecosystems
- Phosphorus in Lakes, Reservoirs, and Large Rivers
- At-Risk Native Species
- Non-Native Species
- Animal Deaths and Deformities
- At-Risk Freshwater Plant Communities

Grasslands and Shrublands
- Land Use
- At-Risk Native Species

Urban and Suburban Areas
- Chemical Contamination

? Indicators needing ADDITIONAL DEVELOPMENT

Core National Indicators
- Fragmentation and Landscape Pattern
- Condition of Plant and Animal Communities
- Natural Ecosystem Services

Coasts and Oceans
- Non-native Species
- Harmful Algal Blooms

Farmlands
- Status of Animal Species in Farmland Areas
- Native Vegetation in Areas Dominated by Croplands
- Stream Habitat Quality

Forests
- None

Fresh Waters
- Stream Habitat Quality

Grasslands and Shrublands
- Riparian Condition

Urban and Suburban Areas
- Suburban/Rural Land Use Change
- Stream Bank Vegetation
- Urban Heat Island
- Natural Ecosystem Services

Indicators with Full, Partial, and Insufficient Data

Throughout the report, we have classified indicators into four categories:

- ● Those with all data required for periodic national reporting
- ◐ Those with some, but not all, of the data needed for national reporting
- ⊖ Indicators with insufficient data for national reporting
- ⓘ Indicators that need further development

Of the 103 indicators included in this report, 58 (56%) are in the first two categories—that is, there are sufficient data to support periodic national-level reporting. Of these, 33 have all the data required and the remaining 25 have some data gaps. These gaps may be regional (i.e., data are available for part but not all of the country) or they may be topical (i.e., data are available on some but not all components of an indicator). An example of the former is the coastal shoreline types indicator (p. 70), where data are available for the Pacific and southern Atlantic coasts, but not for the middle and northern Atlantic or Gulf Coasts. Several of the at-risk species indicators (see the forest and grasslands indicators, pp. 124 and 168) provide examples of the latter. In these cases, we have reported data on the status of native animals, but not plants (such data are available but required additional analysis before they could be used). Table A.1 shows the data availability for all indicators in the report.

Data availability varies by both ecosystem type and indicator category, as shown in Figures A.1 and A.2.

Data are available for more forest indicators than for any other system: there are complete data for about half the forest indictors and some data for another quarter. Full or partial data are available for 50% or fewer of the indicators for farmlands, grasslands and shrublands and urban and suburban areas (although there are more indicators with "full data" for farmlands than for any other system) (see Box A.1). The indicator categories with the highest percentages of data available include those addressing ecosystem extent; contaminants; ecosystem productivity; and food, fiber, and water (i.e., goods provided

Box A.1. Three Systems with Large Data Gaps

Farmlands and **grasslands and shrublands** make up about 60% of the land area of the lower 48 states. **Urban and suburban areas,** which are quite small in comparison, are home to about three-quarters of all Americans. Yet, for these three ecosystems, full or partial data are available for half or fewer of the indicators.

We can report on the acreage of croplands, the food and fiber they produce, and the nutrients and contaminants that occur in farmland streams. Surprisingly, given how important soil is to farming, we cannot report nationally on soil organic matter, soil salinity, or the microscopic animal communities in cropland soils. Finally, we could not report on *any* of the indicators describing biological components in farmland areas, either because data were not available or because the indicators need additional development. (Interestingly, for farmlands, where data do exist, they are more complete than for other systems—there are more indicators with all required data than for any other system.)

Data gaps for grasslands and shrublands include information about how these areas are used, the amount of nitrate in groundwater, the amount of carbon stored in plants and soil, the depth to groundwater, the extent of non-native plants, the frequency of fires, and recreation on these lands. We can report fully on the acreage of grasslands and shrublands and the number of cattle that feed on them, on stream and river flows, on population trends for invasive birds and the number of animal species (but not plants) that are at risk of extinction.

We report data for fewer urban and suburban indicators than for any other ecosystem type—only 6 of 15 indicators. We can report on the extent of urban and suburban areas and on the undeveloped lands they contain. We can also report on nutrients and contaminants in stream water (but not on the degree of contamination in soils). All remaining indicators either have inadequate data or require additional development.

by ecosystems). Indicator categories with the poorest data availability include those addressing landscape patterns, biological communities, and services provided by ecosystems.

No data are included for 45 of the indicators in the report. For 31 of these, the desired indicator is clear, but available data are insufficient for national reporting. For the remaining 14, the indicator itself needs further development. Data gaps and problems with indicator definitions are discussed below.

Trends and Other Context Information

This report does not make normative judgments about whether particular ecosystem conditions are "good" or "bad." Rather, we aim to present the available data in as neutral a form as possible—a "just the facts" approach. However, we also seek to provide information that places current ecosystem conditions in context, to assist the reader in understanding and making his or her own judgments about those conditions.

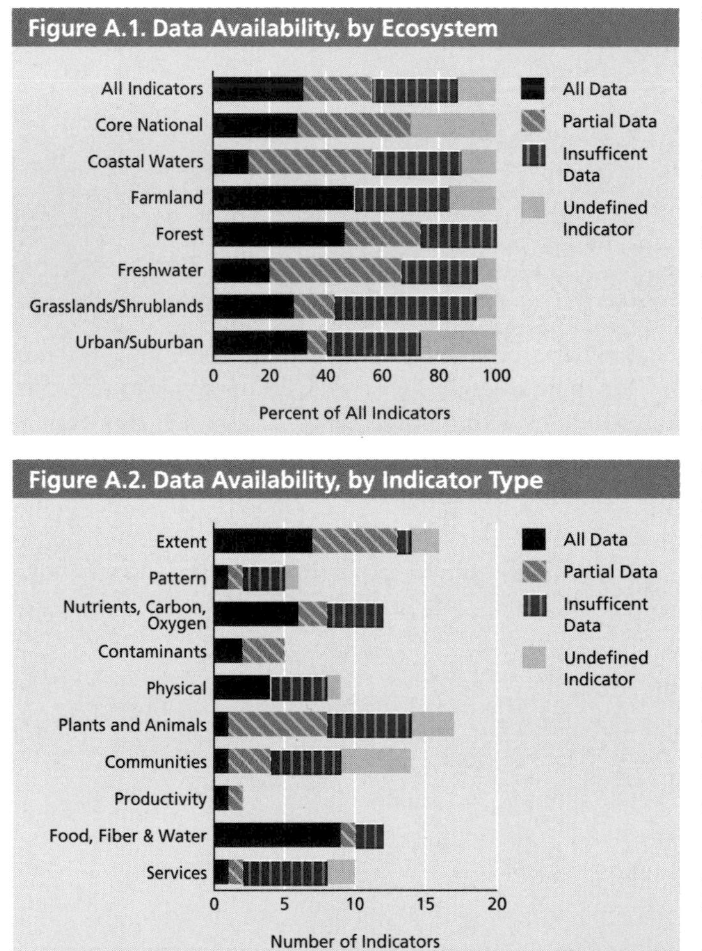

An obvious and (importantly) neutral way to place current conditions into context is to report the value of the indicator over time. Trends provide information both on the direction of change (is the value increasing or decreasing?), but also on the rate of change, which may be useful in determining whether there is reason for concern. In addition, providing information on the geographic distribution of conditions can be useful. So, for example, for some indicators we report whether one region of the country or one ecosystem type (forest, farmlands, etc.) had higher or lower values than other regions or types. A third method for placing information in context is through comparisons to relatively undisturbed "reference" conditions, and a final method is to compare current conditions to broadly accepted reference points, primarily federal limits for the allowable concentration of certain chemicals in the air or water.

As noted above, this report presents full or partial data on 58 indicators (56% of the total). Trends are presented for about half of these (31 indicators). For another 11 indicators, we have provided comparisons against widely accepted standards or against undisturbed or other reference conditions. For the remaining 16, we have information for only one point in time, without useful reference information.

The availability of trends and other reference and comparative information differs according to ecosystem type and indicator category, as shown in Figures A.3 and A.4. For example, as noted above, there are trends for 31 indicators (30%). For urban systems, trends are available for only 7%, while trends are available for about half of the forest indicators.

The situation is even more varied when one considers the availability of trends by indicator category. Trend data are available for more than 80% of indicators describing production and use of food, fiber, and water, and for about 40% of indicators of ecosystem extent. These strong showings are largely a result of the long-standing and well-supported monitoring and reporting programs devoted to accounting for goods of economic interest and the lands used in their production. For indicators of nutrients and related chemicals, contaminants, plant and animal species, and biological communities, trends are considerably less common—20% or less in all four cases. For contaminants and nutrients, however, regulatory standards and nonregulatory guidance levels provide a substantial increase in the amount of context information provided for these indicators.

Data Gaps and Problems in Indicator Definition

Again, of the 103 indicators included in the report, 33 have all data required for periodic national reporting, 25 have partial data available, and the remaining 45 include no data at all. Thus 70 measures (68%) are missing some or all data.

For 7 of these 70 indicators, the data required for national reporting exist (or would have been possible to obtain), but time and/or financial constraints prevented us from assembling them for this report. Three of these involve work with large data sets using geographic information systems (GIS). Another two involve categorizing 16,000 native plants into the ecosystems in which they are typically found. The remaining two would have required addition of questions to a recent recreational survey.

For another 5 of the 70 indicators, we have reason to expect that data adequate for national reporting will become available soon. These include coastal bathing water quality and additional data on shoreline habitat, forest fragmentation, and forest age structure. Table A.2 lists the 12 indicators that are either expected to become available soon or that could be made available now with additional funding.

For 41 indicators, some data are currently collected, but these data are of uncertain quality or comparability. Often the data are not comparable because different agencies or programs use different methods to collect or manage them. The fact that at least some data exist for these indicators means that it may be possible to fill data gaps relatively easily, through collation and aggregation of data from existing programs. However, detailed analyses would be required to determine the quality, coverage, and comparability of the various data sets. Such analyses were beyond the scope of this project, but should be a high priority.

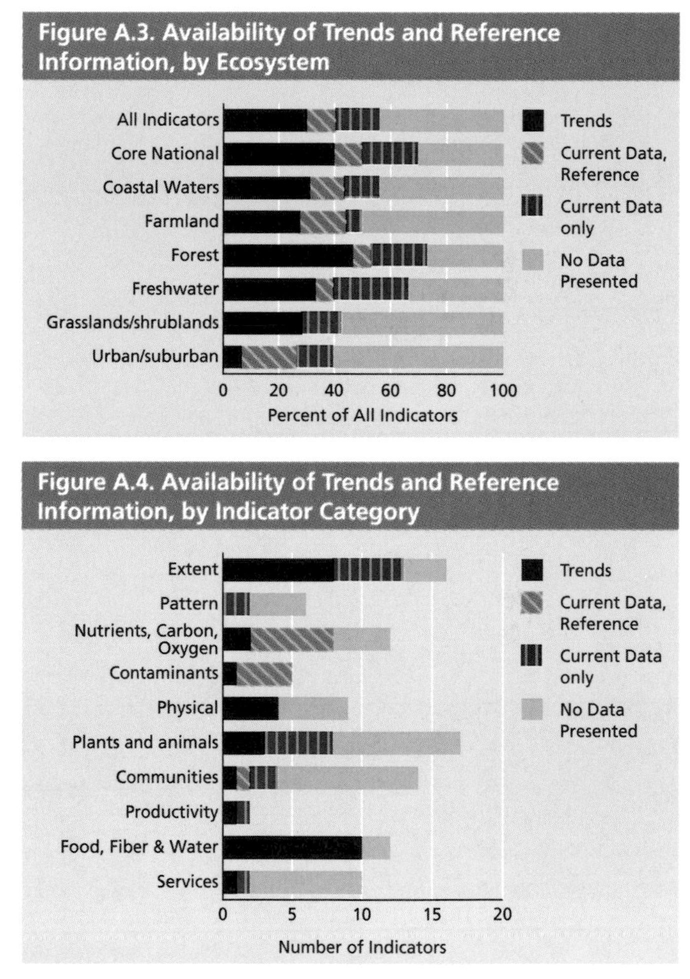

Data are not collected on any significant scale for 10 indicators. For another 7 indicators, the project work groups were not able to agree on a sufficiently well defined measure to even determine whether data are available. Note that 14 of the indicators in the report are marked as needing further development (see Table A.1). However, relevant data are available for almost half of these, so once the additional research is completed to clearly define these measures, a few of these indicators may move to the "with data" category.

Figure A.5 summarizes the status of data collection for the indicators in the report.

Box A.2. Native and Non-native Species

Species are fundamental components of ecosystems, are the subject of many policy and management issues, and are of intense interest to many Americans. We include 16 measures of plant and animal species, about half of which deal with native species and half with non-native or invasive species. We report full data for one of these 16 indicators, partial data for seven, and none for the remaining eight. Only three indicators have trends.

Native species indicators focus on risks of extinction and on unusual mortalities and deformities. Partial data are available for several indicators dealing with the risk of extinction of U.S. plants and animals. However, we could not report on species in farmlands, urban and suburban areas, or coastal waters. For forests, freshwater, and grasslands and shrublands, we are able to report on the status of animal species that depend on these ecosystem types, but not on plants, because the required analyses of the habitats of all 16,000 U.S. plants have not been completed. The only indicators for which trend data are available are those dealing with unusual mortality. However, these indicators are limited because there are data available for only a few groups of species.

The lack of trends is especially troublesome in reporting on the risk of extinction. Interpreting data from a single point in time is complicated because some species are naturally rare. Thus, the rankings are influenced by differences among regions and species groups in the number of naturally rare species, as well as by different types and levels of human activities that can cause species declines. Interpretation will be greatly enhanced when information on population trends for these at-risk species becomes available.

All six ecosystems have indicators that would—if data were available—track key aspects of non-native species, invasive or disruptive species, or both. However, despite the strong attention this subject is receiving in both scientific and management circles, we can report on only two indicators—the number of non-native fish species established in U.S. fresh waters, and population trends in invasive birds in grasslands and shrublands. All other indicators dealing with non-native or invasive species have data that are inadequate for national reporting, require additional development, or both.

Non-native species indicators also vary greatly in their focus. Some focus on plants—the fraction of land area covered by non-native species. Others focus on animals—the number of species in an area, or population trends in those species. Our recommended coastal indicator would integrate both the area covered and the number of non-native species in an area. While these different approaches reflect differences of opinion about what is most important to track, integrated national reporting—across multiple ecosystems—will be made easier if these differences are resolved in the future.

Resources for Filling Data Gaps

Filling the data gaps identified in this report will require a combination of ground-based surveys and information acquired from remote sensing platforms (satellites). Though many hope that remote sensing will form the backbone of future monitoring programs, a preliminary examination reveals that on-the-ground surveys are the method of choice for filling many—and possibly most—of the data gaps.

There are striking differences between the types of monitoring resources likely to be required to fill gaps, depending upon the nature of the indicator. All indicators of chemical conditions (both nutrients and other compounds, and chemical contaminants) and those that track the production and use of food, fiber, and water will probably continue to require on-the-ground data collection, as will a majority of the indicators of species and biological community condition. Measures of landscape pattern (fragmentation

Table A.2. Status of Data Collection for Selected Indicators with Partial or Insufficient Data

SYSTEM	INDICATOR	PAGE
Data available, but time, financial, or other constraints limited access		
Core National	Recreation	60
Farmlands	Fragmentation of Farmland Landscapes by Development	93
Farmlands	Shape of "Natural" Patches	94
Forests	At-Risk Native Species	124
Freshwater	Freshwater Recreational Activities	153
Grassland/shrubland	Area and Size of Grassland and Shrubland Patches	163
Grassland/shrubland	At-Risk Native Species	168
Data expected to be available soon (1–3 years)		
Coastal Waters	Shoreline Types	70
Coastal Waters	Recreational Water Quality	84
Forests	Forest Pattern and Fragmentation	120
Forests	Area Covered by Non-native Plants	125
Forests	Forest Age	126

and related attributes) and measures of extent will likely continue to rely heavily upon remote sensing information. These conclusions are tentative, however, because we have not conducted detailed assessments nor consulted widely about the potential for new and innovative approaches to using remote sensing data to monitor ecosystem attributes that previously required on-the-ground data collection.

In the previous section, we pointed out that for many of the indicators listed as having data insufficient for national reporting, at least some data are available today. The implication for filling the data gaps identified in this project are mixed. On one hand, the fact that data are collected for many measures is heartening, because it means that new monitoring programs may not be needed. However, the disaggregated nature of current monitoring efforts means that simply identifying the nature and scope of the problem will be a painstaking effort. For example, it is a significant undertaking to determine whether data collected by monitoring programs operated by states, federal agencies, research organizations, and others are sufficiently comparable to be collated into a single data set and whether this data set would be appropriate for making regional and national statements. We hope to examine these data sets in detail during the next phase of our research.

Figure A.5. Status of Data Collection

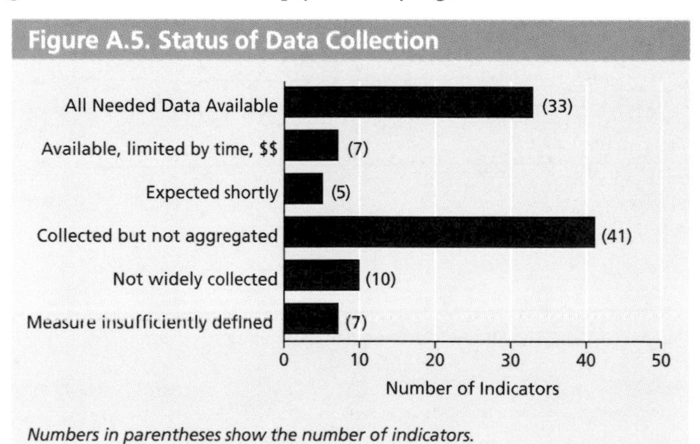

Numbers in parentheses show the number of indicators.

Technical Notes

Core National Indicators

Ecosystem Extent

Note: Several other indicators refer to this technical note for the discussion of remote-sensing data (National Land Cover Dataset) included at the end of this note.

The Indicator

"Coasts and oceans" are indicated by the area of brackish water off U.S. coasts. Brackish water is defined as all waters that have a salinity greater than about 1 part per thousand (ppt) and less than about 30 ppt (measurements are actually made in units called practical salinity units, which are quite close to parts per thousand).

Brackish water systems, including estuaries, are among the most productive ecosystems in the world—before 1985, estuarine-dependent fish species accounted for more than 50% of U.S. fish landings. Brackish water is a mixture of fresh water and seawater, and its distribution is a fundamental parameter of the distribution, abundance, and productivity of estuarine-dependent organisms and of essential fish habitats such as tidal wetlands (mangrove swamps, salt and brackish marshes, and intertidal flats), submerged attached vegetation (macroalgae and vascular plants), and oyster reefs.

Most variability in the salinity of coastal ecosystems is related to freshwater runoff and groundwater discharge. Thus, the areal extent of brackish water is an index of the amount of freshwater that flows from the continent to coastal waters and can be used as a surrogate for nutrient enrichment, sediment loading, and contaminant inputs.

"Croplands" includes the following U.S. Department of Agriculture (USDA) Economic Research Service (ERS) categories: cropland harvested, crop failure, cultivated summer fallow, cropland used only for pasture, and idle cropland; Conservation Reserve Program lands are included. Note: In the Farmlands chapter of this report (see Total Cropland, p. 91), we present multiple estimates of the area of croplands; the ERS was selected for use in this national indicator as illustrative of long-term trends.

"Forests" is defined by the USDA Forest Service as areas of at least one acre with a certain density of trees (at least 10% cover). See also the forest area indicator (p. 117) and its associated technical note (p. 239).

"Fresh waters" includes lakes and streams, as well as wetlands; however, only wetland acreage is reported in this indicator. Wetlands are defined according to the U.S. Fish and Wildlife Service's wetland classification system, which is the national standard, and include the following types of freshwater wetlands: principally palustrine forested wetlands, palustrine scrub-shrub wetlands, and palustrine emergent wetlands.

"Grassland/shrubland areas" for the lower 48 states are defined according to the National Land Cover Dataset (NLCD; see below) and include several land cover categories. Definitions of land cover in Alaska are from a separate study (see below).

"Urban/suburban areas" is generally defined here as land that is substantially covered by one of the following land cover types: low-intensity residential, high-intensity residential, commercial or industrial or transportation lands, and urban and recreational grasses. These categories are based on remote-sensing classification of land cover (see NLCD description below). A series of steps were taken to limit these areas to those thought to be most representative of urban and suburban issues (see the urban/suburban extent technical note for a thorough description, p. 264). There are several other ways that urban areas have been defined by various programs (again, see urban/suburban technical note, p. 264). The approach adopted by the ERS that relies on U.S. Census Bureau data for urban areas is a consistent dataset, however, it is based on different assumptions than the definition of urban/suburban areas in this report. The ERS time series is shown to give a sense of the relative change in urban/suburban areas over the past 50 years.

The land cover and ocean depth (bathymetry) map displays the geographic location of the various ecosystems. Data for forests, grass/shrublands, croplands, and urban/suburban were derived from the definitions in the NLCD. Only those rivers with flow rates exceeding 1000 cubic feet per second (cfs) are shown. Bathymetry data in coastal waters show the depth to the ocean floor in several ranges.

The Data

Coasts and Oceans: Data are not adequate for national reporting. Some data for the salinity of open waters of the U.S. Exclusive Economic Zone are available from the National Oceanographic Data Center (see http://www.nodc.noaa.gov/General/salinity.html). Local and regional data for semi-enclosed bodies of water are collected by a variety of federal and state agencies, but these data have not been compiled into a single source.

Cropland: Data on cropland extent come from the USDA ERS, and are available at http://www.ers.usda.gov/Emphases/Harmony/issues/arei2000/. ERS relies on data provided by the National Agricultural Statistics Service, as well as a variety of other sources. The ERS estimate for croplands is a reasonable estimate; however, it is not the only credible estimate. Specifically, the croplands extent measure (see p. 91) provides estimates of the extent of cropland from other agencies and programs; these estimates of mid-1990s cropland extent range from a low of 431 million acres (USDA Census of Agriculture) to a high of 510 million acres (USDA National Resources Inventory and NLCD). Data from ERS (455 million acres) are used here, but without additional research into which data source is more accurate, it would be equally fair to use any of the other estimates.

Forests: Data on forest extent are from the USDA Forest Service Forest Inventory and Analysis (FIA) program (see http://fia.fs.fed.us). FIA is a survey-based program that has operated since the late 1940s, collecting information on a variety of forest characteristics. See the technical note for the forest area indicator (p. 239) for additional details.

Fresh Waters: Data on freshwater wetlands are from the U.S. Fish and Wildlife Service's National Wetlands Inventory (NWI). See Dahl (2000); data also available at ftp://wetlands.fws.gov/status-trends/SandT2000Report_lowres.pdf. The NWI produces periodic reports on the extent of wetlands in the United States. See also the technical note on freshwater extent (p. 246). River data are from the U.S. EPA River Reach File (see http://www.epa.gov/waterscience/BASINS/metadata/rf3a.htm), which was constrained so that only those rivers with flow rates of at least 1000 cfs were used. Procter & Gamble's Miami Valley Laboratory conducted this analysis for The Heinz Center.

Grasslands and Shrublands: Data on the extent of grasslands and shrublands (lower 48 states) are from the NLCD—see the detailed description below.

Grassland/shrubland data for Alaska are from a vegetation map of Alaska, based on Advanced Very High Resolution Radiometer (AVHRR) remote-sensing images with an approximate resolution of 1 kilometer on a side (see complete description below). The following groupings of classes were used (see http://agdc.usgs.gov/data/projects/fhm/#G, Statewide Vegetation/Land Cover; other classifications listed below): alpine tundra & barrens (#3); dwarf shrub tundra (#4); tussock sedge/dwarf shrub tundra (#5); moist herbaceous/shrub tundra (#6); wet sedge tundra (#7); low shrub/lichen tundra (#8); low & dwarf shrub (#9); tall shrub (#10); and tall & low shrub (#23).

See the Area of Grasslands and Shrublands technical note (p. 256) for information on the pre-settlement estimates of these lands from Klopatek et al. (1979).

Urban and Suburban Areas: Grassland and shrubland data are a relatively straightforward presentation of NLCD vegetation classes, but urban and suburban area data required additional processing. Basically, this involved identification of areas with urban/suburban land cover (using the NLCD classes; see below), then making adjustments to account for the intermixed land use at the edges of urban areas. For example, undeveloped parcels or large parks located within developed areas were included as "urban/suburban" even though they might have been classified as forest or grass/shrub according to the satellite data. See the technical notes on urban/suburban extent (p. 264) for additional information.

The Land Cover and Ocean Bathymetry Map: The map shown in this indicator was constructed from several datasets by USGS's Earth Resources Observations Systems Data Center. These datasets are described below.

Data Quality/Caveats

Because these data are from multiple sources, some caution is appropriate. Different programs use different definitions and may be conducted in different years. Every effort has been made here to identify consistent land cover categories and time periods.

Given the diversity of programs, definitions, techniques, and time periods, there are inevitable conflicts between these various estimates. For example, satellite data (described below) indicate that there are about 690 million acres of forest in the United States (all 50 states), while the USDA Forest Service's FIA program estimates that there are about 747 million acres of forest. Satellite remote sensing, which can provide data on the entire U.S. land surface, may serve as a common reference point, against which other programs that count only forest, for example, or only private lands, or only cropland, may be compared.

The National Land Cover Dataset (NLCD): In the 1990s, a federal interagency consortium was created to coordinate access to and use of land cover data from the Landsat 5 Thematic Mapper. Using Landsat data and a variety of ancillary data, the consortium processed data from a series of 1992 Landsat images, to create the NLCD on a square grid covering the lower 48 states. Each square in the grid, or "pixel," is approximately 100 ft on a side.

Each pixel was assigned one of 21 land cover classes, which are described at http://www.epa.gov/mrlc/classes.html and http://landcover.usgs.gov/classes.html. The steps of this classification process, which can be found in detail elsewhere (see Vogelmann et al. 2001 and Vogelmann et al. 1998), are summarized here. First, an automated process is used to create clusters of pixels for a given regional area. Second, these clusters were interpreted and labeled with the help of aerial photographs. Third, in cases where clusters of pixels included multiple land cover types (i.e., "confused clusters"), models that utilize ancillary data, such as elevation or population density, were used to help assign land cover classes. Finally, lands that are bare—especially clear cuts and quarries—and many grass areas, such as parks, golf courses, and large lawn, are not easily distinguished from other land cover classes during the automated process, so a process of on-screen verifications was used as clarification. These four steps were the general process, and additional steps were taken in certain regions in order to further improve the accuracy of classifications (see http://landcover.usgs.gov/accuracy/ for a discussion of NLCD error analysis).

Note that classification of pixels was based in part on the character of surrounding squares in the grid; thus, a pixel of grass-like land cover surrounded by residential pixels would probably be classified as "urban and recreational grasses" rather than as "pastureland." Where appropriate, the agencies also made use of data from both the Census Bureau and the U.S. Fish and Wildlife Service's National Wetlands Inventory data to help make such distinctions. Satellite data offer an unprecedented opportunity to classify land cover on a consistent basis over very large areas (i.e., the entire country). However, the accuracy of any classification is not perfect. The accuracy of satellite-derived classifications is related to many factors: amount of data available (i.e., many dates of imagery rather than just one), the detail of the required land cover information (i.e., forest vs. deciduous forest vs. sugar maple/beech/yellow birch), classification methods, computing power, and, of course, time and money. Assessments of the NLCD for the eastern United States indicate an accuracy of approximately 80% or higher for general land cover categories (e.g., forest, agriculture, developed). Accuracy assessments for the western United States are currently under way. Improving technology and techniques offer the potential to increase accuracy of the next NLCD (2000) currently being assembled by the Multi-Resolution Land Characterization Consortium. The land cover classes associated with the 30-m (100-foot) square pixels were grouped for the different ecosystems as follows (the number in parenthesis is the NLDC land cover class reference):

- *Forests:* deciduous (#41); evergreen (#42); mixed forest (#43)
- *Croplands:* pasture/hay (#81); rowcrops (#82); small grains (#83); fallow (#84); orchards/vineyards/other (#61)
- *Grass/Shrub:* shrubland (#51); grasslands/herbaceous (#71); bare rock/sand/clay (#31)
- *Water:* open water (#11); wetlands (#91 & #92)
- *Developed:* low-intensity residential (#21); high-intensity residential (#22); commercial/industrial/transportation (#23); urban/recreational grasses (#85)
- *Other:* quarries/strip mines/gravel pits (#32); transitional (#33); perennial ice/snow (#12)

Data Quality/Caveats: The power of satellite-derived classifications is that satellite data can easily cover the entire country and the classification process can be automated (though not completely). This makes it possible to compile a nationally consistent land cover dataset; however, any land cover classification is subject to error. The NLCD for the eastern United States has an accuracy of approximately 80% or higher for the general land cover categories used for our study (see http://landcover.usgs.gov/accuracy). Some of the known misclassifications that occur in the dataset include suburban areas or tree farms classified as forest; grasslands classified as agriculture, or vice versa; and fallow agricultural fields classified as barren lands.

Data Access: NLCD data are available at http://landcover.usgs.gov/mrlcreg.html. Further detail is also available from Vogelmann et al. (2001). Other data can be obtained from the sources cited in this note.

Coastal Bathymetry Data: These data come from the National Geophysical Data Center, and are known as ETOPO5 data. They were generated from a digital database of land and sea-floor elevations on a 5-minute latitude/longitude grid. The resolution of the gridded data varies from true 5-minute for the ocean floors, the United States, Europe, Japan, and Australia to 1 degree in data-deficient parts of Asia, South America, northern Canada, and Africa. Data sources are as follows: Ocean Areas: U.S. Naval Oceanographic Office; United States, W. Europe, Japan/Korea: U.S. Defense Mapping Agency; Australia: Bureau of Mineral Resources, Australia; New Zealand: Department of Industrial and Scientific Research, New Zealand; balance of world land masses: U.S. Navy Fleet Numerical Oceanographic Center. These various databases were originally assembled in 1988 into the worldwide 5-minute grid by Margo Edwards, then at Washington University, St. Louis, Missouri. Data have been described in NOAA (1988). The version of the data making up ETOPO5 is from May 1988, with the exception of a small area in Canada (120-130° W, 65-70° N), which was regridded in 1990; the data are available at: http://www.ngdc.noaa.gov/mgg/global/seltopo.html.

Alaskan Land Cover Data: Data for Alaska are from a vegetation map of Alaska by Flemming (1996), based on AVHRR remote-sensing images with an approximate resolution of 1 kilometer on a side. The following groupings of classes were used (see http://agdc.usgs.gov/data/projects/fhm/#G [Statewide Vegetation/Land Cover]):

- *Freshwater:* water (#1); glaciers and snow (#2)
- *Grass/Shrub:* alpine tundra & barrens (#3); dwarf shrub tundra (#4); tussock sedge/dwarf shrub tundra (#5); moist herbaceous/shrub tundra (#6); wet sedge tundra (#7); low shrub/lichen tundra (#8); low & dwarf shrub (#9); tall shrub (#10); tall & low shrub (#23)
- *Forest:* closed broadleaf & closed mixed forest (#11); closed mixed forest (#12); closed spruce forest (#13); spruce woodland/shrub (#14); open spruce forest/shrub/bog mosaic (#15); spruce & broadleaf forest (#16); open & closed spruce forest (#17); open spruce & closed mixed forest mosaic (#18); closed spruce & hemlock forest (#19)
- *Other:*1991 fires (#21); 1990 fires & gravel bars (#22)

Hawaiian Land Cover Data: These data came from the National Oceanographic and Atmospheric Administration's (NOAA) Coastal Change and Analysis Program (C-CAP), which is a national effort to develop and distribute regional land cover and change analysis data for the coastal zone by using remote-sensing technology. The data used in this program are created from a combination of satellites and fieldwork. C-CAP classifies land cover types into 22 standardized classes that include forested areas, urban areas, and wetlands. C-CAP land cover data are derived from Landsat Thematic Mapper satellite imagery and are available at http://www.csc.noaa.gov/crs/lca/m_eight.html.

References

Dahl, T.E. 2000. Status and trends of wetlands in the conterminous United States 1986 to 1997. Washington, DC: U.S. Department of the Interior, Fish and Wildlife Service.

Flemming, M.D. 1996. A statewide vegetation map of Alaska using a phenological classification of AVHRR data. 1996 Alaska Surveying and Mapping Conference, Anchorage, Alaska.

Klopatek, J.M., R.J. Olson, C.J. Emerson, and J.L. Joness. 1979. Land-use conflicts with natural vegetation in the United States. Environmental Conservation 6:191–199.

NOAA. 1988. Data announcement 88-MGG-02. Digital relief of the surface of the earth. National Geophysical Data Center, Boulder, Colorado.

Vogelmann, J.E., S.M. Howard, L. Yang, C.R. Larson, B.K. Wylie, and N. van Driel. 2001. Completion of the 1990s national land cover data set for the conterminous United States from Landsat Thematic Mapper data and ancillary data sources. Photogrammetric Engineering & Remote Sensing 67:650–662.

Vogelmann, J.E., T.L. Sohl, P.V. Campbell, and D.M. Shaw. 1998. Regional land cover characterization using LANDSAT Thematic Mapper data and ancillary data sources. Environmental Monitoring and Assessments 51: 415–428.

Fragmentation and Landscape Patterns

There is no technical note for this indicator.

The Movement of Nitrogen

The Indicator

This indicator reports both the yield and load of nitrogen from major rivers to the U.S. coastal ocean. The *yield* of nitrogen from major watersheds is defined as the pounds of nitrogen per square mile of watershed area that enters rivers and streams through discharges, runoff, and other sources. The *load* of nitrate, a common form of nitrogen, from major rivers is defined as the tons of nitrate carried to the ocean each year by the four largest U.S. rivers.

Nitrogen can cause significant water-quality problems by stimulating the growth of algae. Two key references provide additional information about how excess nutrients can cause problems in coastal waters. The National Research Council published a study on nutrient pollution in the coastal ocean (NRC 2000) and the National Oceanographic and Atmospheric Administration (NOAA) did a one-time study of actual conditions in the nation's estuaries (National Estuarine Eutrophication Assessment, http://spo.nos.noaa.gov/projects/cads/nees/Eutro_Report.pdf).

The Data

Data Source: Riverine loads of total nitrogen were estimated using streamflow and water-quality data collected by the U.S. Geological Survey (USGS) as part of its National Stream Water Quality Accounting Network (NASQAN), its 1996–1999 National Water Quality Assessment (NAWQA), and its Federal–State Cooperative Program. A few of the stream gauges, most notably those at the mouth of the Mississippi River, are operated by the U.S. Army Corps of Engineers rather than the USGS.

Data Collection Methodology: Stream discharge is estimated by frequent measurement of water depth (stage), which is converted to discharge by use of a rating curve. Data are reported as daily averages. All water-quality samples are representative of the entire river cross-section (depth- and width-integrated) at the time of collection.

At the sites for which data are included in this report, samples were collected at least quarterly over the 4-year period 1996–1999; at most sites, approximately 15 samples were collected each year. A regression model relating concentration to discharge, day-of-year (to capture seasonal effects), and time (to capture any trend over the period) was developed using robust statistical techniques that made no assumption about the underlying statistical distribution of the data. One model was developed for nitrate plus nitrite concentrations (note that nitrite is usually much less abundant than nitrate, so it is normal to discuss the sum of nitrate plus nitrite simply as nitrate); a second model was developed for whole-water organic nitrogen plus ammonia for each station. These models were then used to make daily estimates of concentration, which were multiplied by the daily average discharge to yield the daily load. The daily load of total nitrogen was the sum of predictions of the two models.

Data Manipulation: For the maps, these daily loads were summed over the 4-year period to estimate the load for the entire period and divided by 4 to obtain the average annual load. The coefficient of variation of the average annual load is generally between 20 and 30%. The incremental load was then calculated as the difference between the output load that flowed from the watershed and the input(s) to the watershed. Outputs include the load at the downstream stations and, in the arid western areas, any decrease in runoff, because it was assumed that solutes accompanied any water that was lost to irrigation or transfers to other watersheds (i.e., piping water across watershed boundaries). The incremental yield (shown in the maps) is defined as the incremental load divided by the watershed area. The white areas of the map are areas for which insufficient USGS data exist to calculate loads.

For the time series plots, the daily loads were summed to determine the annual loads shown in the figure. Note that most of the year-to-year variation in the loads is due to differences in runoff, with wet years having higher loads and dry years having lower loads.

Data Access: All USGS data are available at http://water.usgs.gov/nwis. This site includes the discharge and nutrient concentration data used in developing the models that produced the load calculations presented here, but the models themselves are not available. Further information on the NASQAN and NAWQA programs can be found at http:/water.usgs.gov/nasqan and http://water.usgs.gov/nawqa. The NASQAN Web site contains stream discharge data collected by the U.S. Army Corps of Engineers.

References

National Research Council. 2000. Clean coastal waters: Understanding and reducing the effects of nutrient pollution. Washington, DC: National Academy Press.

Chemical Contamination

This technical note also applies to:

- **Coasts and Oceans: Contamination in Bottom Sediments**
- **Farmlands: Pesticides in Streams and Groundwater**
- **Urban/Suburban: Chemical Contamination**

This technical note applies to the core national indicator for chemical contamination, the coastal indicator for sediment contaminants, the farmlands indicator for pesticide, and the urban/suburban indicator for chemical contamination. One technical note applies to these three indicators because they are designed in a very similar fashion. In addition, most of the data (i.e., all freshwater data) for these three indicators are from the same program.

The Indicator—General

In the core national indicator, as well as the indicators for farmlands and urban/suburban areas, a dual approach is used: how frequently compounds are detected, and how often such "occurrences" are at concentrations that are above established human health standards and guidelines and aquatic life guidelines—"exceedances." The coastal sediment contamination indicator presents only data on exceedances of relevant guidelines.

Compounds reported here include many pesticides, polychlorinated biphenyls (PCBs), volatile organic compounds (VOCs), other compounds on the Environmental Protection Agency (EPA) priority pollutant list, potentially toxic trace elements, and a number of pesticide degradation products. The suite of compounds that were measured in different media varied depending on the use of the compounds in a particular area and the chemistry of the compounds. For example, many VOCs (e.g., solvents and fuel additives) are more heavily used in urban than agricultural settings. Further, because of their volatility, VOCs would be expected to be of greater concern in groundwater than in streams or sediments. In general, the suite of compounds was designed to include compounds that occur frequently in ecosystems and/or have a high potential for adverse effects on people or ecosystems.

In order to understand how frequently compounds from a particular suite of contaminants occur in the environment, the U.S. Geological Survey's National Water Quality Assessment (NAWQA) program analyzes groundwater and water, sediments, and fish tissue from streams. EPA's Environmental Monitoring and Assessment Program (EMAP) analyzes sediments and fish tissue from estuaries. It is important to note that all chemical analyses have "detection limits," meaning that even if a compound is present at a concentration lower than the detection limit, the sample cannot be differentiated from one that completely lacks the compound. Analytical methods used for different environmental media are different (e.g., measurements of contaminant concentrations in stream water and groundwater use different techniques than are used in stream sediment analyses, and techniques used in fresh water differ from those used in salt water). However, within an environmental media (sediment, stream water, etc.), consistent analytical methods were used throughout the program. In addition, as data for this indicator are gathered over time, it will be important to consider the effect of improved detection methods (i.e., allowing contaminants to be detected at lower concentrations) on occurrence data.

The second component of the indicators provides a measure of the frequency (e.g., percentage of stream sites) of contaminants that exceeded established reference criteria for the protection of human health or aquatic life. These two types of reference criteria were established for different purposes and thus are based on different assumptions. Specifically, human health standards and guidelines assume that the water will be consumed daily over a person's lifetime, and that the effects of the contaminant would be cumulative (often referred to as "chronic" exposure). Human health standards and guidelines are not currently applied to stream or estuary sediments.

In comparison, guidelines that are designed for the protection of aquatic life are based on shorter-term (about 4 days) exposure. (This refers to U.S. EPA guidelines; Canadian guidelines are different (see references)). This is because aquatic organisms are generally smaller and they are exposed to contaminants in water in a different way than humans are. Also, in general, different species have different sensitivities to specific contaminants.

Standards and guidelines for the protection of wildlife are used to evaluate whether contaminant levels in prey species (fish, in this case) are sufficiently high to cause adverse effects in predator species (often fish-eating birds such as eagles).

Specific standards and guidelines used in this indicator are listed under the appropriate media description below.

Data Sources—General

The data for freshwater streams and groundwater were collected and analyzed by NAWQA (http://water.usgs.gov/nawqa/) in 36 major river basins and aquifers across the United States during 1992–1998.

The data for sediments and fish contamination in coastal waters were collected and analyzed by EPA's EMAP (http://www.epa.gov/emap/) from 1990 to 1997. The data were collected in a manner that allows conclusions to be drawn concerning the majority (approximately 76%) of the area of estuaries in the United States.

Data on sediment contamination in the Great Lakes are collected by a number of agencies. However, these monitoring programs generally focus on areas with highly polluted sediments. As such, these data are not comparable to the data presented here, in that they do not assess the occurrence of sediment contamination across the range of possible locations in the Great Lakes. The EPA Great Lakes National Program Office provided the Great Lakes fish contamination data that are noted in the text.

Freshwater Data: National Water Quality Assessment Program

Methods: The suite of compounds included in the Core National Indicator account for 75% of currently used agricultural pesticide applications (by amounts used), 90% of the nation's historical use of organochlorine pesticide (most of these compounds are no longer permitted for use in agriculture in the United States), plus PCBs and other industrial compounds, VOCs that are currently or soon may become regulated, and other compounds on the EPA priority pollutant list. A number of pesticide degradation products were also included. Nitrate and ammonium were measured in streams and groundwater. Trace elements were measured in stream sediments and groundwater. Radionuclides were measured only in groundwater. All of these contaminants occur naturally in the environment. Thus, they are included in graphics showing exceedances of human health and/or aquatic life benchmarks, but not in graphics showing the occurrence of contaminants. Human health benchmarks apply to those contaminants listed above that were measured in streams and groundwater. Aquatic life benchmarks apply to the contaminants listed above that were measured in streams and stream sediments.

Additional information about the USGS data used in the Heinz Center report can be found at: http://water.usgs.gov/nawqa/heinz_ctr/

The watersheds studied were selected to be generally representative of conditions in agricultural, urban, and mixed land uses. The national contaminants data are based on water samples collected from 109 stream sites and 3,549 wells, stream sediment from 558 stream sites, and composite whole-fish samples collected from 223 stream sites. The sites sampled are representative of a wide range of stream sizes, types, and agricultural, urban/suburban, and mixed land uses, but the sites were not selected to be a statistically representative sample of the nation's streams.

Data for the urban/suburban indicator come from surface water sites on streams that drain 21 urban/suburban watersheds located across the nation. Note that the sites used in this analysis probably are included with "urban and suburban areas" as defined in this report (see p. 181); however, the selection of the sites for sampling was not based on the definition used in this report.

Data for the farmlands indicator are based on water samples from 50 streams and 1084 monitoring wells.

Benchmarks for protection of human health, wildlife, and aquatic life: A variety of U.S., Canadian, and bi-national (International Joint Commission) standards and guidelines were used to evaluate the significance of the detected contaminants in surface water, groundwater, stream sediment, and whole fish.

In conformance with the way the guidelines are written, a concentration exceeding the aquatic-life guidelines in any single surface water sample was counted as an exceedance of the guideline. For human health standards or guidelines, exceedances were identified when a yearly time-weighted mean concentration exceeded the relevant standard or guideline at a surface water site.

For human health, three types of U.S. EPA standards and guidelines were used to evaluate NAWQA data: (1) Maximum Contaminant Level (MCL), (2) Risk-Specific Dose (RSD), and (3) Lifetime Health Advisory (HA-L). Values for these criteria were obtained by the U.S. Geological Survey (USGS) from U.S. EPA (2000, 2001). In all three cases, the standard and guideline levels are concentrations pertaining to lifetime exposure through drinking water.

The MCL is the maximum permissible annual average concentration of a contaminant in water that is delivered to any user of a public water system. The RSD is a guideline for potential carcinogens based on drinking-water exposure over a 70-year lifetime; an RSD value is always associated with a specified cancer risk. The RSDs presented are associated with a cancer risk of 1 in 100,000. The HA-L is an advisory guideline for drinking-water exposure over a 70-year lifetime, considering noncarcinogenic adverse health effects. More detail on these types of benchmarks, their derivation, and their underlying assumptions is provided in Nowell and Resek (1994). For some constituents, more than one of these three types of benchmarks are available. For these constituents, the MCL was used if available; otherwise, the lowest of the RSD (at 1 in 100,000 cancer risk) and HA-L values selected.

Note that the data on freshwater fish tissue do not include information relative to any human health standards because such standards apply to edible fish tissue (e.g., fillets), whereas entire fish were analyzed for the data reported here.

The three types of aquatic-life guidelines used are U.S. EPA chronic water-quality criteria for protection of aquatic organisms (U.S. Environmental Protection Agency 1999), Canadian water-quality guidelines (Canadian Council of Ministers of the Environment 2001a), and Great Lakes water-quality objectives (International Joint Commission [IJC] 1978). All guideline values used in this report are for freshwater aquatic life. The U.S. EPA chronic water-quality criterion for protection of aquatic organisms is the estimated highest concentration of a constituent that aquatic organisms can be exposed to for a 4-day period, once every 3 years, without deleterious effects. If no U.S. EPA chronic water-quality criterion for protection of aquatic organisms exists for a given constituent, then Canadian water-quality guidelines are used, if available. The older Great Lakes water-quality objectives are used only if neither U.S. EPA chronic water-quality criteria for protection of aquatic organisms nor Canadian water-quality guidelines are available for that constituent. The IJC water-quality objectives and Canadian water-quality guidelines are intended to specify maximum concentrations that should not be exceeded at any time.

For contaminants in sediment, the aquatic-life guideline used was the "probable effect level" from the Canadian Council of Ministers of the Environment (2001b). These guidelines are empirically based; they were derived by compiling data from multiple types of studies in the literature that measured both toxicity and contaminant concentrations in sediment. The Canadian probable effect level defines a concentration above which toxicity to aquatic organisms is likely.

For contaminants in whole fish, the New York fish-flesh criteria for protection of piscivorous (fish-eating) wildlife (Newell et al. 1987) were used. These guidelines are intended to protect target wildlife species from adverse effects other than cancer, such as mortality, reproductive impairment, and organ damage. Wildlife guidelines from the state of New York were used because no comparable national guidelines are available for a large number of contaminants.

Additional information on the standards and guidelines used in this report for pesticides is provided at http://ca.water.usgs.gov/pnsp/source/. Information on the numerical values for the standards and guidelines applied to herbicide, insecticide, and volatile organic compounds can be found at

http://oregon.usgs.gov/sumrpt/Benchmrk.1.html,
http://oregon.usgs.gov/sumrpt/Benchmrk.2.html, and
http://oregon.usgs.gov/sumrpt/Benchmrk.3.html.

Estuarine Data: USEPA Environmental Monitoring and Assessment Program (EMAP)

EMAP conducts annual surveys to measure indicators of the health of plants and animals, the quality of their surroundings, and the presence of pollutants (see http://www.epa.gov/emap/). The program, at present, is developing appropriate designs and sets of indicator requirements to characterize the condition of the nation's aquatic resources. Once these developmental issues are addressed, the goal of the program is long-term monitoring that will provide information on the overall health of the environment and the effectiveness of pollution prevention and control strategies.

EMAP-Estuaries (EMAP-E), implemented through partnerships between EPA, the National Oceanographic and Atmospheric Administration (NOAA), USGS, coastal states, and academia, will provide information on the ecological condition of the nation's estuaries as part of the larger program. The data from the EMAP-E program provided in this report spans the period from 1990 through 1997. Beginning in 2000, the EMAP-E effort expanded into a series of annual national surveys (National Coastal Assessment, or NCA) including all coastal states and Puerto Rico. Ecological health is being assessed by investigating the state, regional, and national distributions of fish and bottom-dwelling organisms (benthos). NCA is determining what portions of estuaries can support these plants and animals and finding out why certain areas do not support them. Data from NCA will be available for the next iteration of this report (1999–2005).

For this report, EMAP-E provided information assessing the contaminant levels in estuarine sediments and the condition of benthic organisms in those sediments. These data were collected from over 2000 sites from Cape Cod, Massachusetts, to Brownsville, Texas, and represent over 70% of the total estuarine acreage of the United States (excluding Alaska).

Data Collection Methodology: Evaluation of the potential effects of contaminated sediments on estuarine organisms is difficult because few applicable state or federal regulatory criteria exist for determining acceptable sediment concentrations of all substances. However, contaminated sediments and their potential toxicity to aquatic life are viewed by the public as a major threat to estuarine ecosystems. All site selections were based on probabilistic designs which permit the extrapolation of the data to the entire area. Using a Young-modified Van Veen grab, 5–10 grabs were collected from each site and homogenized. Separate 100-milliliter samples for organics and metals were retrieved from the homogenate and forwarded for quantification of about 125 different compounds (as outlined below). For this report, information assessing the portion of estuarine area with contaminants above ERL or ERM guidelines (see definitions below; Long et al. 1995; Long et al. 1998a,b) is reported.

Data Access: The data presented here were obtained directly from EPA.

List of Contaminants Targeted in Sediments by EMAP

- Pesticides. Pesticides were chosen because of their current and historic prevalent use in society. Sediments were tested for concentrations of 14 pesticides plus six different forms of DDT, which has been banned in the United States since 1972. These pesticides included Aldrin, Alpha-Chlordane, Dieldrin, Endosulfan I, Endosulfan II, Endosulfan sulfate, Endrin, Heptachlor, Heptachlor epoxide, hexachlorobenzene, Lindane (gamma-BHC), Mirex, Toxaphene, Trans-Nonachlor, 2,4'-DDD, 4,4'-DDD 2,4'-DDE, 4,4'-DDE, 2,4'-DDT, and 4,4'-DDT.
- Polychlorinated biphenyls (PCBs). EPA began to phase out the use and manufacturing of PCBs in the United States in 1976, but they are still found in the environment. Human health effects that have been associated with exposure to PCBs include acne-like skin conditions in adults and neurobehavioral and immunological changes in children. PCBs are known to cause cancer in animals. EMAP targeted 21 different PCB congeners.
- Polycyclic aromatic hydrocarbons (PAHs). A group of over 100 different chemicals that are formed during the incomplete burning of coal, oil and gas, garbage, and other organic substances like tobacco or charbroiled meat, PAHs are usually found as a mixture containing two or more of these compounds, such as soot. Some PAHs are manufactured—they are found in coal tar, crude oil, creosote, and roofing tar, and a few are used in medicines or to make dyes, plastics, and pesticides. PAHs are included because of their role as a suspected carcinogen. The following compounds were targeted (plus several isomers of the listed PAHs): Acenaphthene, Anthracene, Benz(a)anthracene, Benzo(a)pyrene, Biphenyl, Chrysene, Dibenz(a,h)anthracene, Dibenzothiophene, 2,6-dimethylnaphthalene, Fluoranthene, Fluorene, 2-methylnaphthalene, 1-methylnaphthalene, 1-methylphenanthrene, 2,6-dimethylnaphtalene, Naphthalene, Pyrene, Benzo(b)fluoranthene, Acenaphthylene, Benzo(k)fluoranthene, Benzo(g,h,i)perylene, Ideno(1,2,3-c,d)pyrene, and 2,3,5-trimethylnaphthalene.
- Heavy metals. Heavy metals occur naturally in the marine environment; however, their concentrations can be increased by human activities such as discharges from industrial processes, burning of fossil fuels, and runoff from roadways that have had an accumulation of particulates from brake drums, for example. Sediments were tested for a total of 15 trace elements: aluminum, antimony, arsenic, cadmium, chromium, copper, iron, lead, manganese, mercury, nickel, selenium, silver, tin, and zinc. Metal concentrations were normalized using metal:aluminum ratios (see Windom et al. 1989).

Benchmarks for sediment quality: The sediment quality guidelines used in this indicator were developed by NOAA through its National Status and Trends Program (see http://response.restoration.noaa.gov/cpr/sediment/SPQ.pdf). Before these guidelines, there were no national criteria or other widely applicable numerical guidelines for sediment quality. These quality guidelines were developed as informal, interpretive tools to estimate the possible toxicological significance of chemical concentrations in sediments. The guidelines have not been promulgated as regulatory criteria or standards, cleanup or remediation targets, discharge attainment targets, or pass–fail criteria for dredged material disposal decisions, or for any other regulatory purpose. (See http://response.restoration.noaa.gov/cpr/sediment/SQGs.html.)

These guidelines were derived from examination of a large number of individual contamination studies, all in salt water. Data from each study were arranged in order of ascending concentrations. Study endpoints in which adverse effects were reported were identified. From the ascending data tables, the 10th percentile and the 50th percentile (median) of the effects database were identified for each substance. The 10th-percentile values were named the "Effects Range-Low" (ERL), indicative of concentrations below which adverse effects rarely occur. The 50th percentiles were designated the "Effects Range-Median" (ERM) values, representative of concentrations above which effects frequently occur. In this report, the ERL is referred to as the "possible effects" guideline and the ERM as the "probable effects" guideline.

The Data Gap

There are large amounts of data on contaminated sediments in the Great Lakes, but these data have for the most part been collected at sites known or suspected of being contaminated, rather than as part of efforts to determine the extent and severity of contamination. Sediments in the defined Areas of Concern (http://www.epa.gov/glnpo/aoc) are generally the most contaminated. Sediments in the open waters tend to have much lower concentrations, and they tend to migrate to sediment depositional areas. See the following for information on surveys that can identify "toxic substances in toxic amounts," which are found in the tributary mouths and embayments of the Areas of Concern: http://www.epa.gov/glnpo/glindicators/sedqual/sedqualitya.html.

Data are not presently available to compare fish tissue contamination to human health standards and guidelines in a consistent way across the country. See the technical note for Selected Contaminants in Fish and Shellfish (p. 228) for additional discussion.

Data are not presently collected in a consistent manner to allow reporting on soil contamination in urban and suburban areas. Individual studies (see Pouyat et al. 1991) have been conducted to determine the extent and nature of such contamination.

References

Canadian Council of Ministers of the Environment. 2001a. Canadian water quality guidelines for the protection of aquatic life: Summary table. In Canadian environmental quality guidelines, 1999. Winnipeg: Canadian Council of Ministers of the Environment. http://www.ec.gc.ca/ceqg-rcqe/English/Pdf/water_summary_table-aquatic_life.htm. Accessed July 24, 2001.

Canadian Council of Ministers of the Environment. 2001b. Canadian sediment quality guidelines for the protection of aquatic life: Summary tables. Updated. In Canadian environmental quality guidelines, 1999. Winnipeg: Canadian Council of Ministers of the Environment. http://www.ec.gc.ca/ceqg-rcqe/English/Pdf/sediment_summary_table.htm. Accessed July 24, 2001.

International Joint Commission. 1978. Great Lakes Water Quality Agreement of 1978, as amended by Protocol signed November 18, 1987, Annex I—Specific objectives.

International Joint Commission http://www.ijc.org/agree/quality.html#ann1. Accessed July 24, 2001.

Long, E.R., L.J. Field, and D.D. McDonald. 1998a. Predicting toxicity in marine sediments with numerical sediment quality guidelines. Environmental Toxicology and Chemistry 17(4):714–727.

Long, E.R., D.D. McDonald, S.L. Smith, and F.D. Calder. 1995. Incidence of adverse biological effects within ranges of chemical concentrations in marine and estuarine sediments. Environmental Management 19(1):81–87.

Long, E.R., G.I. Scott, J. Kucklick, M. Fulton, B. Thompson, R.S. Carr, K.J. Scott, G.T. Chandler, J.W. Anderson, and G.M. Sloane. 1998b. Magnitude and extent of sediment toxicity in selected estuaries of South Carolina and Georgia. NOAA Tech. Memo. NOS ORCA 128. Silver Spring, MD.

Newell, A.J., D.W. Johnson, and L.K. Allen. 1987. Niagara River Biota Contamination Project: Fish flesh criteria for piscivorous wildlife. New York State Department of Environmental Conservation, Division of Fish and Wildlife, Bureau of Environmental Protection, Technical Report 87-3.

Nowell, L.H., and E.A. Resek. 1994, National standards and guidelines for pesticides in water, sediment, and aquatic organisms: Application to water-quality assessments: Rev. Environ. Contam. Toxicol. v. 140, pp. 1–164.

Pouyat, R.V., and M.J. McDonnell. 1991. Heavy metal accumulation in forest soils along an urban-rural gradient in southeastern New York, USA. Water, Air, and Soil Pollution 57–58:797–807.

U.S. Environmental Protection Agency. April 1999. National recommended water quality criteria—Correction. U.S. Environmental Protection Agency, Office of Water, EPA-822-Z-99-001. http://www.epa.gov/ost/pc/revcom.pdf. Accessed July 24, 2001.

U.S. Environmental Protection Agency. Summer 2000. Drinking water standards and health advisories: U.S. Environmental Protection Agency, Office of Water, EPA-822-B-00-001. http://www.epa.gov/ost/drinking/standards/. Accessed July 24, 2001.

U.S. Environmental Protection Agency. 2001. Integrated Risk Information System (IRIS) database: U.S. Environmental Protection Agency, Office of Research and Development, National Center for Environmental Assessment. http://www.epa.gov/iris. Accessed July 24, 2001.

Windom, H.L., S.J. Scropp, F.D. Calder, J.D. Ryan, R.G. Smith, I.C. Burney, F.G. Lewis, and C.H. Rawlinson. 1989. Neutral trace metal concentrations in estuarine and marine sediments of the southeastern United States. Environmental Science and Technology 3: 314–327.

At-Risk Native Species

This indicator also applies to

- **At-Risk Native Forest Species**
- **At-Risk Native Freshwater Species**
- **At-Risk Native Grasslands and Shrublands Species**

The Indicator

The species reported here are those in groups (such as mammals, birds, and fish) that are considered sufficiently well known that the conservation status, habitat, and location (by state) can be assigned with some degree of confidence for all members of the group. The conservation status assessment for each species is an attempt to determine the relative susceptibility of a species to extinction. The assessment process is based on consideration of up to 12 factors that relate to a species' degree of imperilment or risk of extinction throughout its range. Rare species are particularly vulnerable to extinction and so several aspects of rarity are characterized in the assessment process including population size and number of populations, and range extent and area of occupancy. However, trends in population and range size as well as magnitude and immediacy of threats are also important considerations in assessing a species' overall vulnerability or risk of extinction. Additional information on this ranking process can be found at http://www.natureserve.org/explorer/ranking.htm and in Master (1991).

There is general recognition among experts that both status information (as presented here) and trend information (whether a species is increasing, decreasing, or stable) are critical to understanding the condition of species. If and when trend information for large numbers of species becomes widely available, revising the current measure by incorporating trend information or substituting trend information for status should be considered.

The Data

Data Source: NatureServe (www.natureserve.org) and its member programs in the network of Natural Heritage programs develop and maintain information on each of the species reported here.

Data Collection Methodology: On an ongoing basis, NatureServe research biologists gather, review, integrate, and record available information about species taxonomy, status, and use of different habitats or ecological system types. They are assisted in this work by scientists in the network of Natural Heritage programs as well as by contracted experts for different invertebrate taxa. NatureServe staff and collaborators assign a conservation status by using standard Heritage ranking criteria (see http://www.natureserve.org/explorer/ranking.htm) and by using the best information available to them.

The Heritage ranking process considers five major status ranks: critically imperiled (G1), imperiled (G2), vulnerable (G3), apparently secure (G4), and demonstrably widespread, abundant, and secure (G5). In addition, separate ranks are assigned for species regarded as presumed extinct (GX) or possibly extinct (GH).

Critically imperiled species are often found in five or fewer locations, imperiled species are often found in 20 or fewer locations, and vulnerable species are often found in 80 or fewer locations. Apparently secure species are uncommon but not rare, and secure species are common—meaning they are both abundant and widespread. Presumed extinct species have not been located despite intensive searches, and possibly extinct species are missing and are known only from historic records, although there is some hope of their rediscovery. See Stein (2002) for further details on ranks.

These data are not from a site-based monitoring program, but rather from a census approach that focuses on the location and distribution of at-risk species. For other species, the dataset incorporates information from a wide variety of observations and sources.

Data Manipulation: NatureServe has summarized the actual global ranks into "rounded ranks" for the purposes of presentation and analysis. For example, an actual rank may express the bounds of uncertainty, noting for instance that a given species falls somewhere in the range of "critically imperiled" to "imperiled." In such cases, the rounded rank reflects the *more* imperiled designation, in this instance, critically imperiled. Such rank rounding applies to between 10-20% of species included here. The analysis of the percent of at-risk species by region is based on all species that are known to occur within one or more states in each region.

For the core national indicator, only species groups for which sufficient information is available on the entire group are reported. Thus, mammal status is reported because data are available on the status of all mammals, but the status of mayflies and stoneflies is not included because data on all species in these two groups are not available. Groups reported for the national measure are mammals; birds; reptiles; amphibians; freshwater fishes; freshwater mussels; freshwater snails; crayfishes; fairy, clam, and tadpole shrimp; butterflies and skippers; giant silkworm and royal moths; sphinx moths; underwing moths; papaipema moths; tiger beetles; stoneflies; grasshoppers; mayflies; dragonflies and damselflies; ferns/fern allies; gymnosperms; and flowering plants.

For the forest, freshwater, and grassland/shrubland indicators, species were first identified as "forest species" or "freshwater species" or "grassland/shrubland species." In this process, species were assigned to an ecosystem if they live in that ecosystem during at least part of their life cycle and depend on access to that ecosystem type for their survival. This was a generally conservative approach; in preparing these lists, only species that are strongly associated with a habitat type were included. This means that some species that make frequent use of forest, or grasslands/shrublands, or fresh water may be excluded, butalso that the group of species reported for each of these systems here is quite representative of species that are dependent upon those habitats for their survival. Groups reported for the forest and grassland indicator are mammals, birds, reptiles, amphibians, grasshoppers, and butterflies and skippers. Groups reported for the freshwater indicator are freshwater and anadromous fishes; amphibians; reptiles; butterflies and skippers; freshwater mussels; freshwater snails; crayfishes; fairy, clam, and tadpole shrimp; dragonflies and damselflies; mayflies; stoneflies; and caddis flies.

At present, it is not possible to use the data presented here to distinguish naturally rare species from those that have been depleted in number. Increases in the number of at-risk species over time, however, would generally be interpreted as an increase in the number of depleted species after accounting for changes due to changes in taxonomy or to discovery of new species.

Data Quality/Caveats: Heritage conservation status ranks are updated on an ongoing basis through literature review and feedback from users of the network's databases, and also through periodic review of all statuses. Uncertainty about conservation status of a species is captured in part through the use of range ranks (see "Data Manipulation" above). A species' status may change over time due to several reasons, and not solely due to a species becoming more or less at risk of extinction. For example, more populations may be found than were known to exist, or a species may be split taxonomically into two species, such that the two new species may individually be at greater risk of extinction than their single parent species. Because status may change for reasons other than an actual change in condition, and because a species may experience a significant increase or decrease in population size without an incremental change in status, trend is itself a particularly useful measure to use in addition to conservation status and may be reported in future editions of this report if and when data on trends become available.

These data are not from a site-based monitoring program, but rather from a census approach that focuses on at-risk species; for more common species knowledge has been incorporated from a wide variety of observations and sources.

Data Access: Updated and more detailed information on all species is available at http://www.natureserve.org/explorer. For more customized data requests, contact jason_mcnees@natureserve.org.

The Data Gap

Data are not currently available on the status of most coastal and marine species. However, NatureServe will be incorporating status assessments for marine fishes into their databases. NatureServe already reports the status of nearly 150 coral species found off the coast of southern Florida. NatureServe expects to broaden its coverage of marine species to include many more invertebrates and, hopefully, Hawaiian fishes, which is a large challenge given that these are largely different varieties than those found in coastal regions of the mainland United States.

Data on the status of vascular plants exist (and are reported here), but for the most part, these plants have not been classified according to their habitat associations, in the manner that the animals reported here have been (i.e., animals that are dependent upon forests, or grasslands/shrublands, or fresh water). This is primarily a resource issue—there are far more vascular plants than vertebrate animals, and the size of the workload involved in categorizing them has prevented this work from taking place.

See the indicator for Status of Animal Species in Farmland Areas (page 103) for further discussion of the data gap with respect to species in agricultural landscapes.

See the indicator and technical note for urban/suburban Species Status (pp. 191 and 269) for discussion of the data gap with respect to species in urban/suburban areas.

References

Master, L.L. 1991. Assessing threats and setting priorities for conservation. Conservation Biology 5(4):559–563.

Stein, B.A. 2002. States of the Union: Ranking America's biodiversity. NatureServe, Arlington, VA. 25 p.

Condition of Plant and Animal Communities

The Indicator

The indicator would report separately on the biological community condition of (1) the combined area of land, lakes, and coastal waters and (2) linear features (streams and coastlines). Community condition would first be broken into two major categories: lands and waters under intensive human use and those that are left in semi-natural-to-natural condition. Intensive human use would be divided further into two categories: physically altered and highly managed. Semi-natural-to-natural lands and waters would be divided further into three categories: disturbed, less disturbed, and undisturbed.

Examples of system-specific components and indications of the possible data sources that might be used for reporting on each category follow.

Physically Altered Communities

- Areas covered by 30% or more constructed materials (e.g., asphalt, concrete, buildings), as measured from satellites. Data are available from the National Land Cover Dataset (NLCD; 30-m resolution. These data are derived from the Multi-Resolution Land Characterization Consortium, which is a partnership between the U.S. Geological Survey, the USDA Forest Service, the National Oceanographic and Atmospheric Administration (NOAA), and the Environmental Protection Agency (see the technical note for the national extent indicator, p. 207, for more details).
- Open mines, quarries, and gravel pits, measured from satellite. Data also available from the NLCD.
- Area of road surface (including unpaved roads). Data from the Federal Highway Administration, U.S. Department of Transportation, Highway Statistics 1999, http://www.fhwa.dot.gov/ohim/hs99/.
- Lined and culverted streams. Data not available.
- Hardened coastline miles. Partial data available from NOAA's Environmental Sensitivity Index atlases. (See Shoreline Types, p. 70.)

Highly Managed Communities

- Cropped land (not including interspersed natural area), as measured from satellites. Data from the NLCD.
- Forests planted with nursery stock. Data from Forest Service; see the forest management categories, page 119.
- Intensively grazed grassland. Data not available; also, no specific threshold has been identified that would be used to define which lands are intensively grazed.
- Stream miles impounded into lakes. Data not available.
- Drained or impounded wetlands (areas that remain wetlands but have been highly altered). Data not available.

Disturbed Semi-Natural Communities

- Forests heavily affected by invasive species. Data not available.
- Grasslands and shrublands heavily affected by invasives. Data not available.
- Coastal area heavily affected by invasive species. Data available only for selected estuaries (see p. 222).
- Freshwater rivers and streams with low IBI (Index of Biological Integrity, a species-based measure of disturbance). Data not available. See p. 253.
- Lands and waters with highly altered species mix, such as would be characteristic of altered fire or hydrologic regimes. Neither data nor methods are currently available.

Less Disturbed

- Semi-natural lands and waters that are neither clearly disturbed nor identified as undisturbed.

Undisturbed

- Biological communities with species mix essentially the same as would occur without man's influence. There is no generally accepted method to identify such lands and waters.

Plant Growth Index

The Indicator

The plant growth index is based on data collected by the Advanced Very High Radiation Radiometer (AVHRR) aboard the National Oceanographic and Atmospheric Administration's (NOAA) polar-orbiting satellites. Each 1.1 km^2 mapping area (pixel) has been measured twice a day. Daytime measurements in the visible wavelengths (0.58–0.68 m) and near-infrared wavelengths (0.725–1.1 m) are transformed into a Normalized Difference Vegetation Index (NDVI), which has a near-linear relationship to absorbed photosynthetically active radiation for a given land cover type. NDVI also correlates well with net uptake of carbon dioxide and plant biomass production.

For this indicator, NDVI is calculated at 2-week intervals and summed throughout the growing season; only values that exceed non-growing-season, background NDVI are included. Growing season start dates, end dates, and background NDVI were calculated for each land cover type and region. (For a detailed explanation of calculating growing-season accumulated NDVI, see Reed and Yang 1997).

Because the relationship between NDVI and absorbed photosynthetically active radiation varies by cover type, the growing-season accumulated NDVI was calculated separately for the forest, farmland, and grassland/shrubland areas in each county of the conterminous 48 states, for each year between 1989 and 2000 (except for 1994, when the satellite failed). The 11-year average growing-season accumulated NDVI was also calculated for each of the three land cover types in each county. The values in each county segment for each year were then normalized by using the corresponding 11-year average for that county segment to produce a plant growth index where a value of 1.0 equals the long-term average. Areas with plant growth indices greater than 1.0 have higher-than-average accumulated NDVI; within the same cover type and in an area as small as a county, this implies higher-than-average plant growth for that year. The regional and system-specific plant growth indices are the area-weighted averages of the segments contained within the region and system.

Land cover for each 1.1 km^2 pixel for the growing season was identified from the National Land Cover Dataset. These data are derived from the Multi-Resolution Land Characteristization Consortium, which is a partnership between the U.S. Geological Survey, the U.S. Forest Service, NOAA, and the Environmental Protection Agency. (See the national extent indicator's technical note on p. 207 for more details.)

The Data

Data Sources: Data on accumulated NDVI and analysis of those data are from the USGS's Earth Resources Observations Systems (EROS) Data Center, Sioux Falls, South Dakota (see http://edcwww.cr.usgs.gov/).

Data Quality/Caveats: In 2000, the NOAA-14 orbit drifted to a late afternoon overpass time. The effects of this on the plant growth index have yet to be fully understood. However, because the index is accumulated from the beginning of the growing season—a point that is identified each year from the inherent seasonal patterns in the NDVI data—scientists at the EROS Data Center believe the 2000 estimates are comparable to those of previous years.

The Data Gaps

Data for 1994 are unavailable because of satellite failure. Data are available only for the land area of the conterminous 48 states. The Coasts and Oceans chapter of this report includes an indicator of chlorophyll concentrations in coastal waters, a measure related to algal growth. That measure is based on maximum rather than accumulated concentrations, and thus is not directly comparable.

References

B.C. Reed and L. Yang. 1997. Seasonal vegetation characteristics of the United States. Geocarto International 12(2):65–71.)

Production of Food and Fiber and Use of Water

The Data

Forest Products: Data were obtained directly from the USDA Forest Service. The data used in the graph for the entire United States are the same as those used in the timber harvest indicator (see p. 130). Because no data were available for 1980, the data were divided by the interpolated value for 1980. The same data are not available on a regional basis, so "removals of growing stock" was used instead. This statistic is defined as the net volume of growing-stock trees removed from the inventory during a specified year by harvesting, cultural operations such as timber stand improvement, or land clearing. "Growing stock" is a classification of timber inventory that includes live trees of commercial species meeting specified standards of quality or vigor. Cull trees are excluded. When associated with volume, it includes only trees measuring 5.0 inches (12.7 cm) in diameter or greater at breast height. In general, the trends in "total timber harvest" and "removals of growing stock" are similar. Again, these data were divided by the interpolated value for 1980 in each of the six regions. Data are included from all 50 states.

Marine Fish Landings: Data were obtained directly from the National Oceanographic and Atmospheric Administration (NOAA) and are described in the fish and shellfish landings indicator (see p. 81). Data for Hawaiian waters were included in the Pacific Coast region and those for the Gulf of Mexico were reported in the Southeast region, even though a portion of these off the coast of Texas should be included in the Southwest region. Prior to 1976, much of what is now the Alaskan fishery—as well as portions of the other regional fisheries—was in international waters. These waters came under the control of the United States with the establishment of the U.S. Fishery Conservation Zone in 1976. Thus, the large rise in fish landings visible after 1976 (see the commercial fish and shellfish landings indicator, p. 81) resulted from the acquisition of new territory rather than a jump in the productivity of a given area of ocean. It was possible to include this situation in the indicator in the Coasts and Oceans chapter (p. 81); however, a similar approach was not possible for this indicator. For this reason, data prior to 1978 were not included. Landings were divided by the 1980 value, either nationally or regionally.

Freshwater Withdrawals: Data were derived from the U.S. Geological Survey Circular series "Estimated Use of Water in the United States," which has been published every 5 years since 1950 (note: consistency issues prevented the use of data prior to 1960). More recent compilations (1985–1995) are available electronically at http://water.usgs.gov/public/watuse/ (see the technical note for Water Withdrawals, p. 254). Withdrawals for any given year (and region) were divided by the 1980 value.

Agricultural Products: Data are available online from the U.S. Department of Agriculture's Economic Research Service (ERS). State-by-state data are from Table 7 of *U.S. Agriculture, 1960–96: A Multilateral Comparison of Total Factor Productivity* (Technical Bulletin 1895, available at http://www.ers.usda.gov/data/stateproductivity/). State data were summed for the entire U.S. graph. The data in Table 7 are normalized such that the output for Alabama in 1996 equals 1. These normalized data were summed, either for the nation as a whole or for each region, and then divided by the 1980 value to produce the index values for all other years.

Human Population: Data are available online from the U.S. Census Bureau via the "national" and "state" links at http://eire.census.gov/popest/estimates.php. U.S. or summed regional data were divided by the value for 1980 to produce the index values for all other years.

Outdoor Recreation

The Data

Data Source: Data come from a national survey conducted by phone (National Survey on Recreation and the Environment [NSRE] 1994–95; see http://www.fs.fed.us/research/rvur/recreation/publications/outdoor_recreation/title.htm), in which questions were asked about participation in 68 specific outdoor recreation activities. Similar surveys have been conducted since 1965; however, comparable data on recreation days are not available from them. NSRE 2000, whose data are of the same format as those shown here, is currently under way and will be released soon after this report is scheduled to go to press (see http://www.srs.fs.fed.us/trends/nsre.html).

Data from a total of 17,216 interviews were collected from January 1994 through May 1995. The NSRE survey was composed of two random-digit-dialing (RDD) telephone surveys. During the interviews, which averaged 20 minutes in length, Americans above the age of 15 were asked, among other questions, about participation in activities and the number of days and trips spent in recreation activities.

The NSRE data were grouped into several major activity groups, and the only manipulation of the data here was to re-bundle these groups slightly, as can be seen by comparing Table 4.2 in the NSRE report to the listing below. Note that these categories are compatible with those used in the NSRE 2000.

- *Walking and Biking:* biking, bike touring, walking
- *Viewing Activities:* bird-watching, wildlife viewing, fish viewing, sightseeing, studying nature near water
- *Picnics, Family Activities:* picnicking, family gathering
- *Motor Sports:* off-road driving, snowmobiling
- *Snow Skiing:* downhill skiing, cross-country skiing
- *Hiking, Climbing, Etc.:* hiking, orienteering, backpacking, mountain climbing, rock climbing, caving, horseback riding
- *Camping:* developed area, primitive area
- *Hunting:* big game, small game, migratory bird

- *Fishing:* freshwater, saltwater
- *Swimming and Beachgoing:* surfing, swimming/non-pool, snorkeling/scuba, visiting a beach or waterside
- *Motor Boating and Water Skiing:* motor boating, water skiing, jet skiing
- *Sailing, Floating, Rowing, Etc.:* sailing, canoeing, kayaking, rowing, floating, rafting, sailboarding/windsurfing

Data Limitations/Caveats: The RDD survey approach reaches a random sample of telephone numbers rather than of people. A substantial portion of non-representativeness of some groups is attributable to inability to reach selected households and absence of some households from telephone listings. Affluent families are virtually certain to have telephone numbers and many have more than one. However, many low-income households may not have a telephone. As a result, affluent people may have been over-represented somewhat in the survey sample. On the basis of Census data, differences in age, race, and gender were adjusted for over- or under-representation during data analysis.

Data Availability: Data for the 1994–95 NSRE are freely available on the Internet (see http://www.fs.fed.us/research/rvur/recreation/publications/outdoor_recreation/title.htm) and data for NSRE 2000 will eventually be available at no cost.

The Data Gap

As mentioned in the text, the list of activities for which recreation days are available is not exhaustive, and further distinctions for some activities (e.g., swimming, hunting, viewing) on whether they were conducted in a saltwater or freshwater setting are desirable.

Natural Ecosystem Services

There is no technical note for this indicator.

Coasts and Oceans

Coastal Living Habitats—Coral Reefs, Wetlands, Seagrasses, and Shellfish Beds

The Data: Coastal Wetlands

Data Source: The coastal wetland data came from the U.S. Fish and Wildlife Service's (FWS) National Wetlands Inventory (NWI; Dahl et al. 2000, p. 44). The data presented here include estuarine vegetated wetlands, which are approximately 87% of the total coastal wetlands included in the FWS report. Excluded types include estuarine non-vegetated and marine intertidal, neither of which fall into the category of "biologically structured habitat"—that is, they are not characterized by significant vegetation that is habitat for various animals and plants. Note that no estimate of "presettlement" coastal wetlands is included in this indicator. There are estimates of coastal wetlands prior to major development along the coastline that affected many wetlands. Gosselink and Baumann (1980) estimate that 10 million acres of coastal wetlands existed in 1923, which was prior to most of the coastal development.

Data Collection Methodology: The NWI produces periodic reports of changes in wetland area. The data, summarized in the aforementioned report, are derived from three separate analyses; one covering the 1950s to the 1970s; one covering the 1970s to 1980s, and one covering the 1980s to the 1990s. For this report, decadal estimates are presented as the midpoint of the decade. For example, "1980s" data are presented as "1985." Note that more detailed data are available from the NWI maps and accompanying digital data but that acreage summaries are not compiled for national or regional reporting.

NWI counts all wetlands, regardless of land ownership, but recognizes only wetlands that are at least 3 acres. To ensure adequate coverage of coastal wetlands, supplemental sampling along the Atlantic and Gulf coastal fringes was added.

The patchy distribution of Pacific coast estuarine wetlands precluded gathering statistically valid data on this wetland type. Therefore, consistent with past studies, NWI did not sample Pacific coast estuarine wetlands such as those in San Francisco Bay, California; Coos Bay, Oregon; or Puget Sound, Washington.

A permanent study design is used, based initially on stratification of the 48 conterminous states by state boundaries and 35 physiographic subdivisions. Within these subdivisions are located 4375 randomly selected, 4-square-mile (2560 acres) sample plots. These plots were examined with the use of aerial imagery, ranging in scale and type; most were 1:40,000 scale, color infrared, from the National Aerial Photography Program.

Data Quality/Caveats: Field verifications addressing questions on image interpretation, land use coding, attribution of wetland gains or losses, and plot delineations were made. For example, for the analyses in the 1980s to 1990s, 21% of the sample plots were verified.

Data Access: *The Status and Trend of Wetlands in the Conterminous United States 1986-1997* is available on the Web at http://wetlands.fws.gov/bha/SandT/SandTReport.html.

The Data Gap: Coral Reefs

According to the federal interagency Coral Reef Task Force (CRTF; http://coralreef.gov), accurate geo-referenced information on the exact location of specific natural resources and habitat types is essential for effective management of coral reefs. Comprehensive maps and habitat assessments form the foundation for a variety of reef conservation measures, including creating accurate baselines for long-term monitoring. However, according to the CRTF, most coral reefs in U.S. waters, and particularly those in the Pacific Ocean, have not been accurately mapped with modern techniques and at a scale relevant to emerging conservation issues.

In March 2000, the CRTF released a plan of action (http://coralreef.gov/CRTFAxnPlan9.PDF) committing the agencies to produce comprehensive digital maps of all coral reefs in the United States and trust territories within 5 to 7 years. During this period, the ongoing mapping of the Caribbean region will be completed, and new efforts will begin in the Pacific where critical data gaps presently exist. This interagency effort will produce maps at both low and high resolutions that address locally identified conservation and management needs.

The Data Gap: Shellfish Beds

The National Shellfish Register of Classified Growing Waters has been produced every 5 years since 1966; the most recent is the

1995 Register, released in 1997 (http://spo.nos.noaa.gov/projects/95register/). The Register is a cooperative effort among the nation's shellfish-producing states, federal agencies such as the U.S. Food and Drug Administration and the National Oceanographic and Atmospheric Administration, and the Interstate Shellfish Sanitation Conference (ISSC).

The program's focus is on the number and area of shellfish beds that are classified according to sanitary guidelines administered by the ISSC. States have been encouraged to monitor as broad a range of shellfish beds within their waters as possible, in order to protect public health. However, the resulting increase in area monitored confounds reporting on trends in overall shellfish bed area, making it inappropriate simply to report the acreage trends contained within the 1995 Register.

Note that shellfish beds that are no longer living (i.e., relict beds) continue to provide valuable habitat to marine organisms; over time, these beds will slowly disappear as the shells are dissolved.

The Data Gap: Submerged Aquatic Vegetation

While many programs monitor the extent of submerged aquatic vegetation (seagrasses, kelp, and other similar underwater plants), we are aware of no effort to compile and assess national trends. However, the United Nation's Environment Program is actively gathering all available information on the worldwide distribution of seagrasses.

References

Dahl, T.E., et al. 2000. Status and trends of wetlands in the conterminous United States 1986-1997. Washington, DC: U.S. Department of the Interior, Fish and Wildlife Service.

Gosselink, J.G., and R.H. Baumann. 1980. Wetland inventories: Wetland loss along the United States coast. Z. Geomorph. N.F. Suppl. Bd. 34:173–187.

Shoreline Types

The Indicator

Total miles of shoreline are shown for the three regions grouped together, and the breakdown of different shoreline types is shown as a percentage of each region's total shoreline miles. As discussed below, there is some double counting of shoreline types (e.g., a stretch of shoreline may be both mud flat and wetland); this occurs for only about 10% of the shoreline miles. This double counting has two minor implications for the figures. First, a sum of the miles of different types in the left-hand graph would slightly exceed the actual miles along the shore for these three regions. Second, the values in the right-hand graph represent the percentage of total shoreline length for all the types in a given region, which is somewhat more than the actual miles along the shore because of the double counting.

The Data

Data Source: These data were provided by the National Oceanographic and Atmospheric Administration (NOAA), National Ocean Service, Office of Response and Restoration, Hazardous Materials Response Division. Data on Florida's shoreline were collected by the Florida Marine Institute and processed and interpreted by NOAA.

Data were extracted from Environmental Sensitivity Index (ESI) atlases, a product of NOAA's Office of Response and Restoration. The ESI method provides a standardized mapping approach for coastal geomorphology as well as biological and human use elements. More information is available at http://response.restoration.noaa.gov/esi/esiintro.html. Data from multiple atlases (1984–2001) were aggregated into the regions used in this report. For most of the regions, digital data were unavailable for parts of the coastline. The currency and the completeness of the coverage affect the quality of the shoreline summary. With regard to these issues, information specific to the three regions for which data are available is presented in the "Data Quality" section below. Complete metadata for each atlas, including collection methods and source information, can be viewed at http://response.restoration.noaa.gov/esi/metadata.html.

Data Collection Methodology: ESI shoreline types were classified using a combination of overflight information, aerial photography, local habitat maps, National Wetlands Inventory data, and ground truthing. For more detailed information specific to each atlas, see the individual atlas metadata, specifically sections 5.1—Detailed Description: ESI; and 2.5.1—Source Information: ESI.

Data Manipulation: This indicator presents a simplified summary of shoreline types, by region. It is a summary of the total length of land/water interface for the region, as well as the total length of each of the five shoreline types described below.

The first step in defining the indicator was to consolidate the shorelines from the various atlases for each region and reconcile older terminology with current ESI shoreline type classifications. The next step was to combine the ESI shoreline type classes into the five more general categories, based on substrate and slope, that are used in this report. The five categories and the ESI types that make them up, are as follows:

- *Steep sand, rock, or clay:* ESI categories 1 (unvegetated steep banks, cliffs, and seawalls), 1A (exposed rocky shores), 2A (exposed wave-cut platforms in bedrock, mud, or clay), 2B (exposed scarps and steep slopes in clay), 3B (scarps and steep slopes in sand), and 8A (sheltered rocky shores and sheltered scarps in bedrock, mud, or clay)
- *Mud or sand flats:* ESI categories 7 (exposed tidal flats) and 9A (sheltered tidal flats)
- *Beaches (sand or gravel):* ESI categories 3AF (fine- to medium-grained sand beaches), 4 (coarse-grained sand beaches), 5 (mixed sand and gravel beaches), and 6A (gravel beaches)
- *Wetlands, mangroves, etc.:* Includes grasslands, scrublands/shrublands, and marshes. ESI categories 8D (vegetated, steeply sloping bluffs), 9B (vegetated low banks), 10A (salt- and brackish-water marshes, 10B (freshwater marshes), 10C (swamps), and 10D (scrub–shrub wetlands)
- *Armored:* ESI categories 1B (exposed, solid manmade structures, 6B (riprap), 8B (sheltered, solid manmade structures), and 8C (sheltered riprap)

After the regional shoreline was characterized, tables were generated detailing the length of each shoreline segment and its associated ESI type. These data were input into a series of computer programs to sum the total shoreline length and that of each of the shoreline types. ESI shoreline data are quite complex, however, in that a single shoreline segment may contain up to three ESI classifications. For example, a segment may have a tidal flat on the water side backed by a sand beach, then a marsh on the landward side. If a segment has different shoreline types, that segment is counted multiple times. However, when the total length of land/water interface is calculated, each shoreline segment is counted only once, regardless of the number of shoreline types it contains. Thus, the sum of the lengths of all the shoreline types will be greater than the total shoreline length. This double counting occurs for about 10% of the coastlines characterized.

Data Quality/Caveats: ESI shoreline coverage of the three regions shown was complete; the Pacific Northwest region also included considerable area along the shore of the Columbia River. Some of the older atlases used for this region were compiled more than 15 years ago. Though the West Coast is generally not considered an overly dynamic coastline, some changes may have altered coastline shape or type for a small percentage of this region.

Data Access: The data reported here are the result of an analysis undertaken specifically for this project; however, some data are available at http://response.restoration.noaa.gov/esi/esiintro.html.

The Data Gap

Data are not currently available for the majority of coastal regions; however, the necessary analyses are under way at NOAA.

Areas with Depleted Oxygen

The Indicator

The percentage of brackish water exposed to a range of oxygen concentrations for at least 1 month will be reported as anoxic (no oxygen), hypoxic (>0 and <2 parts per million [ppm]), low (2–4 ppm), or sufficient (>4 ppm). Low oxygen levels for a brief period may do little to disrupt the marine ecosystem; however, when those levels persist, significant effects on the local ecosystem can be expected. The percentage of brackish waters that are hypoxic for at least 1 month will be reported by region. Note that bottom waters are the first to become hypoxic or anoxic because less light is available for the oxygen-producing algae to grow, excess organic matter generally sinks and is decomposed in bottom waters, and little exchange with the atmosphere occurs—a process that can introduce oxygen to surface waters.

Dissolved oxygen is an important habitat parameter for both aerobic and anaerobic organisms. In addition to directly affecting the distribution, abundance, and diversity of aerobic organisms (including commercial and sport fish), oxygen depletion (the development of hypoxia and anoxia) in bottom waters alters biogeochemical reactions involving biologically important elements such as carbon, nitrogen, phosphorus, sulfur, and iron.

The distribution of dissolved oxygen is an integrative measure of the dynamic balance between processes that produce, consume, transport, and exchange dissolved oxygen. Plants (dominated by microscopic algae called phytoplankton) generally produce more oxygen than they consume. The amount of dissolved oxygen in the water at any given time and place reflects the balance between this production and several ways that oxygen is lost from a given location: consumption by respiration, loss from surface waters to the atmosphere because of equilibrium processes (e.g., cold water can hold more oxygen than warmer water), and movement of water masses between regions of the world's ocean.

Anthropogenic nutrient loading is considered by many to be the primary cause of increasing trends in the duration and extent of hypoxia and anoxia. Influxes of nutrients stimulate phytoplankton production which can lead to more bacterial decomposition and rapid increases in biological and chemical oxygen demand when this production is not consumed by zooplankton, fish, or shellfish. Consequently, the occurrence of hypoxic and anoxic bottom waters is an important indicator of coastal eutrophication and of the effects of human activities on biogeochemical cycles. Long-term observations of the extent of oxygen depletion in brackish waters will reflect changes in land cover and land-use patterns in coastal watersheds, including the efficacy of efforts to control nutrient loading to coastal ecosystems (see the National Oceanographic and Atmospheric Administration study, Hypoxia in the Gulf of Mexico, at www.nos.noaa.gov/products/pubs_hypox.html).

The Data Gap

Dissolved oxygen should be measured with a precision of ±0.5 ppm. Measurements should be frequent enough (several times a month) to capture seasonal variability on the spatial scales appropriate to estuaries and coastal areas. In addition, the required data cannot be easily accessed because they reside in a variety of databases that are maintained on an ecosystem-by-ecosystem basis by county, state, and federal agencies and institutions.

Observations with sufficient resolution in time and space to calculate the extent of bottom water hypoxia with known certainty exist for some regions (e.g., the northern Gulf of Mexico, Gulf of Maine, Middle Atlantic Bight, South Atlantic Bight) and for many estuaries and bays (e.g., Long Island Sound, the lower Hudson River estuary, Chesapeake Bay, San Francisco Bay, Puget Sound). Although some of these data are available from the National Oceanographic Data Center (www.nodc.noaa.gov/), the National Estuary Program (http://www.epa.gov/owow/estuaries/), and the National Estuarine Research Reserve System (www.ocrm.nos.noaa.gov/nerr), much of the required data resides with state agencies and other federal programs; and sufficient data on both dissolved oxygen and salinity distributions are not available to calculate this index on regional to national scales.

Contamination in Bottom Sediments

The Indicator

This indicator provides information on the concentration, in coastal bottom sediments, of four major classes of contaminants that can harm fish and other aquatic organisms and can adversely affect human health if ingested while consuming fish or shellfish. Sediment concentration levels will be reported separately for estuaries and the coastal ocean out to 25 miles; currently only data for estuaries are available.

The sediment quality guidelines used in this indicator were developed by the National Oceanographic and Atmospheric Administration, through its National Status and Trends Program (see http://response.restoration.noaa.gov/cpr/sediment/SPQ.pdf). Before these guidelines, there were no national criteria or other widely applicable numerical guidelines for sediment quality. These quality guidelines were developed as informal, interpretive tools to estimate the possible toxicological significance of chemical concentrations in sediments. The guidelines have not been promulgated as regulatory criteria or standards, cleanup or remediation targets, discharge attainment targets, pass–fail criteria for dredged material disposal decisions, or for any other regulatory purpose. See http://response.restoration.noaa.gov/cpr/sediment/SQGs.html.

These guidelines were derived from examination of a large number of individual contamination studies, all in salt water. Data from each study were arranged in order of ascending concentrations. Study endpoints in which adverse effects were reported were identified. From the ascending data tables, the 10th percentile and the 50th percentile (median) of the effects database were identified for each substance. The 10th-percentile values were named the "Effects Range—Low" (ERL), indicative of concentrations below which adverse effects rarely occur. The 50th percentiles were designated the "Effects Range—Median" (ERM) values, representative of concentrations above which effects frequently occur. In this report, ERL is referred to as the "possible effects" guideline and ERM as the "probable effects" guideline.

The Data

For this indicator, The Enironmental Monitoring and Assessment Program for estuaries (EMAP-E) provided information assessing the contaminant levels in estuarine sediments and the condition of benthic organisms in those sediments. These information were collected from over 2000 sites from Cape Cod, Massachusetts, to Brownsville, Texas, and represent over 70% of the total estuarine acreage of the United States (excluding Alaska). These data and EMAP-E are described in more detail in the technical note for the national contaminants indicator, p. 210)

Coastal Erosion

The Indicator

The condition of the U.S. coastline—whether it is managed or natural, and whether it is eroding, accreting, or stable—has become a matter of great concern. Not only can a wide sandy beach or broad expanse of coastal marshland be aesthetically pleasing, but it can also protect coastal homes from hazards such as storms and high tides. An eroding shoreline can translate into hundreds of million of dollars in damage to coastal property and loss of tourism revenues.

Management responses to erosion are also problematic. Replacing sand ("beach nourishment") is costly and may have environmental impacts such as disturbance of fish and wildlife habitat and damage to dunes from heavy equipment. Construction of bulkheads and other structures ("armoring") is generally a longer-term approach, but has very significant effects on fish and wildlife that use the shoreline or beach. Neither nourishment nor armoring necessarily stops erosion; however, armoring typically lasts longer than nourishment.

An accurate assessment of how much of the U.S. shoreline is eroding and how much is accreting or stable is necessary to determine how many coastal homes may be in jeopardy (see The Heinz Center 2000). It will also allow planners and officials to take action to protect existing homes from damage and help them manage future development of the coastal zone. Such assessments are complicated by the fact that erosion is a natural and naturally varying phenomenon. Erosion changes on a seasonal and multi-year basis; there will always be some areas that are eroding and some that are accreting, and these areas will shift over time.

The Data Gap

As discussed in the indicator text, guidelines will be necessary for classifying stretches of coastline as "accreting" or "eroding." It is thought that the associated change in the horizontal movement of the shoreline will be in the range of one-half to several feet per year. In addition, the coastal management community will need to agree on impact to the shoreline of groins, which are erosion control structures typically built perpendicular to the shoreline. It is not a simple matter, as it would be for bulkheads, to assign a length of shoreline affected by a groin.

Most shoreline erosion and beach nourishment data are developed on a short-term, project-specific basis. Few long-term or regional studies have been carried out, and differences in data collection and analysis protocols make it difficult to compare site-specific reports and compile the data for either regional or national reporting. A study by Dolan et al. (1985) contains a compilation of erosion data from the 1960s, 1970s, and 1980s. This one-time study may be a model for future analyses.

Various methods have been used to determine whether shoreline locations are eroding, accreting, or stable. These include shoreline profiles, National Oceanographic and Atmospheric Administration's (NOAA) National Ocean Service Topographic Survey Sheets, and aerial photographs, which can be geo-referenced or orthorectified. In addition, Light Detection and Ranging (LIDAR) has been an effective tool for measuring erosion and has been used in at least two different programs. The Airborne LIDAR Assessment of Coastal Erosion (ALACE) project was a partnership between NOAA, the National Aeronautical and Space Administration (NASA), and the U.S. Geological Survey that utilized LIDAR collected via aircraft to map a good portion of the sandy beaches of the lower 48 states; NOAA continues to utilize LIDAR for site-specific analyses of shorelines rather than broad surveys of the U.S. coastline in its Topographic Change Mapping program. (For further information on ALACE or the Topographic Change Mapping program, see http://www.csc.noaa.gov/crs/tcm/.)

Depending on what methods are used, comparison of site-specific reports may or may not provide an accurate regional or national assessment. For example, shoreline profiles may not be dense enough to provide results that can be compared with those of aerial photography. Another consideration in comparing site-specific erosion studies is the time period over which the change in shoreline condition is measured. Reporting on the extent of erosion nationally will require establishing parameters for comparison between various datasets.

Beach nourishment may be undertaken to control erosion, or it may be the by-product of harbor or inlet construction or maintenance, when the excavated material is placed on an adjacent beach. Nourishment that occurs as a by-product is typically not well documented.

References

Dolan, R., F. Anders, and S. Kimball. 1985. Coastal erosion and accretion. National atlas of the United States of America, Department of Interior, U.S. Geological Survey.

The H. John Heinz III Center for Science, Economics and the Environment. 2000. Evaluation of erosion hazards. Washington, DC: The Heinz Center. (available at http://www.heinzctr.org/publications.htm)

Sea Surface Temperature (SST)

The Indicator

This indicator was calculated as follows: (1) the seasonal average sea surface temperature (SST) of near-shore water (shoreline out to 25 miles) was calculated for the warmest season in each region (termed the "seasonal mean maximum"), which typically occurred during summer or fall; (2) the long-term mean (during the warmest seasons) for the period of observation (1985–1998) was calculated; and (3) the long-term mean was then subtracted from the seasonal mean maxima. Thus, values greater than zero are positive "anomalies" (i.e., deviations from the long-term average), and those less than zero are negative anomalies.

Because of the large heat capacity of the ocean, changes in water temperature on the scales reported here are likely to reveal universal changes, such as those caused by global warming, sooner than will be apparent in air temperature (i.e., changes in water temperature are less susceptible to daily and seasonal variability). Changes in annual cycles of water temperature and the occurrence of interannual to decadal trends not only will affect the kinds of organisms that will thrive in a region, but are thought to be associated with the degradation of coral reefs (bleaching) and may be related to the development of harmful algal blooms and the growth of invasive species. On longer time scales (decades to centuries), such changes may be related to decreases in the supply of nutrients to surface waters from the deep sea and a cascade of effects from decreases in primary production to declines in fish production.

As discussed in the text, there is considerable evidence building that the surface waters of the oceans are warming gradually (e.g., Barry et al. 1995 and Levitus et al. 2000).

The Data

Data Source: Data from 1985 through 1998 were analyzed for The Heinz Center by the National Ocean Service of the National Oceanographic and Atmospheric Administration (NOAA). The NOAA/National Aeronautics and Space Administration (NASA) Oceans Pathfinder SST data were obtained using Advanced Very High Resolution Radiometers onboard several NOAA Polar Orbiting Environmental Satellites. Complete data are not available for 1996 and 1997, but are available for 1998. It is important to note that SST data are available back to 1979; however, these data are not yet comparable to the series beginning in 1985.

Data Manipulation: Data were acquired on a grid of square pixels nominally 10 km (about 6 miles) on a side. Both the day and nighttime data were processed to remove clouds (using an "erosion filter") and then averaged to produce monthly means, which were then averaged to produce seasonal means. See "The Indicator" section above for a description of the calculations necessary to generate the SST anomalies.

Data Quality: Systematic errors are rare in such an analysis, and the data are expected to be within 2°F of actual temperatures measured 3.3 ft below the surface.

Data Availability: Data are available for free on the Web from NASA at http://podaac.jpl.nasa.gov/sst/.

References

Barry, J.P., C.H. Baxter, R.D. Sagarin, and S.E. Gilman. 1995. Climate-related, long-term faunal changes in a California rocky intertidal community. Science 267:672–675.

Levitus, S., J.I. Antonov, T.P. Boyer, and C. Stephens. 2000. Warming of the world ocean. Science 287:2225–2229.

At-Risk Native Marine Species

There is no technical note for this indicator.

Non-native Species

Further refinement is required to produce an indicator that combines both number of species and the area they inhabit. However, even with such an indicator, assessment of the national situation would be impossible without a program of data collection and assessment. Monitoring activities will need to be conducted regularly in estuaries around the country. These activities must be supported by historical research to determine which species are native and which have been introduced.

Two definitions are appropriate for this indicator. "Invasive species" are defined in Executive Order 13112, "Invasive Species" (Feb. 3, 1999), as "alien species whose introduction does or is likely to cause economic or environmental harm or harm to human health" where "alien species" are, "with respect to a particular ecosystem, any species, including its seeds, eggs, spores or other biological material capable of propagating that species, that is not native to that ecosystem."

According to an Office of Technology Assessment study (U.S. Congress 1993), just 79 non-native species had cost the American public some $97 billion in damages to natural resources and lost industrial productivity during the 20th century. Pimentel et al. (2000) recently published a more comprehensive estimate; they found that more than $120 billion is spent every year in the United States to deal with the effects of some 50,000 non-native species. Although these figures are not limited to coastal non-natives, they illustrate the scope of the issue.

The discharge of ballast water by oceangoing vessels is a major source of such introductions in coastal waters. With the high speed of modern vessels, organisms taken in with ballast water at one location have little difficulty surviving the trip to a distant destination, where the ballast water and its associated organisms are discharged. Other mechanisms, such as the escape of fish from aquaculture facilities or the intentional introduction of non-native species of shellfish to supplement dwindling native populations, can also contribute to the introduction and spread of non-native species.

The number of successful new invasions appears to have increased dramatically during the 1970s and 1980s, perhaps as a consequence of nutrient enrichment and over-fishing in coastal ecosystems. The list of recent invaders includes several species of benthic algae, submerged aquatic vegetation, toxic dinoflagellates (e.g., *Alexandrium catenella* in Australia) , bivalves (e.g., the zebra mussel in the Great Lakes and the Chinese clam in San Francisco Bay), polychaetes, ctenophores, copepods, crabs, zooplankton, and fish. Such invasions can profoundly alter the population and trophic dynamics of coastal ecosystems. For example, the introduction of the ctenophore *Mnemiopsis leidyi* caused the collapse of the anchovy fishery in the Black Sea by preying on the

Table 1. Proposed Framework for Non-native Species Indicator

	% of Non-native Species		
% of area inhabited or % of total biomass	<25	25–75	75–100
<25	1 (low)	2 (low)	3 (medium)
25–75	2 (low)	3 (medium)	4 (high)
75–100	3 (medium)	4 (high)	5 (high)

Possible ranking system: 1, 2 = low; 3 = medium; 4, 5 = high

anchovy's preferred food, copepods; the introduction of the macrobenthic green algae, *Caulerpa taxifolia*, displaced a diverse community of sponges, gorgonians, and other seaweeds on more than 10,000 acres of sea floor in the northern Mediterranean.

The Indicator

The indicator will report the degree of influence (low to high) of non-native species in major U.S. estuaries. The proposed approach does not evaluate the significance of non-native species (also called non-indigenous, exotic, introduced, or invasive species) to economic or ecological condition, but rather focuses on the degree to which non-natives occupy the system. It would report an index created by combining the percentage of all species in a region that are non-native with the percentage of habitat they occupy (see Table 1). For purposes of this indicator, non-native species fall into at least five categories: fishes, mollusks, crustaceans, higher aquatic plants, and macroalgae. Non-estuarine areas can also have a significant non-native presence; however, the assumption of this indicator is that estuaries are generally more negatively affected.

Table 1 is a proposed framework for establishing the degree of significance of non-native species in a region. Both the number of species and the area they inhabit (or their biomass) are factors, so this measure proposes a combined ranking approach, in which both factors contribute to an overall score. The values presented in the table are arbitrary and are intended only to illustrate the utility of such a ranking system.

The basis for judging significance will vary somewhat among the different categories of organism—for higher aquatic plants, mollusks, and macroalgae, the percentage of the total potential area inhabited would be measured, while for fish and crabs, the percentage of biomass accounted for by non-native species would be measured. The non-native rankings for selected major estuaries should be calculated periodically.

The Data Gap

The number and distribution of native species—not to mention non-natives—are not well documented in most coastal ecosystems, especially when considering species of bacteria, microalgae, and protozoa.

Species lists for the five categories of organism will need to be developed and maintained for each major U.S. estuary. The lists will be based on existing knowledge of the species in the target groupings in each estuary and on ongoing surveys of biological resources that are conducted in these regions for a variety of purposes. These surveys will also provide the data required to assess the ecological significance of the non-native species found in an estuary.

Species lists, data on ecological significance, and some monitoring data on various species are available from surveys and compilations produced by a variety of sources, including state living resource and environmental protection agencies, environmental impact statements, and academic research projects. However, such data are not available in any consistent fashion for many of the major estuaries, and there is no nationwide compilation of data.

The National Aquatic Nuisance Species Task Force (http://www.anstaskforce.gov/), an interagency group, is working to bring together information and data that will undoubtedly be useful for this indicator in the future. Also, the Smithsonian's Environmental Research Center is actively creating databases on invasive species (see http://invasions.si.edu/).

References

Pimentel, D., L. Lach, R. Zuniga, and D. Morrison. 2000. Environmental and economic costs of nonindigenous species in the United States. BioScience 50:53–65.

U.S. Congress, Office of Technology Assessment. September 1993. Harmful non-indigenous species in the United States, OTA-F-565. Washington, DC: U.S. Government Printing Office.

Unusual Marine Mortalities

The Indicator

A deliberate choice was made to focus this indicator on unusual mortalities rather than all marine mortalities. The latter includes death from old age, predation, and human-related causes such as entanglement with fishing gear. Unusual mortalities were selected so that only extraordinary instances of animal deaths would be included; a death from old age is within normal functioning of an ecosystem and should not be treated as a signal of changing ecosystem status.

In response to a major dolphin die-off during 1987–88, the National Marine Fisheries Service (NMFS) established a Working Group on Unusual Marine Mammal Mortality Events (WGUMMME) to create criteria for determining when an unusual mortality event is occurring and then to direct responses to such events (see http://www.nmfs.noaa.gov/prot_res/PR2/Health_and_Stranding_Response_Program/WGUMMME.html). The Working Group consists of a multidisciplinary team that makes judgments based on the following: (1) there is an increased number of mortalities when compared with historical data; (2) animals are stranding at an unusual time of year; (3) strandings occur in a localized area, throughout the geographical range, or they spread geographically with time; (4) the species, age, or sex composition of the stranded animals is different from what is normally seen; (5) stranded animals exhibit similar or unusual pathological findings or general physical condition; (6) mortality is accompanied by abnormal behavior among living individuals in the wild; and (7) critically endangered species are dying.

The Data

Data Source: U.S. Department of Commerce, National Oceanographic and Atmospheric Administration (NOAA), NMFS, Office of Protected Resources, Marine Mammal Health and Stranding Response Program; and Dierauf and Gulland (2001).

Data Quality/Caveats: The data for 2001 are preliminary as these unusual mortality events (UMEs) have not been officially closed and the total number of affected animals tallied. Also, there was a single UME for gray whales that spanned three years (1999–2001). A total of 678 animals were lost to a UME, with 273 in 1999. NOAA has not yet finalized the accounting for this event, so the remainder was simply split between 2000 and 2001. Given that there is no apparent trend to the data, this gross simplification should not distort the interpretation of this indicator.

Data Access: The 2001 data for two UMEs and the total number of gray whales lost in the 1999–2001 UME were obtained directly from NMFS. All other UME data were obtained from Dierauf and Gulland (2001).

The Data Gap

There is no program for sea birds, fish, and shellfish similar to that in place for marine mammals run under the auspices of the WGUMMME. It will be necessary to establish guidelines for what constitutes a UME for these animals, which typically perish in much larger numbers than mammals. For example, a guideline for fish may be that 1,000 deaths of members of two or three species would be necessary to qualify for a UME. This guideline may vary by species and by location and will certainly be different from the guidelines appropriate for shellfish and sea birds. Because these UMEs would all involve large numbers of animals, only the number of events will be reported. In addition, it is unclear if data exist on UMEs for sea turtles; however, these mortalities would be reported by the number of animals lost as is done for mammals.

References

Dierauf, L.A., and F.M.D. Gulland, eds. 2001. CRC handbook of marine mammal medicine: Health, disease, and rehabilitation. 2nd ed. Boca Raton, FL: CRC Press, Inc.

Harmful Algal Blooms

The Indicator

For the purposes of this analysis, harmful algal blooms (HABs) are defined as (1) an increase in the abundance of species that are known to produce toxins harmful to marine animals or humans (see Table 2); (2) the occurrence of lesions or mass mortalities of marine animals caused by HAB species; and (3) the occurrence of human pathologies caused by HAB species. A single event counts only once toward the relative intensity scale, even if it produces multiple impacts (e.g., an increase in the abundance of a HAB species that causes mass mortalities and an increased human health risk will be counted as a single event).

There are approximately 5000 species of microalgae in the world. Of these, about 100 are toxic. The scientific community refers to the phenomenon that cause these events as HABs, recognizing that HAB species represent a broad spectrum of taxa (e.g., dinoflagellates, diatoms, cyanobacteria) and trophic levels (e.g., autotrophic, heterotrophic, mixotrophic) and that many HAB species cause problems at low cell densities (i.e., a visible bloom is not necessarily required for a HAB event to occur). A second group of problematic algal blooms is recognized: those that cause problems such as oxygen depletion, habitat loss, starvation, or respiratory or reproductive failure in marine animals by virtue of their high abundance or biomass. These issues are addressed to some extent in other indicators.

Although definitive scientific evidence is lacking, HAB events appear to be increasing in number, extent, and severity (see *National Assessment of Harmful Algal Blooms in US Waters*, http://www.habhrca.noaa.gov/FinalHABreport.pdf). Rapid increases in the number of people living, working, and playing in the coastal zone have increased the input of nutrients to coastal waters, and HAB events may be occurring more frequently as a consequence. In addition, increases in shipping (and the inadvertent transport of non-native species of algae in ballast water) and the transport of shellfish between regions and continents may be increasing the frequency of HAB events by introducing new HAB species to U.S. coastal waters, or moving them to new locations within the United States. A growing human population also increases the demand for food from coastal waters in the form of wild and cultured fish and shellfish. The aquaculture industry is threatened by HAB events and may contribute to their increase.

The Data Gap

Most harmful algal events, such as fish kills, are typically identified after the event occurred or when it is well under way. Systematic monitoring programs that (1) quantify the abundance of harmful algal species, (2) quantify the concentrations of biotoxins or establish unequivocal causal relations between HAB

Table 2. Harmful Algal Species: By Region and Toxic Effect

Region	Effect	Species
Northeast	Paralytic Shellfish Poisoning (PSP)	*Alexandrium tamarense*
	Fish kills	*Gymnodinium mikimotoi*
	Shellfish mortality	*Aurococcus anophagefferens*
Southeast	Neurotoxic Shellfish Poisoning (NSP)	*Gymnodinium breve*
	Fish kills, Human health	*Pfiesteria piscicida*
Gulf of Mexico	Neurotoxic Shellfish Poisoning (NSP)	*Gymnodinium breve*
	Ciguatera Fish Poisoning (CFP)	*Gambierdiscus toxicus*
	Amnesic Shellfish Poisoning (ASP)	*Pseudo-nitzschia spp.*
West Coast	Paralytic Shellfish Poisoning (PSP)	*Alexandrium catenella*
	Amnesic Shellfish Poisoning (ASP)	*Pseudo-nitzschia spp.*
	Diarrhetic Shellfish Poisoning (DSP)	*Dinophysis spp.*

Note: PSP, NSP, CFP, ASP, and DSP all cause human health problems.

species and mortality events, or (3) quantify increases in human health risks are rare. Consequently, the data required to calculate this index on regional or national scales do not exist.

Although efforts to monitor and report these events are increasing nationwide, there is no standard measure of HABs. This is largely because of the heterogeneous nature of HABs (in terms of taxonomy, nutrition, the conditions under which they become toxic, the kinds of toxins produced, and their effects) and the fact that some species cause problems when they bloom while others cause problems at low cell densities.

State, federal, and academic programs collect most existing data, such as those in the database maintained by the Florida Marine Research Institute, for specific purposes (e.g., research or mitigation) or for specific locations (e.g., the west coast of Florida) where HABs have caused problems in the past. Thus, there is little consistency among programs, and there is no mechanism in place to establish regional or national databases. Both the National Oceanographic and Atmospheric Administration (NOAA; http://www.nodc.noaa.gov/col/projects/habs/index.html) and the Environmental Protection Agency (EPA; http://es.epa.gov/ncer/rfa/02ecohab.html) have initiated efforts to address these problems, and the U.S. Global Ocean Observation System Program, in collaboration with the Southern Association of Marine Laboratories and the Gulf of Mexico Program, is developing a prototype system for the northern Gulf of Mexico (http://www.hpl.umces.edu/projects/HABSOS.pdf).

In addition, the Harmful Algal Bloom and Hypoxia Research and Control Act was enacted in 1998 (PL 105-383) in response to concerns that HABs and related environmental events (e.g., hypoxia, fish kills) are increasingly a threat to human and coastal ecosystem health. The act called for the establishment of an interagency task force on HABs and hypoxia; a national assessment of HABs and hypoxia; and an assessment plan for the Gulf of Mexico. A federal interagency task force released the *National Assessment of Harmful Algal Blooms in US Waters* in October 2000 (http://www.habhrca.noaa.gov/FinalHABreport.pdf).

Condition of Bottom-Dwelling Animals

The Indicator

The worms, clams, and crustaceans that inhabit the bottom substrates of estuaries are collectively called benthic macroinvertebrates. These organisms play a vital role in maintaining sediment and water quality and are an important food source for bottom-feeding fish, shrimp, ducks, and marsh birds. Benthos are often used as indicators of disturbances in estuarine environments because they are not very mobile and thus cannot avoid environmental problems.

The Data

Data Source: The U.S. Environmental Protection Agency's (EPA) Environmental Monitoring and Assessment Program (EMAP) (http://www.epa.gov/emap/) collected these data. See the technical note for Contamination in Bottom Sediments for a description of the EMAP program. These data were collected as part of the EMAP for Estuaries (EMAP-E).

Data Collection Methodology: EMAP-E examined benthic samples from over 2000 sites from Cape Cod, Massachusetts, to Brownsville, Texas. All site selections were based on probabilistic designs that permit the extrapolation of the data to the entire area. Using a Young-modified Van Veen grab, three replicate grabs were collected from each site and forwarded for identification and quantification of species. Using an index developed by EMAP-E (Engle and Summers 1999, Engle et al. 1994, Van Dolah et al. 1999, Weisberg et al. 1997), the condition of the benthic community was determined for each replicate sample, each site, and the bottom surface area of U.S. estuaries. The index reflects changes in benthic community diversity and the abundance of pollution-tolerant and pollution-sensitive species. A low benthic index rating indicates that the benthic communities are less diverse than expected, are populated by more than expected pollution-tolerant species, and contain fewer than expected pollution-sensitive species. The data in this report reflect an assessment of benthic communities as "good" (high index score), "fair" (moderate index score), or "poor" (low index score). For this report, these classes were described using the terms "undegraded," "moderate," and "degraded." These terms were chosen to ensure a neutral description of the index information (i.e., whether a site's benthic community indicates that the site is degraded or not in relation to a reference site).

Data Quality/Caveats: The indices used in the three regions were developed independently and may not be comparable. Each has been demonstrated to be accurate in the region in which it was developed, but there is some question about whether they can be combined because of the different procedures used in their development.

The definition of undegraded and degraded areas also varied because the levels and types of stress differ from region to region. As a result, the indices in less disturbed areas, such as those being developed in southern California, are designed to detect smaller levels of perturbation than are indices developed for areas like the Chesapeake Bay, where hypoxia and resulting defaunation are prevalent.

Finally, some indices are closely identified with particular sampling methods, creating challenges for integration of results. For example, Gulf of Mexico and Atlantic coast indices are based on animals held on a 0.5-mm screen, while an index used in southern California is based on samples sieved through a 1.0-mm screen.

Data Access: The data presented here were obtained directly from EPA.

The Data Gap

Benthic infaunal data are available from most areas of the country, but the index tools necessary to conduct regional-scale assessments of benthic condition are available for estuaries in only three areas of the country: the Mid-Atlantic, the South Atlantic, and the Gulf of Mexico. Altogether, these indices cover less than half of the nation's estuarine waters. EPA has recently issued national guidance on index development (EPA 822-B-00-024), which should facilitate development efforts in the remaining areas. The guidance document is available on the Web at http://www.epa.gov/waterscience/biocriteria/States/estuaries/estuaries1.html. In addition, few data are available on benthic community condition in coastal ocean waters (out to 25 miles).

References

Engle, V.D., and J.K. Summers. 1999. Refinement, validation, and application of a benthic condition index for northern Gulf of Mexico estuaries. Estuaries 22(3A):624–635.

Engle, V.D., J.K. Summers, and G.R. Gaston. 1994. A benthic index of environmental condition of Gulf of Mexico estuaries. Estuaries 17:372–384.

Van Dolah, R.F., J.L. Hyland, A.F. Holland, J.S. Rosen, and T.R. Snoots. 1999. A benthic index of biological integrity for assessing habitat quality in estuaries of the southeastern USA. Marine Environmental Research 48(4–5):269–283.

Weisberg, S.B., J.A. Ranasinghe, D.D. Dauer, L.C. Schaffner, R.J. Diaz, and J.B. Frithsen. 1997. An estuarine benthic index of biotic integrity (B-IBI) for Chesapeake Bay. Estuaries 20(1):149–158.

Chlorophyll Concentrations

The Indicator

Coastal Ocean: For each year, the average chlorophyll-a (referred to here as chlorophyll) concentration (parts per billion, or ppb) for the season with the highest average is reported; this is referred to as the "seasonal mean maximum." Data are reported for each region in a band of coastal water extending 25 miles from the shoreline. This boundary was chosen so the index would be more sensitive to changes in nutrients input from terrestrial sources than influences from the deep sea.

Estuaries: It is proposed to report the percentage of U.S. estuary area that has seasonal mean maximum chlorophyll values below 5 ppb, from 5 to 20 ppb, and above 20 ppb.

The Data

Coastal Ocean: Data from the National Aeronautical and Space Administration's (NASA) Sea viewing Wide Field-of-view Sensor (SeaWiFS; see http://seawifs.gsfc.nasa.gov) were analyzed for the nine ocean regions by the National Ocean Service, National Oceanographic and Atmospheric Administration (NOAA). "Water leaving radiance" (reflectance, or light reflected from the sea surface) is used to estimate chlorophyll concentrations at the surface using a series of assumptions accepted by the scientific community. The data utilized for this analysis are termed "level 3." In all cases, seasonal maxima were determined for strips of water 25 miles wide along the coast. These strips were analyzed using square pixels 6 miles on a side. Note that earlier data from the Coastal Zone Color Scanner are available; however, they are not directly comparable to the SeaWiFS data.

The Data Gap

Coastal Ocean: Algorithms used to translate water leaving radiance into chlorophyll concentration currently provide only rough estimates of concentration in those waters where concentrations of suspended sediments and colored dissolved organic matter are high; for example, near-shore waters influenced by surface and groundwater discharges, coastal erosion, and sediment resuspension. A major research effort is currently under way to improve coastal algorithms. Spatial resolution is also a problem. The data presented here are based on a fairly coarse scale (6-mile resolution), but data with 10 times more resolution will soon be available. In order to provide more reliable estimates, satellite data need to be analyzed together with field (in situ measurements) data that typically are not available electronically and, therefore, not easily accessible. In addition, techniques for integrating the two types of data are needed. Currently, data showing relative changes in chlorophyll within a region can be trusted; however, data showing actual concentrations for any given region may be off by a factor of two. Thus, unless differences are large, meaningful comparisons between regions are not yet possible.

Estuaries: As discussed in the text, no regularly reported data are available for this portion of the indicator. Data from NOAA's National Estuarine Eutrophication Assessment (see http://spo.nos.noaa.gov/projects/cads/nees/Eutro_Report.pdf) suggest that 40 percent of the area of major estuaries has "high" chlorophyll levels (>20 ppb), with another 46 percent having "moderate levels" (5–20 ppb). At the extremes, the north and south Atlantic regions had mostly low-to-moderate levels, while three-quarters of the estuary area in the mid-Atlantic had high chlorophyll levels. These results are not based on quantitative data analysis but on the knowledge of scientists familiar with the estuaries in each region. Monitoring data do exist for some estuaries, but need to be assembled into a uniform, national database, and new programs would be required for the remaining estuaries. A combination of aircraft and satellite remote-sensing and in situ measurements will be required to determine the estuarine component of this indicator.

Commercial Fish and Shellfish Landings

The Data

Data Source: Data are from the U.S. Department of Commerce, National Oceanographic and Atmospheric Administration, National Marine Fisheries Service (NMFS), Office of Science and Technology, Fisheries Statistics and Economics Division. In addition, data on foreign and joint-venture landings for Alaska came from Kinoshita et al. (1993 and 1997) and NMFS "blend data" for June 2000 (for a description of "blend data," see http://www.fakr.noaa.gov/sustainablefisheries/blend.htm). The estimates for foreign catches that occurred prior to the establishment of the Fisheries Conservation Zone in 1976 (dotted line in the figure) came from Wise (1991) and are based on NMFS data; however, these data could not be verified.

Data Collection Methodology: Fish landings data for the Pacific Coast were collected by four state fishery agencies, and NMFS reported fish processed at sea by U.S. boats (fish processed by foreign boats and then exported are not reported). On the Atlantic and Gulf coasts, landings data were usually collected cooperatively by the 19 state fishery agencies and NMFS. Some data were also collected by marine fishery commissions.

Atlantic and Gulf commercial fishery data represent a census of landings and were principally reported using seafood dealer weighout slips, while data on the Pacific coast were principally reported using trip ticket reports and observer reports for at-sea processors. Since 1994, an increasing portion of the fishery catch and effort data for federally managed species has been collected using federally mandated logbooks. The use of Vessel Monitoring

Systems and other electronic data collection and reporting methods is relatively recent and is limited to a small sector of U.S. fisheries.

Note that NMFS has historically included all commercial landings of mollusks in these data, in part because it is not provided with information that will allow it to separate wild-caught from cultured mollusks. In terms of finfish (simply termed "fish" here), those raised by aquaculture are not included with the exception of Alaska salmon, which are released at smolt size (2–4 inches in length) and are caught several years later when they return from the ocean to spawn.

Data Manipulation: All finfish landings have been transformed, when necessary, from landed weight (e.g., dressed, filleted) to round (live) weight equivalents. All mollusks have been standardized from the collected landing report format (e.g., bushels, totes, gallons, counts, and dozens) and reported as meat weight (i.e., without shell) landings. The collecting state and federal agencies themselves transform the landings data.

Data Quality/Caveats: The National Research Council conducted a review of NMFS fisheries data and published *Improving the Collection, Management, and Use of Marine Fisheries Data* in 2000. The report made several recommendations for strengthening fishery data collection, such as implementing a national set of standards and protocols under the umbrella of a Fisheries Information System, but no serious flaws in the existing data collection system were noted.

Data Access: Non-confidential commercial fisheries landings data for 1950–2000 are available at no cost from the NMFS Fisheries Statistics and Economics Division Web site (http://www.st.nmfs.gov/st1/commercial/index.html). The Web site allows users to summarize the data by year, region, state, species, fishing gear, pounds, and dollars.

References

Kinoshita, R.K., A. Greig, D. Colpo, and J.M. Terry. 1993. Economic status of the groundfish fisheries off Alaska, 1991. U.S. Dept. Commer., NOAA Tech. Memo. NMFS-AFSC-15.

Kinoshita, R.K., A. Greig, D. Colpo, and J.M. Terry. 1997. Economic status of the groundfish fisheries off Alaska, 1995. U.S. Dept. Commer., NOAA Tech. Memo. NMFS-AFSC-72.

Wise, J.P. 1991. Federal conservation and management of marine fisheries in the United States. Washington, DC: Center for Marine Conservation.

Status of Commercially Important Fish Stocks

The Data

Data Source: The National Oceanographic and Atmospheric Administration's (NOAA) National Marine Fisheries Service (NMFS) Research Centers in Woods Hole, Massachusetts; Miami, Florida; La Jolla, California; and Seattle, Washington (two centers). Natural Resources Consultants, Inc., analyzed the data under contract for The Heinz Center.

Data Manipulations: As reported by NMFS in the recent publication *Our Living Oceans*, there are 203 stocks within federal jurisdiction. Excluded from these analyses are near-shore stocks, many of which are under state management jurisdiction, and anadromous salmon stocks from the Pacific Northwest.

For 158 of these stocks, data are adequate to consider reporting on their status; the remaining 45 stocks constitute a small fraction (2.5%) of recent landings. Analyses presented in this report were limited to those stocks for which at least 10 years of data were available over the 1981–1999 time period. Both spawning stock biomass and total exploitable stock biomass figures were used to track stock trends. This restriction (i.e., 10 years of data) meant that only 49 of the 158 stocks having status data could be used (these 49 stocks represent about 75% of the weight of fish caught in U.S. waters). It should be noted that requiring spawning stock biomass or total exploitable stock biomass figures certainly restricted the number of trackable stocks; other means exist (e.g., catch per unit effort, relative abundance, indices that combine several stocks) to track stocks that were not included in this analysis.

Biomass refers to the total weight of fish, and can change either because there are more or fewer fish or because, on average, fish are larger or smaller, although changes in biomass are generally described as changes in population size. Stock trends (i.e., "increasing," "decreasing" or "no trend") were determined by linear regression. Trends were determined for four overlapping 10-year periods (1981–1990, 1984–1993, 1987–1996, 1990–1999), which reduced the likelihood that normal year-to-year fluctuations would influence the results. (This is analogous to the effect of using a running average.) Two conditions were necessary for a trend to be reported: the regression line had to have a correlation coefficient (R) indicating at least 95% confidence that the slope was different than zero; and the regression line had to indicate a minimum 25% change over the 10-year period (increasing or decreasing). Trends for the following 49 stocks were studied:

- *North/Mid-Atlantic:* Atlantic menhaden, Georges Bank cod, Georges Bank haddock, Georges Bank yellowtail flounder, southern northeast yellowtail flounder, mid-Atlantic summer flounder, Gulf of Maine red fish, Gulf of Maine cod, Gulf of Maine–Georges Bank plaice, Gulf of Maine–Georges Bank witch, Georges Bank winter flounder, and southern New England winter flounder.
- *Gulf of Mexico:* Brown shrimp, white shrimp, and pink shrimp.
- *Southern California:* Mackerel and sardine.
- *Pacific Northwest:* Dark blotched rockfish, lingcod (northern), lingcod (southern), widow rockfish, yellowtail rockfish, bacaccio rockfish, canary rockfish, cow rockfish, Pacific hake, petrale sole, chilipepper rockfish, sablefish, Dover sale, and longspine thorny rockfish.
- *Alaska Region:* West & central Alaska pink salmon, west & central Alaska sockeye salmon, west & central chum salmon, west & central Alaska Chinook salmon, west & central Alaska Coho salmon, Bering Sea pollock, Bering Sea Pacific cod, Bering Sea yellow fin sole, Bering Sea rock sole, Bering Sea sablefish, Gulf of Alaska pollock, Gulf of Alaska sablefish, Pacific halibut, red king crab, blue king crab–Pribilofs, blue king crab–St. Matthews, Tanner crab, and snow crab.

Data Quality/The Remaining Data Gap: Currently, we are able to evaluate trends on only about 25% of the commercially important stocks found in U.S. waters.

Data Access: Stock biomass data are available by contacting the NMFS research centers noted above in "Data Source."

Selected Contaminants in Fish and Shellfish

The Indicator

Mercury, DDT, and polychlorinated biphenyls (PCBs) are the chemical contaminants of most concern with respect to the human health impacts of the consumption of seafood. Many fish consumption advisories have been issued for because of high concentrations of these compounds.

This indicator would report the concentration of DDT, PCBs, and mercury in the edible portion of fish and shellfish. The edible portion is preferred for this indicator because whole-body analyses can overstate the level of risk, as some contaminants concentrate in portions that are not eaten.

The Data Gap

The Food and Drug Administration (FDA), Environmental Protection Agency (EPA), and state governments have a variety of monitoring and reporting programs in place; however, these programs do not provide the basis for national reporting on contaminant concentrations.

The FDA has the power to conduct wharf examinations and collect and analyze fish and shellfish samples for a wide variety of defects including chemical contaminants, decomposition, radionuclides, various microbial pathogens, food and color additives, drugs, filth, and marine toxins such as paralytic shellfish poison and domoic acid. When necessary the FDA has the authority to detain or remove any imported or domestic product from interstate commerce that fails to meet standards. Though these powers are broad, they are not used in a manner that provides periodic national reporting on chemical contamination.

The FDA works with state regulators when commercial fish, caught and sold locally, are found to contain methyl mercury levels exceeding 1 part per million (ppm). The agency also checks imported fish at ports and refuses entry if methyl mercury levels exceed the FDA limit. There is no FDA reporting program based on these inspections, however.

With the cooperation of state, county, regional, and country officials, the FDA has a Pesticide Residue Monitoring program whose emphasis is on the raw agricultural product but also includes some seafood products. In its 1999 Total Diet Study (TDS; http://vm.cfsan.fda.gov/~comm/tds-toc.html), sometimes called the Market Basket Study, the FDA determined intake levels of various pesticide residues, contaminants, and nutrients in foods, in representative diets of specific age-sex groups in the United States. In the 1999 and subsequent studies, a total of 267 different foods were represented in the 26 market baskets analyzed. Of those 267 foods, only seven involved seafood. While this program might provide estimates of consumption of various contaminants in the American diet, it does not provide a consistent means of tracking contaminant concentrations in fish from U.S. waters.

EPA has provided a national guidance manual (www.epa.gov/ost/fish/guidance.html) to states for developing consumption advisories and contaminant monitoring programs, but does not directly conduct such monitoring. The manual, which is not binding upon states, was designed to promote consistency in sampling and analysis methods, risk assessment methods, decision-making procedures, and approaches for communicating risks to the public. In addition, both EPA and FDA have issued action levels for concentration of various contaminants, and states may choose between them when deciding whether to issue fish consumption advisories.

EPA maintains a Listing of Fish and Wildlife Advisories (http://fish.rti.org). This database describes state-, tribal-, and federally issued fish consumption advisories. Information in the database is provided voluntarily by the states and may not include the actual concentration data used to determine an advisory need. Since advisories may be based on different levels in different states; it is not even possible to use this database to determine how many cases exceeded a certain level.

From 1990 to 1995, EPA published the National Survey of Mercury Concentrations in Fish (www.epa.gov/ost/fish/mercurydata.html). In addition to the fact that this survey was discontinued, several factors contribute to the variability of the current database. States collect data for purposes other than mercury analyses, and not all sampling strategies are based on a random sample. For example, data collected for the purpose of annual water-quality monitoring may not produce the same results as a site-specific study of fish tissue mercury concentrations. States use different techniques to sample fish. The sampling techniques used by each state influence sample size, fish size, and fish type. States do not adhere to the same standards for assimilating a composite sample. The absence of a standardized method for grouping fish may result in grouping different species of fish into composites, which can affect both the representativeness of the sample and the results of analyses. States use various analytical procedures to measure the concentration of total mercury in fish. Variations among analytical equipment, use of different protocols and procedures, and different levels of laboratory staff experience can all bias the assessment of mercury concentrations in fish. In addition, mercury analyses reported on a wet weight basis cannot be directly compared to concentrations reported on a dry weight basis.

Recreational Water Quality

The Indicator

The most commonly used indicators of fecal contamination are total coliform bacteria, fecal coliform bacteria, *Escherichia coli*, and *Enterococcus* (the latter two are bacteria as well). Although indicator bacteria do not necessarily cause illness, they are abundant in human waste where pathogenic organisms, such as viruses and parasites, are also likely to exist. Bacterial indicators are currently measured instead of pathogenic organisms because the indicators occur in much larger numbers and can be measured with faster, less expensive methods than the pathogens of concern. However, with advances in biotechnology, it may soon be feasible to monitor pathogens using genetic tests.

This indicator focuses on *Enterococcus*, which was selected over other measures of bacteria because it has been shown to be the most closely correlated with human health effects. The U.S. Environmental Protection Agency (EPA) recommended the use of *Enterococcus* as the fecal-indicator bacteria for recreational water quality standards in 1986, but it is still not as widely used as the coliform measures. The reporting categories for this indicator correspond to the daily (104 cells per milliliter of water) and monthly (35 cells per milliliter) geometric mean thresholds suggested by EPA as national beach water quality standards. It should be noted that the selection of *Enterococcus* is logical today based on EPA guidelines; however, it is possible that new recommendations from EPA and other sources may alter the organism(s) reported in this indicator (for a discussion of a multi-organism indicator, see http://www.healthebay.org/beachreportmethod.asp).

Because some events are short-term but extend over large areas and others are chronic closures in small areas (near a small local source, for example), the indicator is based on the number of beach-mile-days exceeding thresholds of concern, rather than on the number of exceedances or closures. These different scenarios would be weighted inappropriately if the measure were limited to the number of events or to the mileage of beaches that exceeded thresholds at any time during the year.

The indicator is also based on the underlying microbiological data rather than on the number of beach closures or advisories, as is done in EPA's national report (http://www.epa.gov/OST/beaches/); differences in procedures used by local governments in making closure decisions make such reporting less informative. Moreover, the amount of beach monitoring varies dramatically among states, and an indicator based on the number of closures may focus undue concern on states or beach areas that are the most vigilant.

The Data Gap

In 2000, the U.S. Congress passed the Beaches Environmental Assessment and Coastal Health (BEACH) Act. The Act authorizes EPA to award grants to local entities (states, tribes, and territories) to develop and implement monitoring programs at beaches along the coast, including along the Great Lakes. In response to recent legislation, the state of California is moving toward routine reporting of closures in beach-mile-days. Most other states do not summarize their data in this format.

Only one study has ever estimated the number of beach-mile-days exceeding bacteriological thresholds of concern, and that was a one-time research project (Noble et al. 2000).

There are several challenges to reporting this indicator at a national level. These involve, first, the adoption by states and municipalities of the use of *Enterococcus* as an indicator bacteria and adoption of the use of beach-mile-days as the unit of reporting. Second, national reporting will require obtaining the microbiological data from the numerous local governments that collect it. The indicator also requires an assessment of the extent of beach monitoring, which will require three additional types of information: an estimate of the number of miles of publicly accessible beach that is available for water-contact recreation, the spatial extent of beach associated with each water quality measurement (e.g., distance to the next measurement location or to the farthest location that would be closed based on results from that sample site), and the time between samples. This can be complex in practice because some programs measure bacteria sporadically based on events such as spills or citizen complaints, and defining how much beach is represented by a sample can be difficult. Most monitoring uses sampling sites a mile or more apart, while closure decisions typically apply to much smaller areas around any given sampling point.

In addition, many of the agencies and organizations that monitor water quality do not store their data electronically, and even those that do so do not use an agreed-upon storage format. There are also considerable differences in the number, frequency, and degree of coverage of sampling among states and even among beaches within individual states. More consistency among sampling efforts across the nation would enhance the value of the measure. EPA is working to solve the data management problem by collaborating with coastal states to produce an annual report on the national extent of beach closures. While this is a start, EPA's reporting effort focuses only on closures, rather than on the underlying water quality. Since the standards used to determine when a beach is too dirty for swimming vary from place to place, this information cannot provide a consistent picture of water quality nationwide.

References

Noble, R.T., J.H. Dorsey, M. Leecaster, V. Orozco-Borbon, D. Reid, K.C. Schiff, and S.B. Weisberg. 2000. A regional survey of the microbiological water quality along the shoreline of the Southern California Bight. *Environmental Monitoring and Assessment* 64:435–447.

Farmlands

Total Cropland

Note: Other indicators in this chapter will refer to the discussion of the Natural Resources Inventory (NRI) below.

The Indicator

This indicator reports the acreage of cropland in the United States. Included in this category are pastures and haylands. For the purposes of this indicator, lands that have been idled in long-term set-aside programs, such as the Conservation Reserve Program (CRP), are not included. This is in part because of the objective to report on those lands that are in active use, and because not all of the programs distinguish CRP acres equally well (the remote-sensing data from the National Land Cover Dataset [NLCD] do not separate them from active croplands). In theory, the land area utilized for animal feedlots would be included in this indicator. In practice, however, this acreage is certainly quite small and is not isolated by the various programs used in this analysis. (While the indicator definition excludes CRP lands, one of the data sources used [Economic Research Service, or ERS] does not report CRP acreage separately; thus, the ERS numbers include CRP acreage. There are other differences between the datasets; see below.)

In addition, lands used for intensive livestock feeding are included within the ambit of this indicator. However, it was not possible to determine the degree of coverage of these areas for the data sources described below.

The Data—General

Data Sources: These data were obtained from the U.S. Department of Agriculture (USDA) Natural Resources Conservation Service, National Resources Inventory (NRI) program; the USDA National Agricultural Statistical Service, Census of Agriculture; the USDA ERS, Agricultural Resources and Environmental Indicators publications; and the U.S. Geological Survey. (USGS provided access to and processing assistance with the NLCD, originally produced by a federal interagency consortium, the Multi-Resolution Land Characterization [MRLC] Consortium, see p. 207). See details below on each program.

Comparability Among Data Sources: These four data sources are not fully consistent, and comparisons should be made with care. For example, ERS and Census of Agriculture data include croplands in Alaska and Hawaii, while NRI does not, and only the ERS data reported here include acreage in the CRP—these acres were removed from the data for Census of Agriculture and NRI.

The statement that cropland, including pasture and hayland, occupies about one-fourth of the land area of the United States is based on the estimates from the four programs noted above. These estimates range, for 1997, from 431 million acres (NASS) to 496 million acres (NRI). They are compared to the land area of the lower 48 states (derived from the MRLC dataset), 1.891 billion acres. Thus, the percentages range from 22.8% to 26.2%.

USDA National Resources Inventory

Data Source: Data are from USDA, National Resources Conservation Service, Iowa State University Statistical Laboratory, Summary Report 1997 National Resources Inventory (revised December 2000).

Data Collection Methodology: The USDA Natural Resources Conservation Service, in cooperation with the Iowa State University Statistical Laboratory, conducts the NRI survey to capture data on land cover and use, soil erosion, prime farmland soils, wetlands, habitat diversity, selected conservation practices, and related resource attributes. Data are collected every 5 years from the same 800,000 sample sites in the lower 48 states, Puerto Rico, the U.S. Virgin Islands, and some Pacific Basin territories.

The estimated acreage of nonfederal cropland was classified as irrigated, non-irrigated, cultivated or noncultivated acreage. Data are collected for the NRI using a variety of imagery, field office records, historical records and data, ancillary materials, and onsite visits. The data are compiled, verified, and analyzed to provide a comprehensive summary of the state of U.S. non-federal lands. The NRI is a two-stage stratified area sample of the entire country. Stage one is the Primary Sampling Unit (PSU), and it is a sampling of an area/segment of land typically square to rectangular in shape and ranging from 40 to 640 acres but most typically 160 acres in size. Stage two requires the assignment of sampling unit points that are located within the PSU. Cropland includes pasture and areas used for the production of crops for harvest. For the purposes of this indicator, CRP lands were excluded from the NRI data.

Data Quality/Caveats: Statistics derived for the NRI database are estimates and not absolutes, resulting in some amount of uncertainty. These data are reported at the national level; state-level data are available at http://www.nhq.nrcs.usda.gov/NRI/1997/state_info.html.

Data Access: The NRI report is available at http://www.nhq.nrcs.usda.gov/NRI/1997/.

USDA Census of Agriculture

Data Source: Data are from USDA, National Agricultural Statistics Service (NASS), 1997 Census of Agriculture.

Data Collection Methodology: The Census of Agriculture is a comprehensive accounting of agricultural production information for every county in the United States. For 1992 and 1997, the census was conducted by USDA NASS; prior to 1992, the Bureau of the Census was responsible for censuses every 5 years.

The census is conducted using a mailout/mailback form, direct enumeration, telephone, personal interviews, and follow-up surveys. The mailing list, with 3.2 million contacts, is composed of individuals, businesses, and organizations that are associated with agriculture. Report forms for the 1997 Census of Agriculture were mailed to farm and ranch operators in December 1997 to collect data for the 1997 calendar year.

"Total cropland" includes harvested cropland; cropland used only for pasture or grazing; crop failure; cultivated summer fallow; idle cropland; and cropland in cover crops, legumes, and soil-improvement grasses, not harvested and not pastured. Data on CRP lands were excluded for the purposes of this indicator.

Data Quality/Caveats: The data from each report form were subjected to a detailed item-by-item computer edit. Before publication, tabulated totals for each state were reviewed by state statisticians to identify inconsistencies. Comparisons were also made with previous census data, official NASS Agricultural Statistics Board numbers, and other available check data.

Data Access: The 1964–1997 data are available at http://www.nass.usda.gov/census/census97/volume1/us-51/us1_01.pdf. The 1945–1959 data are not available online but can be obtained by e-mail from NASS at nass@nass.usda.gov.

National Land Cover Dataset

These data are derived from the MRLC Consortium, which is a partnership between USGS, the U.S. Forest Service, the National Oceanographic and Atmospheric Administration (NOAA), and EPA. See the explanation of the NLCD in the national extent technical note, page 207.

USDA Economic Research Service

Data Source: Data were acquired from the U.S. Department of Agriculture, ERS, Resource Economic Division, Agricultural Resources and Environmental Indicators (AREI) 2000 and AREI 1996–97.

Data Collection Methodology: ERS provides national economic data and analysis on issues related to agriculture, food, natural resources, and rural development.

The "cropland" category includes cropland harvested, crop failure, cultivated summer fallow, cropland used only for pasture, and idle cropland. CRP lands are included. ERS compiled these data from NASS Principal Crops and Census of Agriculture data. The data used here were compiled from Krupa and Daugherty (1990), Daugherty (1995), and Vesterby and Krupa (2001).

Data Access: The 1996–97 and 2000 reports may be accessed online at http://www.ers.usda.gov/Emphases/Harmony/issues/arei2000/.

References

Daugherty, A.B. 1995. Major uses of land in the United States, 1992, AER-732, U.S. Department of Agriculture, Economic Research Service.

Krupa, K.S., and A.B. Daugherty. 1990. Major land uses: 1945-1987, Electronic Data Product #89003, U.S. Department of Agriculture, Economic Research Service.

Vesterby, M., and K.S. Krupa. 2001. Major uses of land in the United States, 1997. Resource Economics Division, Economic Research Service, U.S. Department of Agriculture. Statistical Bulletin No. 973.

The Farmland Landscape

The Data

Data Collection Methodology: The data presented here are from the National Land Cover Dataset (NLCD); see the technical note for the national extent indicator (p. 207) for a full description.

Data Manipulation: The U.S. Geological Survey (USGS) Earth Resources Observations Systems Data Center aggregated data from the NLCD into squares 1 km on a side (approximately 1000 30-meter by 30-meter "pixels"). Each of these larger squares was analyzed to determine its land cover composition; 1-km squares in which more than 50% of the pixels were croplands were included within the "farmland landscape." In addition, a "buffer" equivalent to a single 1-km square was added to the edge of the farmland landscape defined above, in order to incorporate areas near those with significant concentrations of cropland.

This set of "farmland landscape" squares was analyzed to determine its composition, using the land cover data for the underlying 30-meter pixels. These data are aggregated using standard regions adopted by the Natural Resources Conservation Service (http://www.nhq.nrcs.usda.gov/land/meta/m2140.html). The following land cover types were reported, based on NLCD categories: farmland, forest, grasslands/shrublands, "developed," wetlands, water, other (see the national extent technical note for further details, p. 207).

Data Quality/Caveats: Note that, in some cases, wetlands are found on croplands, and it can therefore be difficult to separate one from the other. This is especially true because these wetlands may only have water for parts of the year and may be farmed for other parts of the year. Thus, the data on wetlands reported in this indicator should be interpreted with some caution.

Fragmentation of Farmland Landscapes by Development

The Indicator

This indicator indirectly measures the fragmentation of farmland by developed or built-up areas. Cropland interspersed with residential subdivisions raises entirely different policy and farmland management implications than cropland interspersed, for example, with patches of "natural" land cover (forest, grasslands/shrublands, or wetlands). Thus, this indicator considers fragmentation to occur when croplands and natural lands are interspersed with development.

This indicator is an index of spatial fragmentation calculated from digital land cover maps classified from remote-sensing data. Land cover data from, for example, the National Land Cover Dataset (NLCD; see technical note for the national extent indicator, p. 207) will be used. This dataset has classified grid cells, or "pixels," which represent areas measuring approximately 100 feet (30 meters) across. This index is computed by analyzing classified land cover "layers" within a raster-based geographic information system (GIS).

The fragmentation index is calculated for each pixel in the farmland landscape (i.e., either cropland or nearby "natural" lands). This value is based on the characteristics of the surrounding pixel "neighborhood." Such neighborhoods are often created as 3 by 3 or 5 by 5 pixels arrangements. The value for the center pixel is based on the character of the surrounding 8 pixels (in a 3 x 3 square) or 24 pixels (in a 5 x 5 square).

Although the fragmentation index will be calculated for individual pixels, index values for pixels will be aggregated at the scale of one-kilometer squares. Rather than directly reporting index values (i.e., 0 to 1), three fragmentation classes will be reported based on a statistical analysis of these aggregated index values. Each one-kilometer square block will be classified as having a high, medium, or low level of fragmentation. The percentage of surface area in each fragmentation class will be reported by region.

A sensitivity analysis should be performed so that the overall results are not an artifact of the neighborhood size (e.g., such as the 9-pixel arrangement discussed above). In addition, by enlarging the size of the pixel neighborhood (such as to 5 x 5 pixel units), the method will be more sensitive to non-adjacent development.

The index will depend not only on the amount of development interspersed within the farmland landscape, but also on how this development is distributed spatially in the landscape. Thus, development could cover, for example, 20% of two farmland landscapes, but these two landscapes would have very different index values. Clustered rural residential development (e.g., conservation subdivisions), surrounded by cropland and natural areas, would result in relatively high fragmentation index values for those developed portions of the farmland landscape. More scattered, lower density rural residential development (e.g., large estates) would result in somewhat lower fragmentation index values for those developed portions of the farmland landscape. Yet if the total gross residential densities (e.g., total number of dwelling units) were equal in both development scenarios, the proportion of the farmland landscape with an elevated fragmentation index would be much greater in the scattered, low-density development scenario.

The index is sensitive to low-density development if this development can be detected using satellite data (i.e., the development must "fill" a major portion of the pixel used in order to be classified "developed" in the land cover dataset).

It should be noted that this is a subject that has garnered considerable attention in the research community. An example of an alternative approach is the one promoted by the U.S. Department of Agriculture (USDA) Economic Research Service (ERS). Its approach identifies farmland that is influenced by nearby development using property values—based on the assumption that farmland priced beyond its agricultural value must be experiencing development pressure. See *Development at the Urban Fringe and Beyond: Impacts on Agriculture and Rural Land*, AER-803; http://www.ers.usda.gov/publications/aer803/. Another approach would use data from the Natural Resources Inventory (NRI; USDA Natural Resources Conservation Service). Specifically, the "segments per unit" metric might be used for the appropriate land cover category and reported on a regional basis.

The Data Gap

Several indices have been developed to quantify various aspects of pattern at the patch, class, and landscape scales. Data appropriate for calculating this indicator are available from the NLCD, which was used to define the "farmland landscape" for this report (see p. 92). Calculating this index requires digital data and specialized software designed to analyze landscape spatial patterns. The most commonly used software for analyzing landscape spatial patterns (Fragstats) is not capable of processing the very large file sizes that would be required to calculate this index for the entire nation. It may be possible to address this analysis using a statistical sampling technique, analytical approaches relying on GIS software, or other analytical approaches; however, the details of this were not resolved in time for production of this report.

Shape of "Natural" Patches in the Farmland Landscape

The Indicator

The size, shape, and juxtaposition of habitat patches within a landscape, in addition to the total extent of the habitat, influence the population size and viability of sensitive species (Meffe and Carroll 1994). Dozens of metrics have been developed to quantify spatial pattern within landscape mosaics. Some metrics quantify the size, shape, or juxtaposition of individual patches of a single land cover type. Others quantify the spatial relationships (e.g., juxtaposition) among patches of different land cover types.

This indicator measures the geometry, or spatial configuration, of "natural" areas in farmland landscapes. Natural areas are native habitats, including forests, grasslands, wetlands, and naturalized habitats, such as land enrolled in the Conservation Reserve Program (CRP) that formerly was cropland and is now reverting to, for example, forest or grassland. For the purposes of this indicator, open water is not included; however, wetlands are included. (Special attention should be paid to the ability of the land cover datasets, which are based largely on satellite measurements, to account for wetlands.)

This indicator will be calculated for individual patches, yet index values will be reported at the national scale. The indicator will be based on perimeter-to-area (P/A) ratios that have been normalized by area (i.e., divided by the patch's area). Rather than directly reported P/A values, however, three size and shape classes will be determined by a statistical analysis of each region's P/A ratios. The shape classes will be compact (e.g., a circle, which has the lowest P/A ratio for a given area), intermediate, and elongated (e.g., a long, narrow rectangle, which has a high P/A ratio for a given area). The surface area within each size and shape class will be reported nationally. This patch-based index should be area-weighted. This weighting ensures that smaller patches (with higher P/A ratios) have less influence on the aggregate index than larger patches do. The indicator will be calculated for the aggregated area of all forests, grasslands, shrublands, and wetlands in the farmland landscape, rather than being calculated for each "natural" land cover type independently.

Landscape structure—or the spatial configuration of patches, corridors, and the intervening "matrix"—influences ecosystem integrity. Yet spatial pattern is a complex phenomenon that cannot be summarized with a single index. The number, size, shape, orientation, and spatial distribution of land cover patches are important landscape attributes. Other ecologically significant aspects of landscape pattern include the proportion and spatial arrangement of different land cover types.

Agricultural activities have extensively changed many landscapes. "Fragmentation," caused by land use changes and other disturbances, may alter landscape structure by changing land cover area and spatial configuration. Within intensively farmed landscapes, natural areas comprise a relatively small percentage of the surface area. Typically, these natural areas include relatively small and isolated remnants of formerly contiguous native vegetation, in addition to restored conservation areas (e.g., CRP land). These native and naturalized areas provide wildlife habitat, control erosion, and perform other important ecological and cultural functions. Patch size and shape influence the differentiation of patches into distinct edge and interior habitats. Small patches typically have a higher ratio of edge to interior habitat than very large patches with the same shape. Conversely, linear patches have a much higher proportion of edge to interior habitat than patches with the same area, but more compact shapes. Small or highly dissected patches may have little or no interior habitat. The functional connectivity among patches of natural areas depends not only on the distances between the patches, but on the intervening land use and land cover conditions. The land covers (e.g., built-up) and land uses (e.g., farming) that separate natural areas can significantly influence biodiversity and species abundance at landscape and regional scales.

The Data Gap

Calculating this index will require digital data and specialized software designed to analyze landscape spatial patterns. Data appropriate for calculating this index are available from the National Land Cover Dataset, which was used to define the "farmland landscape" for this report (see p. 92). However, the most commonly used software (Fragstats) for analyzing landscape spatial patterns is not capable of processing the very large file sizes that would be required to calculate this index for the nation. That said, there may be simpler approaches that would not have such computing demands, although these have not been fully explored.

It might be possible to make use of existing remote-sensing data through a procedure involving random sampling. In such a procedure, rather than processing the entire dataset, samples would be processed, much as a field program such as the USDA National Resources Inventory (http://www.nhq.nrcs.usda.gov/NRI/1997/) collects data from a representative sample of sites. However, the specific approach needed for such a sampling program was not fully explored during the development of this report.

References

Meffe, G.K., and C.R. Carroll, eds. 1994. Principles of conservation biology. Sunderland, MA: Sinauer.

Nitrate in Farmland Streams and Groundwater

This technical note also applies to the following indicators:

- **Farmlands: Phosphorus in Farmland Streams**
- **Forests: Nitrate in Forest Streams**
- **Urban/Suburban: Nitrate in Urban Streams**
- **Urban/Suburban: Phosphorus in Urban Streams**

This technical note supplements the technical note for

- **Fresh Waters: Phosphorus in Lakes, Reservoirs, and Large Rivers**

The Indicators

Nitrogen (N) and phosphorus (P) are chemical elements that serve as essential nutrients for plants and animals, but at excessive concentrations they can contaminate groundwater and streams. In surface waters they can promote excessive growth of algae (nitrogen typically causes blooms in coastal waters, whereas phosphorus more commonly causes blooms in freshwater systems), whose decay removes oxygen and threatens aquatic animals. At high concentrations, some forms of nitrogen (e.g., nitrate and ammonia) can be directly toxic to fish and create health problems for humans. In groundwater, excessive nitrate poses a threat to humans who drink from contaminated wells. Common forms of nitrogen that are readily available to plants for growth include nitrate and ammonia, and phosphate is the plant-available form of phosphorus. Sources include precipitation, dissolved natural minerals, farm and domestic fertilizers, discharges from septic systems, and effluents from sewage treatment plants.

Graphs for stream sites show mean-annual concentrations of dissolved nitrate plus nitrite or total phosphorus. Graphs for groundwater data are based on nitrate concentration in one sampling of each well. Data are reported as either parts per million (milligrams per liter) as nitrogen or parts per million (milligrams per liter) as phosphorus. The data are labeled "mean total nitrate" although the analytical method actually reports nitrate plus nitrite. This reporting convention is reasonable because except in highly polluted waters, nitrite levels are only a very small fraction of the total and can, therefore, be considered insignificant.

The Data

Data Source: The data were collected and analyzed by the U.S. Geological Survey (USGS) National Water Quality Assessment (NAWQA) program in 36 major river basins and aquifers distributed across the United States from 1992 to 1998. NAWQA samples watersheds with relatively homogeneous land use/land cover to better illuminate the effect of land use on water quality. For this report, data from watersheds where a single land use typically was predominant were used to characterize water quality conditions in farmlands, forests, and urban settings.

Nutrient data are from 15 to 25 samples collected annually at stream sites draining 105 agricultural, 38 urban and suburban, and 36 forested areas. Nitrate data were from samples collected at 1,190 wells in agricultural, 601 wells in urban and suburban, and 79 wells in forested areas. These data are summarized at http://water.usgs.gov/nawqa. Note that the sites labeled "urban" in this analysis should overlap with the "urban and suburban lands" defined as the subject of this report (see pp. 181), but, since different definitions were used in the two efforts, this might not always be the case.

Information on the drinking water standard for nitrogen can be found at http://www.epa.gov/safewater/mcl.html#inorganic. Information on the 1986 phosphorus recommended goal for preventing excess algae growth can be found in EPA 440/5-86-001 (see references). Information on regional nutrient (phosphorus) criteria can be found at http://www.epa.gov/waterscience/criteria/nutrient/ecoregions.

For farmlands, extensive data have been collected from different farming systems at the watershed-level scale, from 1991 to 2000, that will become available shortly through the National Agricultural Library (http://www.nal.usda.gov/). These data will allow additional investigations of the effect of land use and specific farming practices on water quality.

Data Collection Methodology: All samples were collected and analyzed by USGS according to the overall NAWQA design (Gilliom et al. 1995). Stream water samples were collected using depth and width integrating techniques so that the sample is representative of the water flowing past the sampling point (Shelton 1994). Groundwater samples were collected primarily from monitoring wells and low-capacity domestic wells using procedures that resulted in a sample representative of water in the aquifer (Lapham et al. 1995). Methods employed for random selection of well locations for targeted land use are described by Scott (1989) and Squillace and Price (1996). Methods for sample preservation and processing can be found in Shelton (1994) for stream samples and in Koterba et al. (1995) for groundwater samples. Fishman (1993) and Patton and Truitt (1992) describe analytical methods used for nutrient constituents. Land use in the watersheds upstream of stream sampling points or in the vicinity of wells was characterized according to procedures described in Gilliom and Thelin (1997) and Koterba (1998), respectively.

Data Analysis: The data are highly aggregated and should be interpreted mainly as an indication of general national patterns. The data were collected and analyzed by NAWQA in 36 major river basins and aquifers distributed across the United States from 1993 to 1998. The watersheds and aquifers studied were selected to be generally representative of water and land use in each area. Because this is a national assessment, the percentage of targeted land use varies across the nation. For example, watersheds dominated by agricultural land varied from 10 to 99% as cropland and/or pasture; urban and suburban land varied from 6 to 100%; and forested land ranged from 61 to 100%. Water quality is affected by both the percentage of land use in a watershed and the proximity of that land use (as a source of contamination) to streams and rivers. For example, agricultural or urban/suburban land uses might exert a dominant influence on a stream or river, in spite of occupying a small percentage of land cover in the watershed, if these land uses are located in close proximity to the river or stream.

Data Quality/Caveats: Sampling sites were selected to be representative of specific land use types rather than locations where contamination was known or suspected. All samples were collected, processed, preserved, and analyzed using the same methods. Nutrient data were reviewed to identify outliers and inconsistent results by the teams who collected the samples and by a national team (Mueller 1998). Most data have been published by USGS in a series of technical reports focusing on specific study areas and in national summary results (see http://water.usgs.gov/nawqa for a list of reports).

Data Access: All data used in this document are summarized at http://water.usgs.gov/nawqa.

References

Fishman, M.J. 1993. Methods of analysis by the U.S. Geological Survey National Water Quality Laboratory—Determination of inorganic and organic constituents in water and fluvial sediments. U.S. Geological Survey Open-File Report 93-125.

Gilliom, R.J., W.M. Alley, and M.E. Gurtz. 1995. Design of the National Water-Quality Assessment Program: Occurrence and distribution of water-quality conditions. U.S. Geological Survey Circular 1112.

Gilliom, R.J., and G.P. Thelin. 1997. Classification and mapping of agricultural land for National Water-Quality Assessment. U.S. Geological Survey Circular 1131.

Koterba, M.T. 1998. Ground-water data-collection protocols and procedures for the National Water-Quality Assessment Program: Collection, documentation, and compilation of required site, well, subsurface, and landscape data for wells. U.S. Geological Survey Water-Resources Investigations Report 98-4107.

Koterba, M.T., F.D. Wilde, and W.W. Lapham. 1995. Groundwater data-collection protocols and procedures for the National Water-Quality Assessment Program: Collection and documentation of water-quality samples and related data. U.S. Geological Survey Open-File Report 95-399.

Lapham, W.W., F.D. Wilde, and M.T. Koterba. 1995. Groundwater data-collection protocols and procedures for the National Water-Quality Assessment Program: Selection, installation, and documentation of wells, and collection of related data. U.S. Geological Survey Open-File Report 95-398.

Mueller, D.K. 1998. Quality of nutrient data from streams and ground water sampled during 1993–95—National Water-Quality Assessment Program. U.S. Geological Survey Open-File Report 98-276.

Mueller, D.K., J.D. Martin, and T.J. Lopes. 1997. Quality-control design for surface water sampling in the National Water-Quality Assessment Program. U.S. Geological Survey Open-File Report 97-223.

Patton, C.J., and E.P. Truitt. 1992. Methods of analysis by the U.S. Geological Survey National Water Quality Laboratory—Determination of total phosphorus by Kjeldahl digestion method and an automated colorimetric finish that includes dialysis. U.S. Geological Survey Open-File Report 92-146.

Scott, J.C. 1989. A computerized data-base system for land-use and land-cover data collected at ground-water sampling sites in the pilot National Water-Quality Assessment Program. U.S. Geological Survey Water-Resources Investigations Report 89-4172.

Shelton, L.R. 1994. Field guide for collecting and processing stream-water samples for the National Water-Quality Assessment Program. U.S. Geological Survey Open-File Report 94-455.

Squillace, P.J., and C.V. Price.1996. Urban land-use study plan for the National Water-Quality Assessment Program. U.S. Geological Survey Open-File Report 96-217.

U.S. Environmental Protection Agency. 1986. Quality criteria for water-1986. U.S. Environmental Protection Agency Report EPA 440/5-86-001.

Phosphorus in Farmland Streams

See the technical note for Nitrate in Farmland Streams, p. 232.

Pesticides in Farmland Streams and Groundwater

See the technical note for the core national contaminants indicator, p. 210.

Soil Organic Matter

The Indicator

Soil organic matter would be reported as the percentage of organic matter (dry weight) in the upper soil profile (4–6 inches). The data would be presented as a percentage of all croplands having several ranges of percent organic matter, on a national basis and on a regional basis for the latest year for which data are available. It should be noted that it may prove difficult to discern trends in organic matter using the coarse ranges chosen (less than 2%, 2 to 4%, and greater than 4%); an approach addressing change on the regional or local level may be necessary.

Soil organic matter content in the upper soil profile (4–6 inches) was chosen because human activity, particularly management practices, has had its greatest impact here. Soil organic matter content is related to the cation exchange capacity of the soil, soil water-holding capacity, nitrogen mineralization rates, and microbial activity.

Soil organic matter content is also related to biogeochemical processes, and the cycling of carbon and nitrogen within the upper soil profile is related to soil carbon content. Measurement of changes in the soil organic matter content over time provides a quantitative assessment of the soil capacity to support crops and other plant and animal life.

Soil organic matter content is a critical component of soil structure and is vital to all soil processes. Soil organic matter provides the chemical and biological basis for soil components (sand, silt, and clay) to form soil aggregates and is critical in key physical processes (such as water and gas exchange, penetration resistance, and compaction). Differences in climate, parent material, and management history have produced large regional differences in soil organic matter content.

In addition, since soil organic matter is about 60% carbon, the amount of organic matter is a predictor of the amount of carbon in soils. Storage of carbon in soils has become important in international negotiations on the management of greenhouse gas emissions, as increased carbon storage can be useful in offsetting emissions of carbon from fuel burning and other sources. In order to be of use to such negotiations, this indicator would probably need to measure carbon in the upper 3 feet of soil, not just the upper 4–6 inches. While this is not the current focus of the indicator, such a presentation would make this indicator analogous to those in the forest and grasslands and shrublands chapters (pp. 123 and 165).

The Data Gap

U.S. Department of Agriculture soil survey data (contained within the State Soil Geographic Database [STATSGO] and Soil Survey Geographic [SSURGO] datasets) provide an initial county-level estimate of soil organic matter content, but there are no programs in place to monitor and report soil organic matter content on a national basis. Universities and other research institutions have carried out observations of the changes in soil organic matter content under different management practices, but the results of these investigations do not provide national coverage. The STATSGO database is available at http://www.ftw.nrcs.usda.gov/stat_data.html and SSURGO is available at http://www.ftw.nrcs.usda.gov/ssur_data.html.

Soil Erosion

The Indicator

This indicator presents the percentage of U.S. cropland (minus pastures, but including Conservation Reserve Program [CRP] acreage) in each of three categories of land condition (least prone, moderately prone, and most prone to erosion), based on both inherent soil properties and management practices, for 1982, 1992, and 1997, for both wind and water erosion. Also, those lands most prone to wind and water erosion are mapped.

Soil erosion is affected both by the inherent properties of the soil, landscape, and region (e.g., slope, soil type, rainfall) and by management factors that may change more rapidly (specifically, the use of terracing, wind barriers, and the type, amount, and duration of vegetative cover). Soils with higher inherent likelihood of eroding and with high vulnerability due to the way they are managed are likely to erode the most. (Enrollment of these acres into the CRP, which requires steps toward reducing erosion (e.g., planting perennial grasses), will lead to improvement of this indicator.) Conversely, soils with low inherent likelihood of eroding and low vulnerability because of good management are likely to erode least.

Categories for this indicator were developed using parameters measured for use in the Universal Soil Loss Equation (USLE) and Wind Erosion Equation (WEQ). These equations were developed to predict long-term average erosion based on measurements of the inherent soil and plot features and management and surface treatment factors. For water erosion (USLE), inherent soil and plot factors are R, rainfall and runoff; K, soil erodibility; and L and S, topographic factors related to slope steepness and length of slope. Management and surface treatment factors included C, cover management, which essentially measures whether and how much vegetative cover is left on the soil surface, and P, support practice factor, which measures whether there are features such as terraces. The equation form is A (annual soil erosion per unit area) = C*P*R*K*L*S. For wind, the inherent soil and plot factors are I, soil erodibility index, and C, climatic factor. Management and surface treatment factors are K, ridge roughness; L, unsheltered distance along the prevailing wind direction; and V, vegetative cover. Wind erosion, E (annual soil erosion per unit area), is a function of I, K, C, L, and V (see references for more details).

This report uses the underlying principles of these equations to identify cropland area with combinations of inherent soil properties and management practices that are likely to erode most and least. Though inherent soil properties change slowly or not at all, management practices can significantly reduce erosion. Thus, reductions in acreage with high propensity to erode result primarily from application of management practices that reduce erosion, including removal of acreage from cultivation, such as CRP.

Areas with the least susceptibility to both wind and water erosion ("least prone") are generally those with a predicted erosion rate of less than 1 ton per acre per year. Areas with the greatest susceptibility to erosion ("most prone") are those with a predicted erosion rate of 3 tons per acre or more. Areas with moderate susceptibility to erosion have predicted values between about 1 and 3 tons per acre per year.

Standard application of both USLE and WEQ uses the equations to predict total erosion, in tons per acre. In this report, we have chosen not to take this last step in the process. We do so because we believe taking this step overstates actual erosion, as the USLE does not account for deposition, only the initiation of soil movement. Some soil particles move only very short distances, and when erosion is reported in units of "tons per acre" there is a strong implication (and sometimes an explicit statement) that these tons of soil are lost from the farm field.

The WEQ estimates how much eroding soil leaves the downwind edge of the field, in tons per acre per year.

The Data

Data Source: Acreage estimates for lands in each of the three categories were developed using data provided by USDA's Natural Resources Conservation Service, from the National Resources Inventory (NRI). For information on NRI methods, applicability of results, and access to information, see the technical note for Total Cropland, page 229.

References

Bondy, E., Lyles, L., and Hayes, W.A. 1980. Computing soil erosion by periods using wind energy distribution. Jour. Soil and Water Conserv. 35(4):173–176.

Skidmore, E.L., and N.P. Woodruff. 1968. Wind erosion forces in the United States and their use in predicting soil loss. Agriculture Handbook No. 346.

Woodruff, N.P., and Siddoway, F.H. 1965. A wind erosion equation. Soil Sci. Soc. Amer. Proc. 29(5):602–608.

Soil Salinity

The Indicator

This indicator would be reported as the percentage of croplands nationally having one of three salinity levels (less than 2 decisiemens per meter [dS/M], 2 to 4 dS/m, and greater than 4 dS/M; see below for discussion/description). In addition, the percentage of croplands with elevated soil salinity (over 4 dS/m) would be mapped on a Major Land Resource Area (MLRA) basis. (MLRAs are aggregations of geographic areas, usually many thousand acres in extent, which are characterized by a particular pattern of soils, climate, water resources, and land use. See http://www.statlab.iastate.edu/soils/MLRAweb/mlra/ for a discussion and map.)

Salinization is the process by which salts accumulate in the soil. Soil salinity hinders the growth of plants by limiting their ability to take up water. Soluble salts, particularly sodium salts, may also harm soils by reducing soil structure, tillage properties, and permeability to water.

Soil salinization is most often associated with irrigated agriculture because when water is applied to the land to nourish crops, much of it is taken up by plants (or evaporating directly from the soil surface) and is returned to the atmosphere. Since only pure water evaporates from the soil surface or transpires from the plant surfaces, the salts are left behind in the soil. Thus, irrigation has the potential to lead to excess accumulation of salts in the soil. The occurrence of saline soils, however, is not restricted to irrigated soils. The same processes of mineral weathering or dissolution and subsequent concentration because of water evaporation often lead to high salt levels in soils of arid and semiarid regions. The scarcity of rain that makes these areas arid restricts the possibility of leaching and thus leads to salt accumulation. A special case of dryland salinity of particular concern to the northern Great Plains is that of saline seeps. A saline seep occurs when water in excess of that required by plants percolates below the root zone and, upon encountering some type of barrier or restricting layer, moves laterally downhill and emerges in a seepage area, having picked up dissolved salts in transit. Saline seeps are often encountered where farmers practice a wheat-fallow rotation; during dry periods, such a rotation may serve to conserve some water during the non-cropped period to aid the following crop, but in somewhat wetter years, the precipitation in excess of that required by plants initiates the process that leads to a seep. Drainage from saline seeps is estimated to affect about 2.5 million acres in the northern Great Plains.

Soluble salts in soils are measured by determining the electrical conductivity of a saturated paste extract; the units of conductance are reported as dS/m. Few plants are affected when the extract conductivity is below 2 dS/m, while some sensitive plants are affected when values are between 2 and 4 dS/m. Many plants are affected when values are above 4, and few plants can survive at values greater than 16 dS/m. Salts are usually most damaging to young plants, but not necessarily at the time of germination, although high salt concentrations can slow or inhibit seed germination. Most plants are least affected by soil salts when in their mature stages.

Reduced permeability to water is a common problem with salt-affected soil. Soil porosity becomes gradually altered and some soils can become completely impermeable. The mechanisms responsible are swelling of clays, which reduces pore sizes, and dispersion of the soil, so that aggregates break down, and smaller mineral and organic particles move with water and begin to fill smaller pore spaces. Dispersion is the most frequent cause of reduced infiltration. The measurement that most accurately determines whether the soil is affected by soluble salts is the exchangeable sodium percentage, which expresses the portion of the total exchangeable cations that are sodium. An exchangeable sodium percentage value equal to or greater than 15 indicates a sodic soil.

The Data Gap

Soil salinity measurements are needed on dominant soils, on cropping patterns, and particularly on water management practices under both irrigated and non-irrigated conditions in arid and semiarid regions. Salinity measurements are often included in routine soil tests. However, there is no unified effort in place to collect and analyze the results over uniform regions. A program that can monitor changes over time as a function of soils and management practices is vitally needed.

Soil salinity measurements should include data on dominant soils, cropping patterns, and, particularly, water management practices such as irrigation and drainage. Gathering together the existing but fragmented data, collecting new data, and analyzing the results to ensure national coverage require a coordinated effort. Satellite-based technologies, while promising, are able to detect only visible salt deposits. Since visible surface salts are incorporated into the soil by tilling, these approaches may be of use primarily to complement soil testing.

Soil Biological Condition

The Indicator

The Nematode Maturity Index (NMI) is a weighted mean frequency of taxa assigned weights ranging from 1 to 5, with a smaller weight being assigned to taxa with relative tolerance to disturbance and a larger weight to taxa that are more sensitive to disturbance. The index combines both free-living and plant-parasitic nematodes but excludes taxa that simply respond ephemerally to added nutrients. This index can detect differences among fields in a regional survey more reliably than one that measures only free-living nematodes (Neher and Campbell 1996). (See references for a variety of publications that support the use of soil organisms, particularly nematodes, as indicators of soil quality.)

This index is based on the principle that different taxa have different sensitivities to stress or disruption of the successional sequence because of differences in their life history characteristics. Because succession may be disrupted at various stages by common agricultural practices, such as cultivation and applications of fertilizer and pesticides, the successional status of a soil community may reflect the history of disturbance. However, although a disturbance, such as the addition of animal manure to soil, initially produces a predominance of nematodes with smaller values, the abundance of nematodes with large maturity index values soon increases.

Maturity indices have the strength of responding to a variety of land management practices across plant species, soil types, and seasons (Neher et al. 1995). Nematode community structure and function are known to change in response to land management practices such as nutrient enrichment through fertilization by organic or inorganic nitrogen, cultivation, liming, and drainage, as well as to changes in plant community composition and age and to toxic substances such as heavy metals, pesticides, and petroleum products.

The Data Gap

Sampling should be carried out in autumn after cultivation of fields harvested in the fall; this will minimize within-field sampling variation. Free-living nematode populations are generally at their peak at this time because crop residues are incorporated into soil by cultivation and temperatures are moderate.

Cobb's sieving and sugar centrifugal-flotation methods are recommended to optimize recovery of entire nematode communities from soil (Neher et al. 1995). Neher et al. (1998) suggest that it is unnecessary to calibrate indices of nematode community structure at a scale finer than the USDA's Land Resource Regions.

References

Bernard, E.C. 1992. Soil nematode biodiversity. Biology and Fertility of Soils 14:99–103.

Blair, J.M., P.J. Bohlen, and D.W. Freckman. 1996. Soil invertebrates as indicators of soil quality, pp. 273–291. In Methods for assessing soil quality. Soil Science Society of America Special Publication 49, Madison, WI.

Freckman, D.W. 1998. Bacterivorous nematodes and organic-matter decomposition. Agric. Ecosystems Environ. 24:195–217.

Gupta, V.V.S.R., and G.W. Yeates. 1997. Soil microfauna as bioindicators of soil health, pp. 201–203. In C. Pankhurst, B.M. Doube, and V.V.S.R. Gupta (eds.), Biological indicators of soil health. New York: CAB International.

Hendrix, P.F., R.W. Parmelee, D.A. Crossley, Jr., D.C. Coleman, E.P. Odum, and P.M. Groffmann. 1986. Detritus food webs in conventional and no-tillage agroecosystems. BioScience 36:374–380.

Hunt, H.W., D.C. Coleman, E.R. Ingham, R.E. Ingham, E.T. Elliot, J.C. Moore, S.L. Rose, C.P.P. Reid, and C.R. Morley. 1987. The detrital food web in shortgrass prairie. Biology and Fertility of Soils 3:57–68.

Neher, D.A., and C. L. Campbell. 1996. Sampling for regional monitoring of nematode communities in agricultural soils. Journal of Nematology 28:196–208.
Neher, D.A., K.N. Easterling, D. Fiscus, and C.L. Campbell. 1998. Comparison of nematode communities in agricultural soils of North Carolina and Nebraska. Ecological Applications 8:213–223.
Neher, D.A., S.L. Peck, J.O. Rawlings, and C.L. Campbell. 1995. Measures of nematode community structure for an agroecosystem monitoring program and sources of variability among and within agricultural fields. Plant and Soil 170:167–181.
Yeates, G.W., and T. Bongers. 1999. Nematode diversity in agroecosystems. Agriculture Ecosystems & Environment 74:113–135.

Status of Animal Species in Farmland Areas

The Indicator

As discussed on the indicator page, there are multiple approaches to reporting on animal species in farmland areas. One might be to report on the status of species that favor those habitats that existed before farmland was created in an area. Such an approach would, for example, focus on grassland birds in areas of the Great Plains—species that inhabited prairies that have now been converted to farmland. Another approach might be to focus on species that are able to take advantage of farmland landscapes—many game birds and small mammals, for example. Both of these approaches would be useful, but by themselves would be incomplete.

A more appropriate approach, recommended here, would be to focus on the full breadth of species that might inhabit farmlands. To follow the examples above, this would include both grassland birds and game birds and small mammals. Such an approach has been suggested, based on expectations that one might encounter a variety of birds in different regions of the nation. An index could be developed based on comparing this expectation with data on the presence of birds on farmlands in that region—data that may already be available for a significant percentage of farmlands (Breeding Bird Survey, http://www.mbr-pwrc.usgs.gov/bbs/bbs.html; additional information on such an approach toward determining an index of bird "integrity" can be found at htpp://landscape.forest.wisc.edu/LandscapeEcology/Articles/v7i2p137.pdf).

Several reviewers of this report recommended that this indicator focus on domestic animals—their numbers, condition, diversity, and the like. The Farmlands Work Group determined that it was appropriate to focus on the status and trends in wild species as part of this measure (which is intended to describe ecosystem conditions). A measure describing domestic animals would have been appropriate as part of the "human use" set of indicators, but was determined not to be of sufficiently high priority for inclusion.

The Data Gap

There are two major national-scale sources of information on species population status and trends. These include NatureServe's compilation of information from state-based Heritage programs, which provides status information on a global, national, and state basis (www.natureserve.org) for a large number of species, and the U.S. Geological Survey's Breeding Bird Survey (http://www.mbr-pwrc.usgs.gov/bbs/bbs.html), which provides population trend information for a large number of resident birds of North America.

Both programs provide information on a geographic scale that is usually larger than and is not limited to farmlands. Thus, it is likely that it would be necessary to undertake additional work to target these data only to farmlands.

Native Vegetation in Areas Dominated by Croplands

The Data Gap

The technical note for the grasslands/shrublands indicator on non-native plant cover (see p. 261) describes some of the public and private efforts under way to determine the extent of non-native plant cover, which would be useful for inferring the coverage of native species.

The technical note for the farmlands extent indicator (p. 229) describes sources of information on the extent and location of croplands.

Stream Habitat Quality

This technical note is also for the stream habitat indicator for the freshwater system.

The Indicator

The habitat quality of streams and rivers is dependent upon the presence of an appropriate, but changing, mix of habitat features. Key among these are the presence of riffles and pools, size distribution of streambed sediments and embeddedness (degree to which larger gravel and cobbles are buried in silt), amount of large woody debris, and bank stability, although different stream habitat rating methods may measure additional characteristics.

In addition, habitat quality is a relative value, meaning that it must be evaluated in relation to the habitat needs of the native flora and fauna in a region. Therefore, protocols to measure stream habitat quality generally provide for calibration according to a regional reference—that is, stream habitat quality is measured against the values that would be found in a relatively undisturbed or "natural" reference stream in that region. Finally, all stream habitat quality measurement protocols measure a variety of parameters, but not all combine these parameters into a single overall index.

Stream habitat quality measurement is an area of significant current research work. Following are references for four efforts that have or are developing regional or national protocols for this purpose.

References

Maryland Department of Natural Resources; Maryland Biological Stream Survey. http://www.dnr.state.md.us/streams/mbss
U.S. Environmental Protection Agency. Rapid bioassessment protocol. Developed guidance to state agencies. http://www.epa.gov/owow/monitoring/rbp/
U.S. Environmental Protection Agency, Office of Research and Development, National Health & Environmental Effects Research Laboratory. May 2000. Biological indicators for monitoring riparian forest condition. Proceedings of a workshop. Corvallis, OR. EPA-600-R-00-048. COR-00-077.

U.S. Geological Survey. National Water Quality Assessment Program. Stream habitat monitoring protocol. http://water.usgs.gov/nawqa/protocols/OFR-93-408/habit1.html

Major Crop Yields

The Data

Data Source: Data for 1950 through 1998 are from the U.S. Department of Agriculture National Agricultural Statistics Service (NASS), Historical Track Records, United States Crop Production, May 2001. The historical data can be located by using the Crop Production Historical Records link at http://www.usda.gov/nass/pubs/histdata.htm. Data for 1999 and 2000 are from USDA-NASS, Agricultural Statistics 2001; http://www.usda.gov/nass/pubs/agr01/acro01.htm.

Data Collection Methodology: State offices collect and estimate crop yield data from sample surveys of farmers and their business associates (farm service agencies, cotton gins, marketing associates). NASS obtains the yield estimates, which are verified and analyzed on a national level. Survey data are supplemented by information from the Census of Agriculture, which is carried out every 5 years.

Data Manipulation: Yields, which are generally reported as bushels per acre for corn, soybeans, and wheat, and as tons per acre for hay and cotton, were divided by their respective value for 1975. Thus, values above 1.0 indicate higher yields than in 1975, and values lower than 1.0 indicate lower yields than in 1975.

Data Access: See the Web sites listed under "Data Source."

Agricultural Inputs and Outputs

The Indicator

This indicator presents ratios of certain major inputs identified and quantified by the U.S. Department of Agriculture (USDA) to total agricultural outputs, also as quantified by USDA. The intent is to report the amount of inputs needed to produce a unit of output, because the quantities of, and tradeoffs between, individual inputs (such as pesticides and fertilizer) are important. For example, if decreasing amounts of fertilizer are required to produce a unit of output, this has implications for the cost of production (fertilizer is a significant cost) and for off-farm environmental impacts (excess fertilizer can contribute to water pollution).

The Data

Data Source: Data came from *Agricultural Productivity in the United States* published by USDA's Economic Research Service (ERS).

Data Collection Methodology: The output data represent all agricultural outputs, including animals and animal products (meat animals, dairy products, poultry, and eggs) and crops (food grains, feed crops, oilseed crops, sugar crops, cotton and cotton seed, vegetables and melons, and fruit and tree nuts). Aggregation of multiple outputs or inputs into a single index often requires assumptions about the comparability of unlike things—adding tons of corn to tons of strawberries would be nonsensical. USDA economists use an approach that involves determining the adjusted price of a given output, which is multiplied by the output quantity, so that all outputs can be added together into the single value shown here. ERS developed a similar scheme for adding inputs together; however, because the focus in this indicator is on changes in different inputs as well as the overall amount of inputs, the individual inputs are presented here. The yearly quantity of each input has been adjusted to some extent by ERS to reflect the changes in quality. For example, similar results can now be achieved with smaller quantities of pesticides. Thus, a larger quantity of less effective pesticide might be treated as equal to a smaller quantity of a more effective pesticide. The same is true for the other inputs, such as labor, whose quantities have been quality-adjusted over time.

Data Manipulation: Each input has been divided by the total farm output for that year. The data from ERS are all relative to a given year (1948) and are not reported as actual quantities. Because the focus of this report is on 1950–2000, we chose the midpoint (1975) as a more appropriate index year. Because of this, data (inputs and outputs) were simply divided by the 1975 value. All input data were then divided by the value of total outputs for any given year to produce the data shown in the figure.

Data Access: The data are available at http://www.ers.usda.gov/publications/aib740/; a more detailed version of the data is available at http://www.ers.usda.gov/data/sdp/view.asp?f=inputs/98003/.

Monetary Value of Agricultural Production

The Indicator

The gross value of agricultural production is a measure of the physical output of major crops and livestock multiplied by price (in dollars) received by producers. (The values have all been converted to 1999 dollars.)

The geographic distribution of agricultural sales is a measure of gross sales by crop and livestock producers per square mile. These data do not reflect payments received by producers through government income support, commodity, or conservation programs, nor do they reflect economic activity associated with food processing and distribution or off-farm service and supply businesses.

The Data

Data Source: Data on the dollar value of agricultural sales are from the U.S. Department of Agriculture (USDA) Economic Research Service (ERS), which reports farm income and farm cash receipts.

Data for agricultural sales per square mile are from the U.S. Department of Commerce, Bureau of Economic Analysis (BEA), Regional Economic Information System branch (http://www.bea.doc.gov/bea/regional/reis/), which calculates county cash receipts.

Data Collection and Manipulation (Dollar Value of Agricultural Sales): The USDA National Agricultural Statistics Service (NASS) conducts national surveys that measure acres planted and harvested, yields, production, and market prices. The estimates include cash receipts from the marketing of about 150 crop and livestock commodities.

ERS uses NASS-published, calendar-year cash receipts for major livestock and commodity-producing states. ERS develops indexes to indicate direction and magnitude of changes in monthly sales quantities and multiplies them by NASS-published monthly prices. Data for other states are developed in cooperation with the NASS state offices, which use all available sources, including informed opinions, often corroborated by data from state survey programs, producer associations, and the state's extension service. California data come from state-conducted surveys.

ERS adjusts NASS quantity and value of production data for major crop commodities in major producing states to adjust for production of feed used on farms for livestock and for Commodity Credit Corporation sales and to account for the fact that some sales do not take place in the same year as the crop is harvested. Data from NASS that cannot be released to the public because of confidentiality constraints are included in the overall ERS dataset.

Data were adjusted for inflation using the Gross Domestic Product Implicit Price Deflator (IPD) provided by the Economic Research Service. All data were adjusted to the average level of prices that existed in 1999. The following formula was used to convert each figure in the series from current dollars to constant dollars (available at http://www.owlriver.com/pic.mhsc.org/DataPages/sd-079.htm).

$$\text{Year Z constant dollar value} = \frac{\text{(Year Z current dollar value)*(Base year IPD index number)}}{\text{Year Z IPD index number}}$$

Data Collection and Manipulation (Agricultural Sales per Square Mile): The U.S. Department of Commerce's BEA uses a variety of data sources to develop county-level estimates of farm receipts. For 16 major producing states, NASS-affiliated state offices prepare annual county estimates of farm cash receipts. For other states, state-level cash receipts estimates produced by NASS are allocated by BEA to counties in proportion to the corresponding Census of Agriculture data for the relevant year.

These county-level data were used to produce county-level estimates of cash receipts per square mile by dividing total cash receipts by the number of square miles in a county. County area data are from a standard dataset produced by Environmental Systems Research Institute, Inc. (ESRI; http://www.esri.com), a maker of geographic information system software and data products.

Data Access: Data on U.S. national farm cash receipts for 1924–1999 are available online at http://www.ers.usda.gov/data/farmincome/finfidmu.htm.

The U.S. county cash receipts data can be requested through BEA, Regional Economic Information System branch (http://www.bea.doc.gov/bea/regional/reis/).

Recreation on Farmlands

There is no technical note for this indicator.

Forests

Forest Area and Ownership

The Data

Data Collection Methodology: The USDA Forest Service's Forest Inventory and Analysis (FIA) program authority is mandated under the Forest and Rangeland Renewable Resources Research Act of 1978 (PL 95–307). Since the late 1940s, FIA has used a two-phase sample (generally, double sampling for stratification) to collect information on the nation's forests. Phase one establishes a large number of samples (more than 4 million, roughly every 0.6 miles). These are selected using aerial photographs or other remote-sensing images, which are then interpreted for various forest attributes. Phase two establishes a subset of approximately 450,000 phase-one points (roughly every 3 miles) for ground sampling. About 125,000 of these samples are permanently established on forest land. The forest characteristics measured include ownership, protection status, species composition, stand age and structure, tree growth, occurrences of mortality and removals, tree biomass, incidences of pathogens, natural and human-caused disturbances, and soil descriptors.

Forest land is any land that is at least 10% stocked by forest trees of any size, including land that formerly had such tree cover and that will be naturally or artificially regenerated. The minimum area for classification of forest land is 1 acre. For the forest area and ownership indicator, public forests include those owned by federal, state, and local governments, as well as other public entities such as the Tennessee Valley Authority. Private lands include those owned by individuals, corporations, nongovernmental organizations, and tribes. The Forest Service's FIA program derived estimates of historic forest area from a wide variety of sources. For example, the sources included forest-clearing data collected during the 1870 and later decennial censuses, limited state and regional surveys, and the expert opinion of resource professionals.

Data Manipulation: Raw data from the 125,000 field samples are processed and merged with information from the remote-sensing phase of the sampling procedure to provide statistically reliable estimates of area and ownership.

Data Quality/Caveats: FIA surveys provide forest area data with a reliability of ±3–10% per 1 million acres (67% confidence limit). This standard applies to all data reported for 1953 and later. Regional totals generally have errors of less than ±2%. No error estimate is provided for data from before 1953. Note also that data collected before 1953 come from a wide variety of sources (see above).

Data Access: All data are available free of charge except for products that require special processing or shipping fees. Electronic databases are unavailable at the national level prior to 1987, and most regional data from before 1977 are not available electronically. Forest statistics, online databases, and a map of U.S. forest distributions are on the Web at http://fia.fs.fed.us. The data provided here also are available in Smith et al. (2001).

References

Smith, W.B., J. Vissage,D. Darr, and R. Sheffield. 2001. Forest statistics of the United States, 1997. Gen. Tech. Rep. NC-219. St. Paul, MN: U.S. Department of Agriculture, Forest Service. 191p.

Forest Types

Please see the technical note for Forest Area and Ownership (see p. 239), which also serves as the technical note for the Forest Types indicator.

Forest Management Categories

The Data

The data for this indicator were collected by the Forest Service's Forest Inventory and Analysis (FIA) program, which is described in the technical note for Forest Area and Ownership (p. 239). These data do not include information on private lands that are legally reserved from harvest, such as lands held by private groups like The Nature Conservancy. In addition, many "natural" and "semi-natural" lands are at times reserved from harvest because of administrative or other restrictions.

We hope that, in future reports, it will be possible to report on the existence of protected or reserved areas on a broader range of land ownerships. One dataset being developed for this purpose will report the acreage of lands according to a system of categorizing management intensity developed by the U.S. Geological Survey Gap Analysis Program (see http://www.gap.uidaho.edu/handbook/Stewardship/default.htm). This database is currently under development by the Conservation Biology Institute in conjunction with the USDA Forest Service; see http://www.consbio.org/cbi/what/pad.htm.

Note that "interior Alaska" includes all forests except the Southeast Coast area up to and including the Kenai Peninsula. Thus, "interior" includes areas that may not be thought of as part of Alaska's interior, yet they are included because of their remoteness. The acreage shown here for interior Alaska (about 113 million acres) does not include the Tongas National Forest (about 12 million acres). Note also that there is an apparent drop in interior Alaska acres in 1997; however, the 2-million-acre decrease came about from a reclassification, not a true loss of forest.

Forest Pattern and Fragmentation

The Indicator

As a means of illustrating the amount of forest providing different degrees of distance from non-forest cover, this indicator provides information on the percentage of forest surrounded by small, medium, and larger "neighborhoods" (defined below) containing at least 90% forest.

The "percentage of forest" that meets a certain set of criteria is calculated by determining what fraction of "pixels" (squares of forest 30 meters, or about 100 feet, on a side) is in the center of a "window" that meets the criteria. Thus, the percentage of forest that has 90% or more forest cover within a radius of about 250 feet (the "immediate neighborhood," about 5 acres) is determined by counting the number of pixels that are in the center of a 5-acre window that contains at least 90% forest.

The Data

Data Source/Collection Methodology: Data for this indicator were prepared by Kurt Riitters, USDA Forest Service (see http://www.srs.fs.fed.us/4803/landscapes/). The data are based on the National Land Cover Dataset, which is described in more detail in the technical note for the national extent indicator, p. 207). This is a 30-meter resolution remote-sensing-based dataset that provides, among other things, forest/non-forest cover information for the lower 48 states. The unit of data is the pixel, which is a square approximately 30 meters on a side.

Data Manipulation: The data presented here are from a "moving window" analysis. In this approach, the algorithm describes many successive, overlapping "windows" of a certain size, making it possible to characterize the area surrounding each individual forest pixel, in addition to knowing its forest/non-forest status. As the window "moves" across the dataset, each pixel is used as the center of a window; thus, it is possible to determine how many forest pixels are surrounded by different amounts of forest.

Five window sizes were used for this analysis but only three are reported here. The three reported sizes are 2.25 hectares, referred to here as the "immediate neighborhood," 5 acres, or "within a radius of about 250 feet"; 65.61 hectares, referred to here as the "local neighborhood," 160 acres, or "with a radius of about one-quarter mile"; and 5314. 41 hectares, referred to here as the "larger neighborhood," 13,000 acres, or "within a radius of about one and a half miles." These sizes correspond to 25 pixels (a square of 5 x 5 pixels); 729 pixels (a square of 27 x 27 pixels) and 59,049 pixels (a square of 243 x 243 pixels). The other two window sizes were 7.29 hectares and 590.5 hectares. (Note: This analysis uses a square window, since each remote sensing pixel is square. Thus, the page text description of the "radius" of the "neighborhood" is an approximation to make the presentation clearer to a non-technical audience, and is written as if the window were round.)

The analysis on which the data presented here was based determines, for each pixel and window size, whether it is surrounded by at least 60% forest, at least 90% forest, or exactly 100% forest. For this report, the 90% criterion was chosen. The 90% criteri-

Table 3. Degrees of Forest Cover for Different "Window" Sizes

Window area (hectares)	West			East			East & West		
	% core	% interior	% connected	% core	% interior	% connected	% core	% interior	% connected
2.3	51.0	*64.1	85.4	59.0	*70.7	88.6	56.5	68.7	87.6
7.3	32.7	55.7	83.3	40.7	62.4	86.3	38.2	60.4	85.4
66	7.1	*44.0	78.5	11.1	*48.2	80.0	9.9	46.9	79.6
590	0.2	33.1	75.0	0.9	34.8	75.2	0.7	34.3	75.1
5310	0.0	*22.4	70.5	0.0	*25.1	72.6	0.0	24.2	71.9

on was selected based on considerations of data quality and previous experience with this analytical approach. The alternate interpretations, along with a detailed description of the methodology, are described in detail in K.H. Riitters et al. (submitted).

Table 3 presents the results of the full analysis, including all window sizes and all three degrees of forest cover. As in the original publication, the table uses the term "core" to refer to areas surrounded by 100% forest cover for the indicated window size, "interior" to refer to areas surrounded by at least 90% forest cover for the indicated window size, and "connected" for areas surrounded by at least 60% forest cover for the indicated window size. Data presented in the body of the report are indicated with an asterisk.

The satellite remote-sensing data presented here can, in theory, distinguish non-forest areas as small as 100 feet on a side (10,000 square feet) from adjacent forest pixels. In practice, the accuracy of doing this depends on the contrast between forest and non-forest land cover, which is, in general, quite good. In addition, geometry plays an important role in distinguishing non-forest land cover. For example, a clearing that fills several 100-foot by 100-foot pixels would probably be more easily detected than a winding road that may fill some pixels and only partially fill others.

For further reading on habitat fragmentation, see other related indicators in this document and also Noss and Csuti (1997) and Wilcove et al. (1986).

References

Noss, R.F., and B. Csuti. 1997. Habitat fragmentation, pp. 269–304. In G.K. Meffe and R.C. Carroll (eds.), Principles of conservation biology. Second edition. Sunderland, MA: Sinauer Associates.

Riitters, K.H., et al. Fragmentation of continental United States forests. Submitted to Ecosystems.

Wilcove, D.S., C.H. McLellan, and A.P. Dobson. 1986. Habitat fragmentation in the temperate zone, pp. 237–256. In M.E. Soulé (ed.), Conservation biology: The science of scarcity and diversity. Sunderland, MA: Sinauer Associates.

Nitrate in Forest Streams

See the technical note for Nitrate in Farmland Streams, p. 232.

Carbon Storage

The Indicator

Metric tons of carbon are measured for the following components: biomass (total live tree material above ground plus coarse roots—"trees" in the figure), soil (soil organic matter), and dead plant material and coarse woody debris on the forest floor (forest floor litter). The weight of organic materials in plants is approximately 50% carbon.

Determining the amount of carbon stored in a forest can change dramatically within a few days, such as following a fire or timber harvest. Such fluctuations are natural in most forests and do not provide a useful indication of forest condition. Carbon lost during long-term agricultural use of soils can be restored by reforestation. In fact, the steady increase of carbon stored in eastern forests is a reflection of the re-establishment of forests on abandoned agricultural lands.

Carbon storage does not necessarily end when harvest occurs. Some wood products are used in long-term applications such as housing. Other products (e.g., newspapers) may end up in landfills, thus storing carbon for long periods of time. Landfills also generate methane (a carbon-rich greenhouse gas), so they both store and release carbon.

The Data

Data Sources: The information presented here is from the USDA Forest Service Forest Inventory and Analysis (FIA) program (see p. 239) and is based on field estimates of the size of trees of various species, along with statistical models of the relationships between tree stem volume and the other components of carbon storage. Carbon contained in branches, leaves, the forest floor, and soil are estimated from, and are therefore less precise than, data for harvestable wood. Although extensive, the field measurements used as the basis for this indicator do not include national parks and wilderness areas or slower-growing forests. Expansion to these areas is currently planned.

Data Collection Methodology and Data Manipulation: Carbon storage is estimated by the FIA program using on-the-ground measurements of tree trunk size from many forest sites and statistical models that show the relationship between trunk size and the weight of branches, leaves, coarse roots (>0.1 inch in diameter), and forest floor litter. Such data are combined with estimates of forest land area obtained from aerial photographs and satellite imagery. Forest floor litter includes all dead organic matter above the mineral soil horizons, including litter, humus, small twigs, and coarse woody debris (branches and logs greater than 1.0 inches in diameter lying on the forest floor). Data for Alaska and Hawaii are not included in this data series. Note that there are 1.1 English tons per metric ton. In most international discussions, carbon storage is reported in metric tons.

Data Access: Data for 1950 through 1987 are from Birdsey (1996), and data for 1992 are from Birdsey and Heath (1995). Additional information about carbon storage in forests can be obtained at http://www.fs.fed.us/ne/global/research/carbon/forcarb.html.

The Data Gap

Data on soil carbon are scarce, and the influences of management activities on soil carbon are still poorly known. More intensive measurements of soil carbon are planned by both the Forest Service and the USDA Natural Resources Conservation Service (NRCS).

Some forests have not yet been fully inventoried, notably in parts of Alaska and for pinyon-juniper forests throughout the western United States. Where data were available, they were employed in the estimation process; where data were not available, assumptions were used. Data on these areas are now being collected and will be incorporated into future estimates.

Estimates of carbon storage in the soil and forest floor litter were developed using models based on data from specific forest ecosystem studies. There are no inventories specifically designed to estimate carbon storage over large regions in ecosystem components other than wood, although the Forest Service and the NRCS are currently testing protocols for measuring total carbon in a forest ecosystem. Measurement protocols for forest floor litter and soil carbon are being developed and are being implemented as funds become available.

The amount of product in landfills is based on studies conducted by the Forest Service Forest Products Laboratory and other sources. The Forest Service developed conversion factors to translate products in use and materials in landfills to carbon-equivalents. These conversion models account for all steps in the transformation of cut timber into products and through use to disposal. The models are run separately for each region of the United States and for different kinds of harvest (e.g., pulpwood, sawtimber). More information on these models can be found in Row and Phelps (1991).

References

Birdsey, R.A. 1996. Carbon storage for major forest types and regions in the conterminous United States, pp. 1–25. In R.N. Sampson and D. Hair (eds.), Forests and Global Change, Vol. 2: Forest Management Opportunities for Mitigating Carbon Emissions. Washington, DC: American Forests.

Birdsey, R.A., and L.S. Heath. 1995. Carbon changes in U.S. forests. In Joyce, L. A. (ed.), Productivity of America's forests and climate change. U.S. Department of Agriculture, Forest Service, General Technical Report RM-271, Rocky Mountain Forest and Range Experiment Station, Ft. Collins, CO.

Row, C., and Phelps, R.B. 1991. Carbon cycle impacts of future forest products utilization and recycling trends. In Agriculture in a world of change, Proceedings of Outlook '91, U.S. Department of Agriculture, Washington, DC.

At-Risk Native Forest Species

See the technical note for the core national at-risk species indicator (p. 214).

Non-native Plants

The Indicator

The term "non-native" generally refers to species found in the United States whose native range is outside North America. More recently, this term has also been applied to species that are native to North America, but which are now found outside their historic range. Other terms for non-native species include "alien," "non-indigenous," or "introduced." The term "invasive" is also applied to many non-native species; invasive species spread aggressively into areas occupied by native species. Clearly, not all non-native species are invasive; nor are all invasive species from outside North America.

This indicator will report total area covered by non-native species. In some cases, the total area covered by any single species may be relatively low, but total area covered by all non-natives may be larger.

A useful introduction to the issue of non-native species can be found in the Office of Technology Assessment publication *Harmful Non-Indigenous Species in the United States* (1993; http://www.wws.princeton.edu/~ota/disk1/1993/9325_n.html).

A more recent, policy-oriented view of non-native species issues can be found in the Congressional Research Service report *Harmful Non-Native Species: Issues for Congress* (1999; http://cnie.org/NLE/CRSreports/Biodiversity/biodv-26.cfm).

Two state-based surveys of the kinds of non-native species and their impacts and controls can be found at http://www.ct.nrcs.usda.gov/landscp/invasive/problems.htm (Connecticut) and http://www.mdflora.org/publications/invasives.htm (Maryland).

Forest Age

The Indicator

It is important to note that the age of a tree does not necessarily convey information about the size of the tree. Fast-growing species attain sizes comparable to much older trees of another species, and trees of the same species and age growing in different locations may be very different in size. In addition, processes such as forest fires and hurricanes can act to limit the age of trees in a region (e.g., hurricanes are more prominent in the eastern United States).

The Data

This indicator presents data for a subset of all forests in the United States—those defined by the USDA Forest Service as "timberlands." Timberlands is a designation that covers lands on which harvesting is not prohibited by law and which grow an average of 20 cubic feet of wood per acre per year. Thus, the data presented here do not include national parks and wilderness areas and other natural and semi-natural forestland not classified as timberlands and thus not included in previous inventories. As a result, these data describe nearly all eastern forests, but only about 40% of western forests. Data on slow-growing forests and those in parks and wilderness areas are being collected, but they are not yet available.

Data Source: Data for this indicator were collected by the USDA Forest Service Forest Inventory and Analysis (FIA) program, which is described in the Forest Area and Ownership technical note (p. 239).

Data Collection Methodology: The age of a stand of trees is a classification based on the mean age of trees with dominant or codominant crown positions in the stand. Dominant/codominant crowns are those tree crowns dominating or sharing space in the upper layer of the tree canopy. The age of these trees is generally determined using tree cores from which annual growth increments were counted.

Forest Disturbance: Fire, Insects, and Disease

The Data

Data Source: Data reported here are from the USDA Forest Service Forest Health Monitoring (FHM) program. FHM, a component of the Forest Service's Forest Health Protection program, is a national program designed to determine the status, changes, and trends in indicators of forest condition on an annual basis. The program uses data from ground plots and surveys, aerial surveys, and other data sources and develops analytical approaches. See http://www.na.fs.fed.us/spfo/fhm/index.htm.

Data on forest fire acreage in national forests (referenced in text) was included in the 1999 Heinz Center prototype for this report (Heinz, 1999), and is from the General Accounting Office (GAO, 1999).

Data on acreage affected by diseases/parasites were obtained from "Forest Insect and Disease Conditions in the United States" for the years 1999, 1998, and 1997 (available at http://www.fs.fed.us/foresthealth/annual_i_d_conditions/index.html). These reports provide data on recent (i.e., past 5 years) acreage affected by the five major insects reported here. Data on historical acreage affected by these insects were provided by the Forest Inventory and Analysis program.

Insect data are collected using aerial surveys, implemented using a nationally standardized program, addressing both public and private forests. Disease data are collected using ground surveys and are considered to be less reliable.

Forest fire data were provided by the USDA Forest Service National Forest System, but are not limited to national forests. These data do not presently distinguish between forest fires and fires on other land cover types.

References:

GAO. 1999. Western National Forests: Nearby communities are increasing threatened by catastrophic wildfires. United States General Accounting Office. GAO/T-RCED-99-79.

The Heinz Center. 1999. Designing a Report on the State of the Nation's Ecosystems: Selected Measurements for Croplands, Forests, and Coasts & Oceans. The H. John Heinz III Center for Science, Economics and the Environment. Washington, D.C. http://www.heinzctr.org/publications.htm.

Fire Frequency

Note: This serves as the technical note for the Grassland/Shrubland fire frequency indicator.

The Data

The USDA Forest Service has an active program of research into fire and fuels management, including development of tools for assessing fire risk due to changes in fire frequency. In particular, the Fire Regimes for Fuels Management and Fire Use project, which began in 1997, involves mapping and characterization of presettlement natural fire regimes and current vegetation conditions and development of an index of departure for use in national-level fire management planning.

As part of this program, the Forest Service has developed estimates of presettlement fire frequency, using biophysical information, preexisting remote-sensing products, and expert knowledge about disturbance and successional processes and developed stylized successional pathways for unique combinations of presettlement fire regime and potential natural vegetation. These estimates can be found at http://www.fs.fed.us/fire/fuelman/firereg.htm.

Additional information on this procedure may be found in Schmidt et al. (in press).

However, current fire return intervals, based on tree ring scars and similar site measurements, have not been determined for the majority of the United States. The research project described above has developed estimates of fire return intervals by inference from existing vegetation. Essentially this involves assumptions about the fire return interval required to permit a certain vegetation type to develop. While these are valuable estimates, they are based on a significant amount of expert knowledge and modeling, rather than being relatively direct measurements of fire return frequency, and thus were not appropriate for inclusion in this report.

References

Cissel, J.H., F.J. Swanson, and P.J. Weisburg. 1999. Landscape management using historical fire regimes: Blue River, Oregon. Ecological Applications 9:1217–1231.

Knapp, P.A. 1997. Spatial characteristics of regional wildfire frequencies in intermountain west grass-dominated communities. Professional Geographer 49:39–51.

Knapp, A.K., J.M. Briggs, D.C. Hartnett, and S.C. Collins, eds. 2000. Grassland dynamics. New York: Oxford Univ. Press.

Sauer, C.O. 1950. Grassland climax, fire and man. Journal of Range Management 3:16–21.

Schmidt, K.M., J.P. Menakis, C.C. Hardy, D.L. Bunnell, N. Sampson, J. Cohen, and L. Bradshaw. In press. Development of coarse-scale spatial data for wildland fire and fuel management. General Technical Report RMRS-GTR- CD-XXX. Ogden, UT: U.S. Department of Agriculture, Forest Service, Rocky Mountain Research Station.

Swanson, F.J., J.A. Jones, D.O. Wallin, and J.H. Cissel. 1993. Natural variability—implications for ecosystem management. In M.E. Jensen and P.S. Bourgeron (eds.), Eastside Forest Ecosystem Health Assessment. Vol. II: Ecosystem management: Principles and applications. Portland OR: U.S. Forest Service.

Wallin, D.O., F.J. Swanson, and B. Marks. 1994. Landscape pattern response to changes in pattern generalization rules: Land use legacies in forestry. Ecological Applications 4:569–580.

Forest Community Types with Significantly Reduced Area

The Indicator

Rarity of species and ecosystems is a common conservation criterion, but in measuring risk, rarity may be less relevant than extent of historical decline or potential for further decline. Many rare species and communities have apparently always been rare and may not be highly vulnerable to extinction. On the other hand, a major decline in a once-dominant or widespread species or ecosystem type may have ecological consequences far more severe than the loss of the last few individuals of a chronically rare species or the loss of a plant community that never covered more than a small area.

This indicator will be based on an identification of forest community types that occupy at least 70% fewer acres than at presettlement. Note that the "forest community types" described in this indicator are more specific than the groupings described in Forest Types, p. 118. The "forest types" reported in that indicator are broad classifications, each of which would include many "forest community types."

The indicator will report the number of these community types and the present acreage of the suite of significantly reduced community types. It will also report the change in area of these community types from one reporting period to the next, allowing readers to understand whether reductions in the area of these already-reduced types is continuing or has been stopped or reversed.

Note that use of a presettlement baseline is not intended to imply that forest community types were "pristine" or completely unaffected by human activity. It is clear that Indians exerted influence over the presettlement landscape, although the extent of that influence is currently under debate and is likely to have differed by region. The use of a presettlement baseline is also not intended to serve as a goal for action or policy. It is rather intended as a relatively long-term reference point, against which to compare current conditions.

A recent review of threats to imperiled species in the United States found that 85% of all imperiled species were threatened by habitat degradation or loss (including 92% of vertebrates, 87% of invertebrates, and 81% of plants) (Wilcove et al. 1998). A separate study tallied species that were listed or were candidates for listing under the Endangered Species Act (ESA) for three major endangered ecosystems. As an example, in 1993 the longleaf pine–wiregrass ecosystem, which has declined by nearly 99% since presettlement times, contained 27 ESA-listed species and 99 species that were candidates or proposed for listing under ESA (Noss et al. 1995).

Forest community types for this indicator are defined at the "alliance" level of the National Vegetation Classification System (Grossman et al. 1998). An alliance is a group of plant associations that share a similar architecture and one or more diagnostic species, which are generally the dominants in the primary canopy. In some cases, aggregations of ecologically related alliances may be tracked. The alliance level of classification is roughly equivalent to "covertype" as defined by the Society of American Foresters.

Ecosystems can decline in area through outright conversion to another land cover or through gradual changes, like those that accompany fire suppression, which allows other species to take over a forest. For this indicator, as long as an area has the characteristics of a specific forest community type, it would be counted as part of that type. If, for example, significant vegetation changes occurred as a result of fire suppression, the forest may eventually be classified as a different type.

The Data Gap

The Association for Biodiversity Information and the USDA Forest Service Forest Inventory and Analysis (FIA) program are collaborating on development of methods that would allow estimation of the area of alliances (or in certain cases, aggregations of alliances) from existing FIA data. This would provide a recent historical perspective on changes in alliance area, and would allow the area of these community types to be tracked in the future.

Many scientists recognize the value of developing a national map of presettlement vegetation at the alliance or comparable level to provide a more quantitative basis for the assessment of forest cover change. A preliminary approach to this analysis could be done by crosswalking alliances to the Kuchler Potential Natural Vegetation types (Kuchler 1964). The Association for Biodiversity Information is seeking funding to complete this work.

Specific data in the indicator writeup are from sources as follows. Data on redwood acreage and Great Lakes pine forest are from Klopatek et al. 1979 and Powell et al. 1993. It is important to note that other estimates exist for the reduction in acreage of redwood (see Noss 1995) and Great Lakes pine forest (see Frelich 1995). Data on oak savanna are from Nuzzo 1986.

References

Eyre F. H. et al. 1980. Forest cover types of the United States. Washington, DC. Society of American Foresters.

Frelich, L. 1995. Old forest in the Lake states today and before European settlement. Natural Areas Journal 15:157–167.

Grossman, D.H., et al. 1998. International classification of ecological communities: Terrestrial vegetation of the United States. Volume I: The national vegetation classification standard. Arlington, VA. The Nature Conservancy. http://www.natureserve.org/publications/icec/toc1.html.

Klopatek, J.M., R.J. Olson, C.J. Emerson, and J.L. Jones. 1979. Land-use conflicts with natural vegetation in the United States. Environmental Conservation 6:191–199.

Kuchler, A.W. 1964. Manual to accompany the map: Potential natural vegetation of the conterminous Unitied States. Special Publication 36, American Geographical Society, New York.

Noss, R.F., E.T. LaRoe, and J.M. Scott. 1995. Endangered ecosystems of the United States: A preliminary assessment of loss and degradation. U.S. Geological Survey, Biological Report No. 28.

Nuzzo, V.A. 1986. Extent and status of Midwest oaks savanna: Presettlement and 1985. Natural Areas Journal 6(2):6–36.

Powell, D.S., J.L. Faulkner, D.R. Darr, Z. Zhu, and D.W. MacCleery. 1993. Forest resources of the United States, 1992. General Technical Report RM-234. Fort Collins, CO: USDA Forest Service, Rocky Mountain Forest and Range Experiment Station. Revised, June 1994.

Wilcove, D.S., et al. 1998. Quantifying threats to imperiled species in the United States. BioScience 48:607–615.

Timber Harvest

The Indicator

Sawlogs are logs that are at least 8 feet long, with minimum defects or bends, and that are at least 6 inches in diameter (measured inside the bark) for softwoods and 8 inches for hardwoods. *Pulpwood* includes trees, chips, or logging residues used to produce wood pulp, from which products such as paper are made. *Fuelwood* is cut as a source of energy and is used primarily for residential firewood. *Veneer* logs are trees from which veneer is sliced for plywood and other veneer products. *Logging residues/other* describes parts of trees that are cut or otherwise killed in the harvesting process (e.g., for road building), but that are not removed to make products. *Other products* is a miscellaneous category of products from trees, including pilings, poles, shingles, and charcoal.

The Data

The data presented in this indicator are not directly comparable with the data presented in the growth and harvest indicator, because that indicator reports only the volume of "growing stock," an inventory category that excludes certain trees and parts of trees.

Data Source: Data on forest products and their source were collected by the USDA Forest Service Forest Inventory and Analysis (FIA) program and the Forest Service Forest Products Laboratory, which also supplements these data with information from U.S. Department of Commerce published reports and industry trade association sources.

Data Collection Methodology: The FIA collects data through a large-scale field sampling program, described in the technical note on forest area and ownership (p. 239). Also included here are data from periodic Forest Service wood facility surveys, residential fuelwood surveys, studies of active logging operations, and field inventories of harvested trees.

Data Manipulation: FIA field data are used to estimate harvest distributions by ownership based on trees harvested for products. The Forest Service Forest Products Laboratory also conducts utilization studies on active logging operations to estimate wood usage for products and residues left in the woods. These data are merged with log receipt data from wood-using facilities to produce estimates of timber and other material cut to deliver those logs to the facility. Ancillary data from the Department of Commerce on wood use and industry association data are used to validate information on the volume of trees cut annually to produce primary wood products such as sawlogs, pulpwood, veneer logs, fuelwood, and other wood products.

Data Quality /Caveats: Non-fuelwood product totals shown would generally have errors of less than ±10 percent. Data are from FIA wood facility surveys, which are full industry canvasses and are thus assumed to have negligible sampling error. Periodic residential fuelwood studies generally have errors of ±15%.

These data are not directly comparable with the data presented in the "Growth and Harvest" indicator, because that indicator reports only the volume of "growing stock," an inventory category that excludes "trees of poor form or quality and the upper central stem" (U.S. Department of Agriculture definition).

Data Access: All data are available free of charge, except for products that require special processing or shipping fees. Electronic databases are unavailable at the national level before 1987, and most regional data from before 1977 are not available electronically. Forest statistics, online databases, and a map of U.S. forest distributions are on the Web at http://fia.fs.fed.us. Forest Products Laboratory data synthesizing Department of Commerce and industry trade association data are available at http://www.fpl.fs.fed.us. Additional data on wood products use may be found at http://www.fpl.fs.fed.us/.

Timber Growth and Harvest

The Indicator

Definitions for the terms "growth," "harvest," and "timberlands," as used in this indicator, are those used by the USDA Forest Service. Growth is the net annual increase in the volume of living tree stems between inventories after accounting for effects of mortality but before accounting for the effects of harvest. Harvest is a measure of the average annual volume of living trees harvested between inventories. Timberland is the subset of forest land on which harvesting is not prohibited by law and potential wood growth rates are greater than 20 cubic feet per acre per year. Growth is a rough measure of the rate at which forests are converting solar energy into tree biomass. Comparing growth with harvest is a frequently used method of assessing whether wood harvesting is reducing the volume of tree biomass in a forest.

The Data

The data presented in this indicator are not directly comparable with the data presented in the timber harvest indicator, because the data presented here report only the volume of "growing stock," an inventory category that excludes certain trees and parts of trees (these data—defined below—are used for both the growth and harvest categories presented here). The harvest data presented in Timber Harvest (p. 130) encompass a broader suite of timber products, including "growing stock" and other harvested materials.

Data Source: Data for this indicator were collected under the USDA Forest Service Forest Inventory and Analysis (FIA) program, which is described on page 239. FIA data are from national compilations of periodic statewide survey data.

Data Quality/Caveats: The data for this indicator are limited to "growing stock" trees. Growing stock is a Forest Service inventory category that includes live trees of commercial species meeting specified standards of quality or vigor. When used in calculating volume, this category includes only trees 5.0 inches d.b.h. ("diameter at breast height" a common measurement of tree size) and larger, and which have no obvious characteristics that would make them unusable for industrial use (e.g., rot, unusual shape). In addition, volume is computed for the central stem from a 1-foot stump to a minimum 4-inch top diameter outside bark, or to the point where the central stem breaks into limbs. Noncommercial species are species that normally do not develop into trees suitable for industrial wood products. Since many forest products are made from trees and parts of trees that are not counted as "growing stock" for this indicator, the amounts and trends shown here may differ from those shown in the harvest and use indicator.

This indicator does not provide data on the species, age, quality, or other attributes of the trees being harvested or of trees whose growth is measured. General trends in growth and harvest in the East and West do not reflect some important trends that are occurring at smaller scales. Factors influencing trends in growth and removals vary substantially among and within regions. Subregions where growth/harvest ratios are similar today may have very different growth/harvest ratios in the future. In the West, growth/harvest ratio on timberland may be a poor indicator of change in forest biomass because timberland accounts for only 40% of total forest area in the region. These data exclude forest areas in parks and wilderness, where timber harvesting is prohibited, as well as slow-growing forests.

Current inventory practices limit the data shown here to the main trunk of trees of a certain size, shape, and species. Therefore, the data presented are not directly comparable with the figures presented for Timber Harvest and Use, which account for products made from all parts of all species of trees.

Because this indicator does not include information on growth in slow-growing forests and those in parks and wilderness, which make up 60% of western forests, it may not reflect significant changes in forest growth in that region.

Data Access: see the technical note for Forest Area and Ownership (p. 239); additional data for this indicator were taken from the publications listed in the references.

References

USDA Forest Service. 1958. Timber resource for America's future. Forest Resource Report No. 14. Washington, DC.

USDA Forest Service. 1965. Timber trends in the United States. Forest Resource Report No. 17. Washington, DC.

USDA Forest Service. 1982. An analysis of the timber situation in the United States 1952–2030. Forest Resource Report No. 23. Washington, DC.

Waddell, K.L., D.D. Oswald, and D.S. Powell. 1989. Forest Statistics of the United States, 1987. Resource Bulletin PNW-RB-168. Portland, OR: USDA Forest Service

Recreation in Forests

There is no technical note for this indicator.

Fresh Waters

Extent of Freshwater Ecosystems

The Data

Wetlands, Lakes, Reservoirs, and Ponds

Data Source: Data for wetlands are from Dahl (2000). Data used here are from Figure 35 (p. 56) and Figure 42 A–C (p. 62). Data for lakes, reservoirs, and ponds come from Dahl (2000) and also from Frayer et al. (1983); Dahl and Johnson (1991); and unpublished data from the U.S. Fish and Wildlife Service.

Data Collection Methodology/Definitions: The data shown here are derived from the U.S. Fish and Wildlife Service's National Wetlands Inventory (NWI), which produces periodic reports of changes in wetland area. For this report, decadal estimates are presented as the midpoint of the decade. For example, "1980s" data are presented as "1985." The historic estimate for 1780 is based on the estimate of 221 million acres of coastal and freshwater wetlands at that time (see Dahl 1990) minus an estimate of 10 million acres of coastal wetlands in 1922, which should approximate the historical area of coastal wetlands because most of these were converted to other land cover types after World War II (see Gosselink and Baumann 1980). Estimates of wetland extent in the 1780s are based on colonial or state historical records plus land use records, drainage statistics, and information on the extent of hydric soils (i.e., drained and undrained).

NWI counts all wetlands, lakes, reservoirs, and ponds, regardless of land ownership, but recognizes only wetlands that are at least 3 acres, and ponds that are at least 1 acre. A permanent study design is used, based initially on stratification of the 48 conterminous states by state boundaries and 35 physiographic subdivisions. Within these subdivisions are 4375 randomly selected 4-mi^2 (2,560-acre) sample plots. These plots were examined with the use of aerial imagery of varying scale and type; most images were 1:40,000-scale, color infrared, from the National Aerial Photography Program.

The wetland types selected for reporting here were recommended as the most relevant and most reliable for long-term reporting by the NWI (see Dahl 2000, p. 62). For wetlands, they include forested, shrub, and emergent wetlands. Ponds include the category of open-water ponds and non-vegetated palustrine wetlands (i.e., palustrine unconsolidated shore, which are mud flats and the shorelines of ponds); ponds are generally less than 6 feet (2 m) deep and less than 20 acres in size. Lakes and reservoirs are generally larger than 20 acres and deeper than 6 feet, although smaller bodies are included if they are deeper than 6 feet or have a wave-formed or bedrock shoreline.

Data Quality/Caveats: Field verification was conducted to address questions of image interpretation, land use coding, and attribution of wetland gains or losses, and plot delineations were completed. For example, for the 1980s-to-1990s analysis, 21% of the sample plots were verified.

Ephemeral wetlands and effectively drained palustrine wetlands observed in farm production are not recognized as a wetland type and are not included. Wetlands that are farmed during dry years but that normally support hydrophytic vegetation were classified as freshwater emergent wetlands.

The U.S. Geological Survey's (USGS) National Hydrography Dataset (NHD) also has information on lake, reservoir, and pond area (at least 6 acres in size). Considerably higher total acreage (26.8 million acres) is found using this resource. NWI was used because time trends are possible; the cause of the disparity between datasets is not known.

Data Access: *The Status and Trend of Wetlands in the Conterminous United States 1986 to 1997* is available on the Web at http://wetlands.fws.gov/bha/SandT/SandTReport.html.

Riparian Areas

Note: This indicator uses a distance of roughly 100 feet from the edge of a stream to define its "riparian" area. This is based on the availability of remote-sensing data, as described below. We are cognizant that the definition of riparian areas is a complex one, and that no single value for the width of this feature will be appropriate in all situations.

Data Source: Data reported here for the classification of riparian areas along streams and rivers were provided by the U.S. Environmental Protection Agency's National Exposure Research Laboratory, Environmental Sciences Division, and are based on the NHD. The NHD is a comprehensive set of digital spatial data that encodes information about naturally occurring and constructed bodies of water (see http://nhd.usgs.gov/). The NHD was developed based on EPA's River Reach File 3 (RF3), which itself was based on digitization of streams from USGS topographic quadrangle maps. The dataset does not provide information on very small streams, and the lower limit of stream size that is reported in the database is unclear. Data on the vegetation cover within 100 feet of streams and rivers were produced by EPA from remote-sensing imagery and the NHD. The remote-sensing imagery is from the National Land Cover Dataset (NLCD; see the technical note for the national extent indicator for further details, p. 207).

Data Manipulation: For this study, EPA combined these datasets to identify the land cover along streams and rivers (and the shores of ponds, lakes, and reservoirs—see the altered freshwater ecosystems indicator). For each stream reach described in the NHD, land cover was characterized, using the NLCD, in a band approximately 100 feet wide on either side of the stream. NLCD land cover classes were aggregated to produce four general categories (forested; agricultural; urban; and grasslands, shrublands, and woody and emergent wetlands). In one instance, the text describes this latter category as "other natural vegetation," despite the fact that some of these land cover types may not be the historical (i.e., natural) vegetation for that site, or may have been altered in other ways. This terminology is used to highlight the contrast with the highly altered land covers (urban, agricultural). Estimates of the riparian area in each of these different land cover classifications were derived by overlaying stream reaches and land cover.

Data Caveats/Limitations: The NLCD and the NHD are currently the most comprehensive datasets available for land cover and freshwater resources, respectively. However, both of these contain inaccuracies that could affect the calculations presented here. The NLCD is known to contain approximately 20% error in land cover classification; some of the known misclassifications that occur randomly in the dataset include suburban areas or tree farms classified as forest; grasslands classified as agriculture, or vice versa; and fallow agricultural fields classified as barren lands. The NHD is a relatively new dataset and is known to contain numerous errors and inconsistencies. Strahler first- and second-order streams (a method for ranking stream order, which is related to size) are poorly represented in the NHD as well as in the RF3 that serve as the base data. It appears that dry lake beds in the west may have occasionally been included as lakes in the NHD. Additionally, the architecture of the NHD results in some lakes being represented by numerous polygons with different identifications, thus being counted as separate lakes in this analysis. Numerous inconsistencies exist in the NHD attribute data. The designation of stream segments as perennial or intermittent is particularly problematic; in at least one case, this designation can be shown to follow USGS topographic quadrangle boundaries. In addition, many errors can be found in the attribution of ponds, lakes, and reservoirs. Although these inconsistencies were noted, it was not possible given the scope and scale of this analysis to provide across-the-board corrections, nor was it possible to coregister the datasets for all locations. Therefore, the most current versions of both datasets were used as is.

The Data Gap

Information on the number of small, medium, and large streams and rivers is not available. In general, the number of stream miles can be derived from sources such as the NHD; however, there is no universally accepted approach for categorizing streams and rivers based on size (i.e., small, medium, and large). Potential approaches include basing categories on flow rate, drainage area size, or stream order. USGS will soon incorporate a tool within the NHD dataset to allow determination of stream order, which can be determined from maps. Flow rate is a much more difficult parameter to determine.

In addition, there is concern that use of the NHD may understate the extent of small streams. Since the NHD is based upon historic mapping conducted for the USGS, there may be inconsistencies in the degree to which small streams were mapped. Since the rate of conversion and alteration of small streams is believed to be higher than for larger streams, it is important to ensure as great a coverage of small streams as is feasible.

For a discussion of the effects of human activities on small streams, see Meyer and Wallace (2001).

References

Dahl, T.E. 1990. Wetland losses in the United States 1780's to 1980's. Washington, DC: U.S. Department of the Interior, Fish and Wildlife Service.

Dahl, T.E. 2000. Status and trends of wetlands in the conterminous United States 1986 to 1997. Washington, DC: U.S. Department of the Interior, Fish and Wildlife Service.

Dahl, T.E., and C.E. Johnson. 1991. Status and trends of wetlands in the conterminous United States, mid-1970's to mid-1980's. Washington, DC: U.S. Department of the Interior, Fish and Wildlife Service.

Frayer, W.E., T.J. Monahan, D.C. Bowden, and F.A. Graybill. 1983. Status and trends of wetlands and deepwater habitats in the conterminous United States, 1950's to 1970's. Ft. Collins, CO: Dept. of Forest and Wood Sciences, Colorado State University.

Gosselink, J.G., and R.H. Baumann. 1980. Wetland inventories: Wetland loss along the United States Coast. Zoological Geomorphology NF Supplement 34:173-187.

Meyer, J.L., and Wallace, J.B. 2001. Lost linkage and lotic ecology: Rediscovering small streams. In M.C. Press, N.J. Huntley, and S. Levin (eds), Ecology: Achievement and challenge (41st Symposium of the British Ecological Society). Blackwell Science.

Altered Freshwater Ecosystems

The Indicator

This indicator would report the percentage of each of the major freshwater ecosystems (rivers and streams, riparian areas, wetlands, and lakes, ponds, and reservoirs) that are altered. "Altered" is defined differently for each of the following:

- *Rivers and streams* (all flowing surface waters) are altered if they are leveed, channelized, or impounded behind a dam. There are other types of alterations to streams that may be important; these include changes in sedimentation and temperature, and barriers to movement between stream reaches. Such changes can be caused by dams or other alterations to the river or its surroundings. As monitoring and reporting technology and understanding evolve, it may be possible to report on these and other alterations. At present, identifying such changes requires detailed site-specific analyses, which have not been done on a widespread basis (see also The Heinz Center 2002). Both the stream habitat quality and changing stream flows indicators provide important complementary information on stream conditions.
- *Riparian areas* along rivers and streams are considered altered if they have a predominance of urban or agricultural land use.
- *Lakes and ponds* are considered altered if the area immediately adjacent to the shoreline has land cover that is predominantly urban or agricultural. Since there is no agreed-upon proportion of shoreline that must be in these land use categories in order for individual lakes to be classified as "altered," this indicator reports the overall percentage of lake shoreline in agricultural or urban use. This indicator focuses on "natural" waterbodies, that is, those that are not created by impoundment behind a dam. While reservoirs provide habitat, the prevalence of large and frequent fluctuations and associated poor development of the riparian/littoral zone reduces this value. In this case, the number or percentage of natural lakes whose waterflow has been altered by damming would also be reported. Some

impounded lakes are not subject to such fluctuations, but until it is possible to distinguish between different impoundment types, this indicator will be limited to natural waterbodies.

- *Wetlands* are considered altered if they are excavated, impounded, diked, partially drained, or farmed. These categories are used by the U.S. Fish and Wildlife Service's National Wetlands Inventory; they are defined in Cowardin et al. (1979). Wetlands fragmentation (subdivision into smaller and more isolated patches by filling, roads, or other alterations) is also important, but measurement of this change requires detailed site-specific information.

The Data

The methods used to produce the data reported here for altered riparian areas are described in the technical note for the Extent of Freshwater Ecosystems, which immediately precedes this one. The extent indicator describes methods used to characterize riparian areas; the same method could be used to classify the shorelines of ponds and lakes, but the relevant database does not distinguish between natural and impounded lakes/reservoirs.

The Data Gap

There is no nationally aggregated database of the number of impounded river miles or the number of leveed river miles. There is also no method for calculating the extent of downstream effects of dams, other than by conducting site-specific investigations for each dam.

No nationally aggregated database distinguishes impounded waterbodies from natural ones, or identifies which natural lakes are dammed at their outlets. It is possible that existing databases on dam locations, such as those maintained by the U.S. Army Corps of Engineers, could be merged with other datasets, such as the National Hydrography Dataset, to derive this information.

Data on altered wetlands are available through the U.S. Fish and Wildlife Service's National Wetlands Inventory (see http://www.nwi.fws.gov/). At present, these data are not available in electronic form for the entire United States. Further, these data are available only on a quad-sheet-by-quad-sheet basis. The Fish and Wildlife Service is in the process of integrating these data more fully, and it is likely that they will be available in the near future. However, they will be from different time periods in different states, and there is no plan for periodic updating. In addition, there are no plans to produce regional or national reports comparing any updates with past data.

References

Cowardin, L.M., V. Carter, F.C. Golet, and E.T. LaRoe. 1979. Classification of wetlands and deepwater habitats of the United States, FW/OBS-79/31. Washington, DC: U.S. Fish and Wildlife Service.

The H. John Heinz III Center for Science, Economics and the Environment. 2002. Dam removal: Science and decision making. Washington, DC: The Heinz Center. http://www.heinzctr.org/publications.htm.

Phosphorus in Lakes, Reservoirs, and Large Rivers

The Indicator

A variety of nutrients are needed for plant growth in aquatic systems: nitrogen, phosphorus, carbon, sulfur, iron, manganese, and various trace metals (e.g., copper, cobalt, molybdenum, and zinc). Silica is required by some kinds of algae (e.g., diatoms) because it is the main component of the shells that surround the cells. However, nitrogen (N) and phosphorus (P) are by far the most common nutrient elements that limit or control the amount and rate of plant growth in aquatic systems and, thus, define their trophic status and corresponding "water quality." Of these two elements, phosphorus is widely considered to be the element that most commonly limits aquatic plant growth in fresh waters under natural conditions (i.e., minimal impacts from human activity). Total phosphorus (TP) includes all forms of phosphorus present in a water sample—dissolved and particulate, inorganic and organic; adsorbed onto suspended clays and hydrous oxides; present in planktonic organisms and in organic detritus; and phosphorus in dissolved natural organic matter. Phosphorus in macrophytes, fish, and bottom sediments generally is not included.

TP was selected for reporting because it is a comprehensive measure of the many operationally defined and chemical forms of phosphorus, most of which are directly or indirectly available for plant growth. Excess phosphorus can contribute to algal blooms, poor water clarity, and other symptoms of eutrophication.

TP levels are a measure of trophic state (Carlson 1977) and general water quality in lakes, reservoirs, and large rivers. (Large rivers typically behave as lakes; water residence times in stretches of large rivers are sufficiently long that substantial phytoplankton growth can occur in them.) The concentrations of TP that contribute to symptoms of eutrophication are poorly understood for flowing waters, but generally they are thought to be higher than the critical levels in lakes. Consequently, TP is reported separately for lakes and rivers. (The effects of phosphorus enrichment are different for lakes and rivers in tropical areas than they are for temperate zones; this discussion relates to temperate zones only.)

TP measurements are straightforward; TP in lakes should be reported as an average over the growing season (e.g., April to September), which will require several (e.g., 4–6) samples over the course of the period. Consideration was given to the appropriate number of samples each year (e.g., Knowlton et al. 1984), and complications of sampling in areas with minimal seasonal influence, such as Florida (Brown et al. 1988).

TP measurements in rivers are restricted to those large rivers with flows exceeding 1000 cubic feet per second (cfs). To ensure proper characterization of average values for each river, only sites that had at least 30 samples over the course of 2 years were included.

Information on the 1986 phosphorus recommended goal for preventing excess algae growth can be found in EPA 440/5-86-001 (see references). Information on regional nutrient (phosphorus) criteria can be found at http://www.epa.gov/waterscience/criteria/nutrient/ecoregions.

The Data

Data for river phosphorus are from sites operated by the U.S. Geological Survey (USGS) National Water Quality Assessment (NAWQA) and National Stream Water Quality Accounting Network (NASQAN). Data were available from 140 sites, with data collection from 1992 to 1998; 116 of these sites were either NAWQA or NAWQA and NASQAN joint sites.

NAWQA is described generally in the technical notes for the core national indicator for contaminants (p. 210) and for nitrate in farmland streams (p. 232). While that note describes data collection from streams with relatively homogenous land cover (and often relatively low discharge volumes), the data used in this indicator are from larger rivers, with both larger discharge volumes and watersheds with generally more diverse land uses. Thus, these samples represent the integrating influences of many different land uses. The methods for processing and summarizing these data for large rivers, such as computing annual-weighted discharge concentrations, also have been described in the technical note for the Farmlands nitrate indicator.

NASQAN is a USGS program that is focused on four major river basins: the Mississippi, the Rio Grande, the Colorado, and the Columbia River. NASQAN stations are located on major tributaries in the four river basins, along the mainstem of rivers where there is a large increase in flow, and upstream and downstream from large reservoirs. The program generally measures both stream flow and a broad range of chemical constituents. An extensive quality-assurance/quality-control program enables constituents present in very low concentrations (micrograms per liter, roughly parts per billion) to be measured with definable accuracy and precision. See http://water.usgs.gov/nasqan/progdocs/index.html.

Because there was concern over the use of STORET data for this indicator (see below) with respect to the possibility that sampling locations might be strongly influenced by virtue of being located near outfalls from wastewater treatment plants, this question was also raised with respect to the NAWQA/NASQAN data. These programs collect data using procedures that ensure that the sample is representative of the entire stream cross-section. So, even if the stream at the point of collection were not well mixed, the samples would still be representative of the entire stream flow. In addition, the measure that is being reported—annual discharge-weighted average concentrations—addresses the potential concern that samples might be overly representative of summer low flows when wastewater effluent can comprise a large fraction of the flow in some rivers.

The Data Gap

In assessing the availability of data for reporting on phosphorus in lakes and rivers, we reviewed two major datasets in addition to the one reported here (NAWQA/NASQAN). These were STORET, maintained as a data repository by the Environmental Protection Agency (http://www.epa.gov/storet/), and within STORET, data from the National Water Information System (NWIS), a USGS-maintained data system (http://water.usgs.gov/nwis/).

Under contract to The Heinz Center, Procter & Gamble's Miami Valley Laboratory undertook an assessment of the quality and spatial and temporal variability of the data from these two sources. They concluded that phosphorus data were likely to be comparable in terms of reporting thresholds; that is, there were few if any problems related to the use of different reporting thresholds in different states or jurisdictions.

The second step was to determine whether either data system had sufficient numbers and geographic distribution of sampling sites. It was apparent from inspection of a map of lake phosphorus sampling sites that neither NWIS nor STORET as a whole has sufficient coverage across the country. STORET has phosphorus concentration data from a large number of river sampling sites, and this record extends into the 1980s.

However, there was significant concern among workgroup members regarding the fact that STORET data are derived from studies undertaken for many reasons and using many methods for selecting sampling sites. For example, some sampling was undertaken specifically as part of before-and-after effectiveness studies relating to phosphorus removal in publicly owned sewage treatment works (POTWs). Other studies may have been undertaken to determine the nature and extent of known phosphorus contamination problems, while others may have been located randomly as part of efforts to characterize nutrient concentrations in both "clean" and "dirty" areas.

STORET has very little information that can be used to determine the rationale for sampling-site selection. Thus any determination of the appropriate subset of STORET results to use would have to be based on a complex analysis of the proximity of sampling sites to POTWs, urban areas, and the like, which could be used to determine if the sampling was biased to inclusion or exclusion of such sites. Unfortunately, this analysis has not been done and could not be accomplished within the time and resources of this project. Therefore, given the significant potential for STORET data to be unrepresentative, we have decided that it is inappropriate to rely on it for this indicator until such studies can be completed.

References

Brown, C.D., D.E. Canfield, Jr., R.W. Bachmann, and M.V. Hoyer. 1998. Seasonal patterns of chlorophyll, nutrient concentrations and Secchi disk transparency in Florida lakes. Lake and Reserv. Manage. 14:60–76.

Carlson, R.E. 1977. A trophic state index for lakes. Limnol. Oceanogr. 22:361–369.

Knowlton, M.F., M.V. Hoyer, and J.R. Jones. 1984. Sources of variability in phosphorus and chlorophyll and their effects on use of lake survey data. Water Resour. Bull. 20:397–407.

U.S. Environmental Protection Agency. 1986. Quality criteria for water—1986, U.S. Environmental Protection Agency report, EPA 440/5-86-001.

Changing Stream Flows

The Indicator

This analysis is based on changes between flow characteristics of a 20-year period beginning about 1930 and three 10-year periods (1970s, 1980s, and 1990s). All stream gauges used here had a 20-year record for the reference period and a 10-year record for the later comparison period. Some of these 20-year records began in 1930 and ended in 1949, while some began in 1931, 1932, and 1933, and ended correspondingly later. Twenty years was selected as a reasonable period that would allow characterization of hydrologic regimes, and 10 years as the minimum period to use to determine changes.

Data from the earlier period are being used here as a practical baseline for historical comparison, even though many dams and other waterworks had already been constructed by this time, and even though this period was characterized by low rainfall in some parts of the country. This decision means that it is more useful to focus on decade-to-decade changes in the number of streams with major changes in flow, rather than on the number or percentage of streams with such changes, compared to the baseline period.

The indicator assesses changes in magnitude and timing of low-flow events and high flows, extreme events that are ecologically important in riverine ecosystems. Four subindicators were included in the analysis:

- ***Average 7-day low flow (% change):*** assesses the degree of alteration in low-flow magnitude, a parameter of importance to aquatic life. Minimum flows determine habitat availability for aquatic organisms and can influence condition of riparian vegetation. Regulated streams are often required to maintain a minimum flow for aquatic life.
- ***Timing of the 7-day low flow (Julian day):*** describes how timing of low-flow conditions may have changed. A substantial change in seasonal timing of low flow can influence many ecological processes.
- ***Average 1-day high flow (% change):*** assesses the degree of alteration of the average annual peak flow. High flows are significant ecological and geomorphic events for streams and rivers, and a large change in the 1-day high flow is expected to have important ecological consequences.
- ***Timing of the 1-day high flow (Julian day):*** assesses the change in the timing of maximum annual high flow, an event of substantial ecological relevance. A substantial change in seasonal timing of peak flow can influence many ecological processes.

The Data

Data Source: Data reported here are from the U.S. Geological Survey (USGS) stream gauge network. USGS has placed stream gauges and maintained flow rate records throughout the United States since the end of the 19th century. These records are available on the Internet in the form of daily streamflow values reported as the average volume of water per second over a 24-hour period (http://water.usgs.gov/nwis/discharge).

Table 4. Values for Minimal, Moderate, and Large Change

	Minimal Change	Moderate Change	Large Change
Percent Change	<25%	25-75%	>75%
Time	<30 days	30-60 days	>60 days

Data Collection Methodology: Stream gauging data are collected using standard USGS protocols.

Data Manipulation: Queries of the USGS Web site were used to identify 867 sites that had 20-year continuous records within 4 years of the target dates of 1930–1949 and 10-year continuous records for the decades of the 1970s, 1980s, and 1990s. The data for these 867 sites were then put into a format compatible with the Indicators of Hydrologic Alteration (IHA) software package produced by The Nature Conservancy with Smythe Scientific Software (http://www.freshwaters.org/iha.html), which was used to perform all subsequent analyses. The IHA software package compares the values for each subindicator (see list above) for the early 20-year period and the three later 10-year periods for each gauge. Each gauge is classified according to the degree of change of each of the four subindicators (see Table 4). Data analysis was conducted by David Raff, Department of Civil Engineering, Colorado State University.

Data Quality/Caveats: Although the sites analyzed here are spread widely throughout the United States, gauge placement by the USGS is not a random process. Gauges are generally placed on larger, perennial streams and rivers, and changes seen in these larger systems may differ from those seen in smaller streams and rivers. In addition, the USGS gauge network does not represent the full set of operating streamflow gauges in the United States. The U.S. Army Corps of Engineers, for example, operates gauges, and those data are not available through the USGS; they were not used in this analysis.

Data Access: Stream gauge data are available through the USGS Web site at http://water.usgs.gov/nwis/discharge. Analysis results are available through The Heinz Center.

Water Clarity

Two approaches for measuring water clarity are measurements of Secchi depth and satellite-based estimates. Since 1994, the U.S. Environmental Protection Agency (EPA) has supported an impressive program that aggregates Secchi disk measurements made by volunteers during July across parts of the United States and Canada (The Great North American Secchi Dip-In; see http://dipin.kent.edu). In 2000, lakes in 43 states were sampled, but the coverage varied considerably from state to state—in Minnesota, Michigan, and Maine, large numbers of lakes were tested, while in West Virginia and Wyoming, no lakes were sampled, and in states such as Pennsylvania and the Dakotas, relatively few lakes were sampled. In order to make the data nationally representative, this program should be expanded to include more lakes in more states. Because clarity is greatly affected by algal blooms, measurements of clarity should be carried out at the height of the growing season (mid-July to mid-September) in each ecoregion, which may or may not fit with the July observations of the Dip-In program. In addition, scientists are developing ways to measure water clarity from satellite data, which could greatly improve our understanding of how water clarity varies across the country and over time.

The Indicator

This discussion assumes that water clarity will be measured in lakes and reservoirs by the Secchi-disk method, although a satellite-based method may become the preferred approach. Secchi depth measurements of water clarity (or transparency) will be reported in three ranges: low (<3 ft), medium (3–10 ft), and high (>10 ft). The Secchi disk is a white plate with a diameter of 8 inches with black lines radiating from the center. The disk is lowered into the water until it can no longer be seen. The depth at which this occurs is called the Secchi disk transparency or Secchi depth (SD). It is a simple but effective way to measure water clarity.

Water clarity values for lakes and reservoirs will be reported in two ways: by lake area falling into the low, medium, and high categories, and as averages for freshwater ecoregions. (Ecoregions are areas that are similar in climate, geography, and ecological conditions and are defined in Ricketts et al. 1997.) Measurements should be made annually during an "index" period near the height of the algal growing season, which generally corresponds with the height of the recreational use season. In lakes of the Upper Midwest, for example, the index period is mid-July to mid-September, when Secchi-disk transparency is relatively constant and at annual minimum values. The appropriate length of the index period in other parts of the country needs to be determined, but the mid-July to mid-September period should be suitable for all lakes in temperate climate zones. One measurement during this period should be adequate to define ecoregional growing-season minimum values, although one measurement is not sufficient to define the minimum transparency for an individual lake.

Humic-colored lakes and reservoirs are found in many areas of the country (e.g., in northern forests of Minnesota, Wisconsin, Michigan, New York, and New England and in wetland forests throughout the Southeast, from Virginia to Florida). Clay turbidity is a dominant factor in water clarity in lakes and reservoirs of the central plains and the Piedmont region of the Southeast. Humic color and clay turbidity tend not to have a strong seasonal pattern in lakes, so a mid- to late-summer sampling period designed to capture the peak influence of algal growth on transparency should also be appropriate for these lakes and reservoirs.

Ponds have been excluded from this indicator, mostly because the hydraulic properties of ponds are quite different from those of lakes. Because of their shallow nature (typically less than 2 meters, or 6.5 feet), ponds can readily be completely mixed by strong winds. Such mixing can suspend sediments in the water column, which would decrease clarity. Lakes (and reservoirs) typically have a warm layer of water at the surface (epilimnion) that does not easily mix with deeper, colder waters (hypolimnion). Full wind-driven mixing of lakes typically occurs only during the fall and spring when temperatures are fairly uniform across all depths.

The Data Gap

The Great North American Secchi Dip-In program has been evolving since 1994. Supported by the EPA in cooperation with the North American Lake Management Society, the Dip-In is the largest-scale program for collecting SD data in the United States. The program relies upon volunteers who measure the Secchi depth of lakes in their area over a 2-week period in the beginning of July. Data are collected and maintained at http://dipin.kent.edu. While the data do not cover the whole country, they are substantial. In 2000, lakes in 43 states were sampled, but coverage varies considerably from state to state. Several states (Minnesota and Wisconsin in particular) have extensive volunteer monitoring programs coordinated by state agencies, and some state agencies have extensive collections of historical data.

Using satellite imagery is promising as a way of obtaining essentially complete coverage of lake water clarity. This approach is being tested by a NASA-funded consortium involving the Universities of Minnesota, Wisconsin, and Michigan. The consortium is applying a recently developed protocol using Landsat satellite images from the early 1990s and from 1999 to all lakes over 50 acres in the three-state region (see resac.gis.umn.edu/lakeweb/index.htm and Kloiber et al. 2000).

References

Kloiber, S.M., T. Anderle, P.L. Brezonik, L. Olmanson, M.E. Bauer, and D.A. Brown. 2000. Trophic state assessment of lakes in the Twin Cities (Minnesota, USA) region by satellite imagery. Arch. Hydrobiol. Ergebn. Limnol. 85:1–15

Ricketts, T.H., et al. 1997. A conservation assessment of the terrestrial ecoregions of North America. Volume 1: The United States and Canada. Washington, DC: Island Press.

At-Risk Native Species

See the technical note for the core national at-risk species indicator, (p. 214).

Non-Native Species

The Indicator

This indicator reports the percentage of all hydrologic units (simplified here to represent watersheds; see below) having one of several ranges of established non-native species. Introduced species are those that are not native to the watershed in which they are found. These species may be from outside North America, or they may be from another part of this continent. Established species are those that have established persistent breeding colonies. In general, watersheds with higher numbers of non-native species are subject to higher levels of ecological and economic disruption.

Some non-native species become established at low population levels; other species are "invasive"—that is, they spread aggressively, creating ecological and economic disruption. Ideally, this indicator would track only invasive species, perhaps by reporting on a selected group of problematic or potentially problematic species, as identified by recognized experts. However, it is not now possible to identify potentially problematic species, and thus we have chosen to report on all non-native species. But changes can signal the emergence of an invasive species. Some become invasive quickly; others do so only after long lag times.

It is important to note that hydrologic units, which are represented by hydrologic unit codes (HUCs), can be loosely thought of as watersheds. However, only at the finest resolution is this accurate. Thus, the HUCs shown in the figure may include multiple watersheds in whole or in part, or they may actually represent a single watershed.

The Data

Data Source: Nonindigenous Aquatic Species Database, Biological Resources Division (BRD), U.S. Geological Survey (USGS). Roughly 90% of the data are derived from the published literature. Data are collected for the most part by federal and state biologists, although the public does contribute by reporting sightings.

Data Manipulation: Data for introduced species are maintained in a database whose units are 6-digit HUCs (there are 352 6-digit HUCs across the 50 states). The only necessary manipulation was to compute the indices as described above.

Data Quality/Caveats: Although the BRD database (Web site listed below) is widely known about throughout the professional community, in some cases new discoveries are not reported by state and federal biologists.

Data Access: While these types of data are available on BRD's Nonindigenous Aquatic Species (NAS) Web site (http://nas.er.usgs.gov/), the actual data presented here were prepared for this report by USGS.

The Data Gap

NAS includes information on a host of vertebrates, invertebrates, algae, and plants. At this time, however, the database managers do not feel that these data have matured adequately to be presented at the national level.

Animal Deaths and Deformities

The Indicator

This indicator describes unusual mortality among fish, aquatic mammals (such as otter or beaver), waterfowl (i.e., ducks, geese, and swans), and amphibians, along with the incidence of deformities among amphibians. Unusual mortality generally involves the death of multiple animals in a relatively small area over a relatively short period of time. That is, one dead bird would not be considered an "unusual mortality event," but if one dead bird was found every day for a week, in the same location, it might be. In addition, a single death might be considered for inclusion here if the particular circumstances warranted it—for example, if the bird was part of a flock that was known to have fed at a contaminated site.

This indicator reports mortality events according to the number of individuals killed. When data for different species groups become available, it may be necessary to use categories (such as serious, severe, catastrophic) rather than numbers of individuals. This would facilitate comparison of mortality events affecting different species. For example, an event affecting 100 individuals would be viewed with different levels of seriousness if it affected 100 waterfowl, 100 fish, or 100 mammals such as otters.

The Data

Data Source: Data on waterfowl are collected by the Department of the Interior, U.S. Geological Survey, Biological Resource Division, National Wildlife Health Center (NWHC). They were supplied especially for this report.

Data Collection Methodology: NWHC is a research and diagnostic laboratory, with a primary focus on disease prevention, detection, and control in free-ranging wildlife. NWHC maintains a database of outbreaks of wildlife disease and unusual mortalities, usually affecting multiple animals at the same time. The database covers all 50 states, Puerto Rico, and the U.S. Virgin Islands, and covers wildlife disease and mortality events over the past 25 years. The database contains information on avian, mammalian, and amphibian mortality events. Information in the database is provided by various sources, such as state and federal personnel, diagnostic laboratories, wildlife refuges, and published reports.

Data Quality/Caveats: As noted, the NWHC database covers mammalian and amphibian mortality events, as well as avian events. For freshwater reporting, the avian component was selected as the most complete and most likely to provide representative information at this time. Even for birds, however, the database may not accurately reflect all causes or cases of mortality since NWHC is not informed of every mortality event. Smaller events, in particular, may be handled locally and may not be reported to NWHC. The decision whether or not to include a reported event in the database is made by NWHC specialists. The data reported mortality events primarily affecting anseriformes (ducks, geese, and swans); however, other types of birds that died in an event would have been counted. In addition, the database was not developed as a tool for reporting on national trends; it was intended for use by NWHC as a tool for tracking epidemiological information over time. The information is generally not from specifically defined surveillance and monitoring systems; rather, information is provided as events are discovered or reported.

Data Access: Data are reported quarterly in NWHC online reports. See http://www.nwhc.usgs.gov/pub_metadata/qrt_mortality_report.html. These reports also include information on mammal and amphibian mortality. Data reported here were prepared by NWHC staff specifically for this project.

The Data Gap

Mammal and Amphibian Mortality: As noted, the NWHC collects data on amphibian and mammal mortality. These data are less complete than for waterfowl. Reporting on these groups would be possible if additional resources were available to ensure that reports of amphibian and mammal deaths were reported to NWHC on a regular basis from all regions of the country.

Fish: There is no program in place to collect information about freshwater fish die-offs.

Amphibian Deformities: The North American Reporting Center for Amphibian Malformations (NARCAM; see http://www.npwrc.usgs.gov/narcam/) is a project of the U.S. Geological Survey's Northern Prairie Wildlife Research Center. The NARCAM database receives data from a wide variety of sources. NARCAM is not part of a structured monitoring system, but it cooperates with and receives information from several such monitoring programs, among them NAAMP (North American Amphibian Monitoring Project), Frogwatch USA, ARMI (Amphibian Research and Monitoring Initiative), and A Thousand Friends of Frogs. Wildlife refuge personnel, state fish and game agency staff, university students and researchers, and others who have conducted field surveys of amphibians also submit reports, as do members of the general public, who are able to use NARCAM's Web site to submit their reports directly online. Unless the reporter is thought to have sufficient expertise, the submission is forwarded to a verifier (a professional herpetologist or other expert) who can go to the original site and confirm the report.

As of July 2001, more than 2,000 verified reports, from 47 states and 4 Canadian provinces, had been included in the NARCAM database (see http://www.npwrc.usgs.gov/narcam/reports/reports.htm and http://www.nwhc.usgs.gov/amph-dc.html). However, reports are not evenly distributed among the states: Minnesota, where large numbers of malformed amphibians were first reported, accounts for 21.7% of all reports, Wisconsin for 12.2, and Vermont for 12.0. Another nine states account for 26% of all verified reports. According to NARCAM, it is often difficult to find trained volunteers (and funds) for amphibian surveying programs.

Status of Freshwater Animal Communities: Fish and Bottom-Dwelling Animals

This technical note also supports the urban/suburban indicator Animal Communities in Urban/Suburban Streams.

The Indicator

Biological integrity has been defined as "the capacity of supporting and maintaining a balanced, integrated, adaptive community of organisms having a species composition, diversity, and functional organization comparable to that of the natural habitat of the region" (Karr et al. 1986).

Ecosystems that are "healthy," or show high integrity, are more likely to withstand disturbances imposed by natural and anthropogenic stressors. Biological integrity is a broad term that typically refers to measures of structural elements, such as genetics, individuals, populations, and assemblages (communities).

Quantitative methods for assessing biological integrity (generally called "indices of biotic integrity") have been developed for fish and benthic macroinvertebrates. Benthic macroinvertebrates comprise a heterogeneous assemblage of animal groups that inhabit the sediment or live in or on other bottom substrates in the aquatic environment. Macroinvertebrates are defined as organisms that cannot pass through a No. 30 sieve (0.6-mm, or 0.023-inch openings). The major taxonomic groups of freshwater benthic macroinvertebrates are the insects, annelids (worms), mollusks, flatworms, and crustaceans. They are important members of food webs, and their well-being affects the well-being of higher forms, such as fish.

The Data Gap

Most methods for assessing biotic integrity were developed for streams and wadeable rivers. A seminal step was the development of the Index of Biotic Integrity (IBI) for fish, described briefly at http://www.epa.gov/bioindicators/html/ibi-hist.html.

IBIs for fish and macroinvertebrates are based on reference conditions, which are usually determined by comparison to undisturbed or relatively undisturbed areas believed to be representative of conditions in an ecoregion (an ecoregion is "a relatively large area of land or water that contains a geographically distinct assemblage of natural communities" [Abell 2000]). Most IBIs consist of several metrics that can be organized under three major groupings: species richness and composition, trophic structure, and abundance and condition. Each metric is scored from low (1) to high (5), with low values corresponding to the worst condition and high values representing the reference condition. This approach means that all IBIs must be tailored to the specific species makeup in a specific region. At present, there are no national criteria for assessing biological integrity, but the U.S. Environmental Protection Agency has published guidelines for the development of such criteria, and methods and criteria for several regions and states are under development (see U.S. Environmental Protection Agency 1996 and 1998).

Thirty-two states are developing quantitative tests for fish or bottom-dwelling animals or both: Alabama, Alaska, Arizona, Arkansas, California, Connecticut, Delaware, Georgia, Hawaii, Idaho, Illinois, Indiana, Iowa, Maryland, Massachusetts, Minnesota, Mississippi, Montana, Nebraska, New York, North Carolina, North Dakota, Oklahoma, Oregon, Pennsylvania, South Carolina, Tennessee, Texas, Washington, West Virginia, Wisconsin, and Wyoming. Five states (Florida, Kentucky, Maine, Ohio, and Vermont) already have active quantitative testing programs in place, and 10 states (Colorado, Kansas, Louisiana, Michigan, Missouri, New Hampshire, New Jersey, New Mexico, Rhode Island, Virginia) and the District of Columbia have or are developing some type of fish or benthic community assessment program (generally not a quantitative test, as is proposed here). Only South Dakota, Nevada, and Utah have no active or planned program (http://www.epa.gov/ost/biocriteria/States/streams/streams.html, 06/28/01).

In order to develop a nationally consistent set of observations, there must be consistency in key aspects of the monitoring in different states. For example, some states currently use an "average" condition for the basis of their reference, whereas others use "minimally impaired" (e.g., closer to "natural" or "undisturbed"). The result is that states using the former approach appear to be in good shape (on average), while those that compare their sites to a "minimally impaired" reference show a wide range of IBIs (exceptional to poor). Without a common reference condition, IBI rankings will not be comparable from state to state.

In addition, comparing testing results from different places requires some consistency in scoring methods. For instance, EPA's current Environmental Monitoring and Assessment Program (EMAP; http://www.epa.gov/emap/) uses an IBI scaled to 100, while some state programs use a scale of 1 to 60. Aggregation will require knowledge of the linearity of the scoring method. That is, is an EPA score of 50 the same as a state score of 30? Clearly, rules for classification to establish "ranks" will need to be developed.

Finally, consistency is important with regard to the intensity of sampling. Regions that are more heavily sampled are more likely to reflect the "true" aggregated condition than areas that are not. Criteria for the number of observations per region should be developed to screen out results that do not adequately describe the condition of a body of water.

References

Abell, R.A., et al. 2000. Freshwater ecoregions of North America: A conservation assessment. Washington, DC: Island Press.

Karr, J.R., K.D. Fausch, P.L. Angermeier, P.R. Yant, and I. J. Schlosser. 1986. Assessing biological integrity in running waters: A method and its rationale. Special publication 5. Champaign: Illinois Natural History Survey.

U.S. Environmental Protection Agency. 1996. Biological criteria: Technical guidance for streams and small rivers. U.S. Environmental Protection Agency, Office of Water, Washington, DC. EPA-822-B-96-001.

U.S. Environmental Protection Agency. 1998. Lake and reservoir bioassessment and biocriteria technical guidance document. U.S. Environmental Protection Agency, Office of Water, Washington, DC. EPA-841-B-98-007. http://www.epa.gov/owow/monitoring/tech/lakes.html.

At-Risk Freshwater Plant Communities

The Indicator

For purposes of this report, wetlands are defined using the dominant vegetation (including all rooted aquatic species) and hydrologic properties of the National Wetlands Inventory (NWI; for information about the NWI program, see http://wetlands.fws.gov/; for information on the wetlands classification system, see http://wetlands.fws.gov/Pubs_Reports/Class_Manual/

class_titlepg.htm). Wetland plant communities are defined according to the association concept, which is a plant community type of a specific floristic composition resulting from certain environmental conditions and displaying relatively uniform physiognomy. These communities form part of the U.S. National Vegetation Classification System (NVCS), which was adopted as the federal standard for vegetation information by the Federal Geographic Data Committee in 1997. The classification covers uplands as well as wetlands (see http://www.fgdc.gov/standards/status/sub2_1.html for information about this classification system). The conservation status assessment for each association is called a global rank and is based on the relative rarity and degree of imperilment of the association across its entire geographic range. Tracking wetland plant communities at the association level is a way of measuring wetland diversity and provides a tool to assess conditions affecting specific types of wetlands across the entire country.

Riparian areas are the margins of streams, rivers, or lakes. Riparian areas include a range of plant communities, including both upland vegetation communities (often thriving on the increased moisture available near the stream or river) and wetland plant communities on the floodplain. Because riparian vegetation is a mixture of upland and wetland habitats, classification is difficult. In 1997, the U.S. Fish and Wildlife Service developed a classification scheme for the western United States (http://wetlands.fws.gov/Pubs_Reports/Riparian/Riparian.htm), but this system has only begun to be used for collecting data on riparian habitats in that region of the country. As the Service uses this classification to expand its natural resource mapping to riparian habitats, it should be possible to use the resultant inventory to document the status of riparian habitats and their trends in the future. Meanwhile, NatureServe (a nonprofit organization; see www.natureserve.org) and the Natural Heritage Network, which provides status information on wetlands (see below), are developing an approach for reporting on riparian area condition (see "Data Quality/Caveats" below).

The Data

Data Source: NatureServe and its Natural Heritage member programs develop and maintain information on each association in the NVCS. The regions were defined by The Heinz Center and collaborators, using vegetation-based and climate-based ecological regions, the regional boundaries developed by federal land and resource management agencies, vegetation data from remote sensing, and a common-sense approach to regional differences and similarities.

Data Collection Methodology: NatureServe ecologists gather, review, and integrate available information about vegetation pattern from Natural Heritage program databases, published and unpublished literature, and ecology experts in each state. They then assess conservation status using standardized Heritage ranking criteria (see http://www.natureserve.org/explorer/ranking.htm). Heritage ranks range from 1 to 5, with 1 meaning critically imperiled; 2, imperiled; 3, vulnerable to extirpation or extinction; 4, apparently secure; and 5, demonstrably widespread, abundant, and secure.

Data Manipulation: The global ranks are summarized into "rounded ranks." For example, an actual rank may express substantial uncertainty about whether the community is "critically imperiled" or "imperiled." In all such cases, the rank has been rounded to the more imperiled one.

Data Quality/Caveats: Conservation status ranks are continually reviewed and revised by Natural Heritage program biologists. In addition, as development of the system of classifying plant communities evolves (http://www.fgdc.gov/standards/status/sub2_1.html), more communities will be recognized in geographic areas that are currently "underclassified." Such revisions could affect the proportion of communities considered at-risk.

Some variability exists across the country in how the wetland plant community types were defined and in the amount of survey work done, and the definitions of community types are still under review by ecologists with the NatureServe and Natural Heritage programs.

Data Access: Detailed, periodically updated information on each wetland plant community type, including its status, is available at http://www.natureserve.org/servlet/NatureServe?init=Ecol.

The Data Gap

In the near future NatureServe hopes to augment the associations used in this analysis with an "ecological systems" approach. Ecological systems are biological communities found within a geographic region that share similar ecological processes and gradients (e.g., fire regime, elevation, climate, hydrologic regime), biological dynamics (e.g., succession), and other driving environmental features (e.g., soils, geology). Wetland areas defined by such an approach will bear a more direct relationship to major ecological settings (e.g., riparian types, peatlands, marshes) and thus may be a better basis for this kind of analysis.

This ecological systems approach may help in dealing with the fact that riparian areas are not specifically described in the NVCS and are not assessed by NatureServe. A holistic approach could include the entire moist upland–wet lowland zone as part of the riparian area, facilitating mapping and documentation of these systems across a region.

Stream Habitat Quality

See the technical note for the Farmlands Stream Habitat Quality indicator, p. 237.

Water Withdrawals

The Indicator

Five mutually exclusive categories of water use are reported: "Municipal" supply is water withdrawn by public and private water suppliers and delivered to homes and businesses for drinking, commercial, and industrial uses. "Rural" water use is self-supplied water for domestic use and for livestock. Water used for "irrigation" includes application to crops, pastures, and recreational lands such as parks and golf courses. "Thermoelectric" is water used for cooling in the generation of electric power. "Industrial" water use includes self-supplied water (i.e., water not drawn from the municipal supply) for fabrication, processing, cooling, and washing. The industrial category includes commercial and mining uses of water.

The Data

Data Source: Using raw data collected by states and other sources, the U.S. Geological Survey (USGS) compiles estimates of water use for each use category and then aggregates the estimates for each state, Puerto Rico, and the U.S. Virgin Islands and for each of the 21 water-resources regions. The data have been published every 5 years since 1950 in the USGS Circular series "Estimated Use of Water in the United States." More recent compilations are available electronically at http://water.usgs.gov/watuse/. Some state and federal agencies also publish reports on water use for specific states or categories of use.

Data Collection Methodology: Sources of information and accuracy of data vary by state and by water-use category. Most public-supply water withdrawals and deliveries are metered. In some states, large irrigation and industrial users are required to have water meters to measure the amount of water withdrawn. For other categories, such as self-supplied domestic (e.g., "rural") and small industries (e.g., self-supplied commercial), estimates of water use are derived from population or product output. Energy production data obtained from the Department of Energy are used in making water-use estimates for the thermoelectric power category. Information on acres irrigated is obtained from the Department of Agriculture's Census of Agriculture and its Farm and Ranch Irrigation Survey and from state universities. Information on public water supplies is obtained from the Environmental Protection Agency, state agencies, and individual water suppliers.

Data Manipulation: The steps required to transform the raw data into final form vary with the category of use and with the level of detail of the available raw data. Guidelines used for preparing the most recent estimates are available at http://water.usgs.gov/watuse/. In addition, sources of information and accuracy of data are discussed in the USGS circulars published every 5 years.

Data Quality/Caveats: Because the sources of data and the level of detailed information vary for each state, it is difficult to apply an error analysis to the national aggregate water-use estimates. As part of the compilation effort, each USGS compiler is required to provide justification when estimates change by more than 10% from the previous water-use compilation. Once the data are compiled at the state level, they are peer-reviewed by USGS regional water-use specialists and again by USGS national water-use specialists.

Data Access: The data used here are available in the regular USGS Circular series "Estimated Use of Water in the United States" (for historical data) and at http://water.usgs.gov/watuse/ (for more recent data).

Groundwater Levels

The Indicator

This indicator would describe changes in water levels in major regional aquifers by reporting the fraction of the total area of regional aquifers that declined, increased, or remained stable in comparison to a previous period, and would be reported every 5 years. An example of the kind of data that are available for some major aquifers, and which would be used to develop a national indicator, can be seen in a series of maps depicting changes in the High Plains aquifer, which underlies eight states in the central United States (see McGuire et al. 1999).

The Data Gap

This indicator would require significant data on water levels in major regional aquifers (see below). It would also require a scheme for classifying changes in aquifer level as "significant increase," "significant decrease," or "no significant change." Changes in groundwater level have unique levels of significance in different aquifers; a change of a few feet in a shallow coastal aquifer may be quite important in terms of susceptibility to saltwater intrusion, while a change of 10 feet on a very large aquifer may not be as significant. Logically, the values for "stable" will be different in different aquifers (e.g., the High Plains case defined –5 feet to +5 feet as "no significant change"). Therefore, definitions of significant increase or decrease (and thus, no significant change) should be determined on an aquifer-by-aquifer basis.

Water-level data are available for all or parts of every state, but these data cannot be aggregated to provide national coverage because of limited coverage of most aquifer systems and lack of electronic availability of much of the monitoring data. The High Plains aquifer is one of the few multistate aquifers with systematic and coordinated water-level monitoring. States or areas with good water-level-monitoring programs include parts of Florida, Long Island (NY), Pennsylvania, and Utah. To ensure national coverage, the following points must be addressed:

- Data must be collected from areas that represent the full range of topographic, hydrogeologic, climatic, and land use environments within the major aquifers.
- Data must be collected using standardized methods from monitoring wells or other wells not affected by local pumping. Procedures for well selection and data collection are available in Chapter 2 of the USGS's 1980 *National Handbook of Recommended Methods for Water-Data Acquisition.*
- There must be agreement on timing of water-level measurements across the country so that the status of major aquifers in a region or in the entire country can be presented as a snapshot in time.
- Plans must be in place to ensure long-term viability of observation-well networks and data collection programs, including plans for a combination of data collection at long-term monitoring wells and periodic synoptic measurements.
- There must be agreement among the agencies or other sources of data on electronic data storage, access, and dissemination. The agencies that will be responsible for leadership in compiling and publishing the data must be identified.

References

McGuire, V.L., C.P. Stanton, and B.C. Fischer. 1999. Water level changes, 1980 to 1997, and saturated thickness, 1996–97, in the High Plains aquifer. U.S. Geological Survey Fact Sheet. FS-124-99. http://ne.water.usgs.gov/highplains/hp97_web_report/97fs.pdf.

Waterborne Human Disease Outbreaks

The Data

This indicator reports the number of waterborne disease outbreaks (WBDOs) reported to the Centers for Disease Control and Prevention (CDC) through a network of doctors and state and local public health officials. In addition, the U.S. Environmental Protection Agency (EPA) and the Council of State and Territorial

Epidemiologists assist with collection and reporting of WBDOs. CDC generally reports only cases involving at least two individuals with a similar illness, and only where epidemiological evidence implicates water as the probable source of the illness. (Data from 1920 to 1936 include outbreaks that affected more than five people. These early data also include some cases related to contamination of reservoirs and cisterns, which are not included in the 1973–1998 dataset.) This indicator does not report outbreaks due to distribution system problems of unknown origin, nor does it include outbreaks caused by contamination of water or ice at the point of use (e.g., a contaminated water faucet). Outbreaks associated with recreational fresh surface waters are included here; outbreaks associated with marine water, spas, whirlpools, hot tubs, and the like are not reported.

Data Collection Methodology: State and territorial and local public health departments are primarily responsible for detecting and investigating WBDOs and voluntarily reporting them to CDC. CDC requests annual reports from state and territorial epidemiologists or from persons designated as WBDO surveillance coordinators. EPA collects additional information on water quality and treatment as needed from state drinking water agencies.

Data Manipulation: Information from CDC was sorted to identify only those outbreaks that are clearly linked to contamination in lakes, streams, ponds, and the like. Thus, outbreaks linked to contamination at the point of use and those linked to marine waters, hot tubs, spas, and swimming pools were deleted. Outbreaks associated with untreated and inadequately treated drinking water were aggregated.

Data Quality/Caveats: Various factors can affect the chances of an individual illness being linked to a water source. These include public awareness, the likelihood that ill people will consult the same health care provider, availability and extent of laboratory testing, local requirements for reporting cases of particular diseases, and the surveillance and investigative activities of state and local health and environmental agencies. Recognition of WBDOs is also dependent on certain outbreak characteristics; large interstate outbreaks and outbreaks involving serious illness are more likely to receive the attention of health authorities. Outbreaks associated with private water systems that serve a small number of residences or farms are the most likely to be underreported because they generally involve only a few people.

Data Access: Current WBDO data are reported by CDC, Public Health Service, U.S. Department of Health and Human Services, in CDC Surveillance Summaries for Waterborne-Disease Outbreaks, *Morbidity and Mortality Weekly Report.* The 1985–1999 Surveillance Summaries are available at http://www.cdc.gov/mmwr/sursumpv.html; see Volumes 37, 39, 40, 42, 45, 47, and 49. Data from 1978 to 1984 are from CDC's Water-Related Disease Outbreaks Annual Summaries (1980–1985), and data from 1973 to 1977 are from CDC's Foodborne and Waterborne Disease Outbreaks Annual Summaries (1974–1979).

Freshwater Recreational Activities

There is no technical note for this indicator.

Grasslands and Shrublands

Area of Grasslands and Shrublands

The Indicator

This indicator reports the acreage of grasslands and shrublands using land cover data based on satellite measurements from the National Land Cover Dataset (NLCD). For this indicator, pastures and hay-lands were included; however, they were included within farmlands for that system's extent indicator (p. 191) and the national extent indicator (p. 40). Pastures and hay-land are included in this indicator because many fall within the description of grasslands and shrublands given in the introduction of this chapter, and because it is not clear how well the satellite data distinguish them from less-managed grasslands. (Note that in the NLCD the classification "pasture/hay" is defined as areas of grasses, legumes, or grass-legume mixtures planted for livestock grazing or the production of seed or hay crops.)

The U.S. Department of Agriculture Economic Research Service (ERS) has carefully tracked changes in different land uses over the past 50 years in its "Major Uses of Land" series (see http://www.ers.usda.gov/data/majorlanduses/). Its accounting for cropland and forest land is consistent with the approach taken in this report, and its trend in urban area was adopted for comparison purposes (see national extent indicator, p. 40). However, the ERS category that is closest to the definition of grasslands and shrublands used in this report is "grassland, pasture and range." This category, which included 578 million acres for the lower 48 states in 1997, is inconsistent with the definition used in this report because land is included based on grazing activity rather than on the land cover classification; there was no obvious way to reconcile the differences in definition adequately so that the ERS data could be used to track trends in grasslands and shrublands.

The extent of grasslands and shrublands (shown in this indicator) is a key aspect in understanding this ecosystem. Additional indicators in this chapter provide information on other key parameters. In addition, there have been attempts to provide overall ratings of the "ecological condition" or "health" of these lands. One potential measure of ecological condition is seral stage. The concept of rangeland health has been addressed by the National Academy of Sciences (Committee on Rangeland Classification 1994). However, ecological condition is expressed differently at multiple scales, including sites and landscapes, and presently does not lend itself to synthesis at a national scale. Aggregation of site-level rangeland condition data to a national assessment is particularly problematic (Mitchell 2000).

The Data

Data Source: The data for the lower 48 states are from the NLCD, which has a resolution of approximately 30 meters on a side. The NLCD is a product of the interagency Multi-Resolution Land Characterization (MRLC) initiative (see the technical note for the national extent indicator, p. 207).

Data for Alaska are from a vegetation map of Alaska by Flemming (1996), based on Advanced Very High Resolution Radiometer (AVHRR) remote-sensing images with an approximate resolution of 1 kilometer on a side. The following groupings of classes were used (see http://agdc.usgs.gov/data/projects/fhm/#G [Statewide Vegetation/Land Cover] and the technical note for the national extent indicator, p. 207). Briefly, the following are Flemming's (1996) classes that were included within grasslands and shrublands: alpine tundra & barrens (#3); dwarf shrub tundra (#4); tussock sedge/dwarf shrub tundra (#5); moist herbaceous/shrub tundra (#6); wet sedge tundra (#7); low shrub/lichen tundra (#8); low & dwarf shrub (#9); tall shrub (#10); and tall & low shrub (#23).

Presettlement estimates of grass/shrub land cover were derived from data provided by Richard J. Olson, Oak Ridge National Laboratory (personal communication). These data were first published in Klopatek et al. (1979). This dataset provided potential area of Kuchler vegetation types. A set of Kuchler vegetation types provided by the Vegetation/Ecosystem Modeling and Analyis Project (VEMAP) program (http://www.cgd.ucar.edu/vemap/lists/kuchlerTypes.html) was used to select a set of grassland and shrubland vegetation types from Klopatek et al. (1979). While there are minor differences between the Kuchler naming conventions in the Klopatek et al. (1979) and VEMAP lists, the overall suite of vegetation classes matches quite well, and the resulting estimate is considered reasonable.

Data for recent changes in "non-federal grasslands and shrublands" are from the U.S. Department of Agriculture A Natural Resources Conservation Service National Resources Inventory (NRI) program. NRI uses the term "rangelands," which is consistent with our definition of grasslands and shrublands, except that the NRI data used here do not include pasture or lands enrolled in the Conservation Reserve Program. Data from 1982, 1992, and 1997 are derived from the NRI Summary Report (revised December 2000), tables 5 and 8. See http://www.nhq.nrcs.usda.gov/NRI/1997/national_results.html.

Data Limitations/Caveats: In the discussion section of the indicator text, an attempt was made to place bounds on the loss of grasslands and shrublands since the time of European settlement. This was done to give the reader a sense of the change; however, this estimate should be interpreted with caution. There are two caveats in particular. The satellite data used to estimate the acreage of pasture do not indicate whether or not the land is heavily managed (i.e., plowed and seeded). Depending upon the division of pastures between relatively heavily managed and relatively lightly managed (i.e., more natural in character), grasslands converted to pasture could represent a significant addition to the estimates of area converted. Also, a considerable amount of the land that is now classified as pasture is located in the East and was probably forest. Hence, to say that grasslands and shrublands declined 40 to 140 million acres since European settlement ignores the fact that more of the original grasslands and shrublands may have been lost but these losses were offset by gains in eastern pastures.

Data Access: Please see the information contained within the technical note for the national extent indicator (p. 207).

References

Committee on Rangeland Classification. 1994. Rangeland health: New methods to classify, inventory and monitor rangelands. Washington, DC: National Academy Press.

Flemming, M.D. 1996. A statewide vegetation map of Alaska using a phenological classification of AVHRR data. 1996 Alaska Surveying and Mapping Conference, Anchorage, Alaska.

Klopatek, J.M., R.J. Olson, C.J. Emerson, and J.L. Joness. 1979. Land-use conflicts with natural vegetation in the United States. Environmental Conservation 6:191–199.

Mitchell, J.E. 2000. Rangeland resource trends in the United States: A technical document supporting the 2000 USDA Forest Service RPA Assessment. Gen. Tech. Rep. RMRS-GTR-68. Rocky Mountain Research Station, Fort Collins, CO.

Land Use

The Data

Data Source: Data on Conservation Reserve Program (CRP) lands are from the USDA Farm Services Agency (FSA), which manages CRP signups and contracts.

Data Manipulation: Reported here are lands in the following "practice" categories: Introduced Grasses (CP1), Native Grasses (CP2), Wildlife Habitat (CP4), Grass Waterways (CP8), Established Grass (CP10), Wildlife Food Plots (CP12), Filter Strips (CP13), Contour Grass (CP15), Snow Fences (CP17), Salt Tolerant Vegetation (CP18), Alternative Perennials (CP20), Filter Strips (CP21), Cross Wind Strips (CP24), and Declining Habitat (CP25). The figure of 29 million acres (29.37 million acres) reported here is based on data reported at http://www.fsa.usda.gov/crpstorpt/04approved/r1pracyr/r1pracyr2.htm (report MEPRTK-R1, April 30, 2001, last accessed 10/15/01). Data from each of these practice categories were summed for all contracts active at the time the report was prepared.

This report provides information on cover practices for contracts beginning in each program year since CRP was implemented (1987). However, the report generally provides information only for contracts that begin in a specific year, not those that are active in a specific year. Therefore, it is not possible to develop time trend information for cover practices active in any given year. The exception to this statement is that the report does provide a summary of cover practices for all currently active contracts, and this summary provided the figure of 29.37 million acres.

The Data Gap

For this indicator to be reported on effectively at a national level, a standardized set of definitions and criteria for classifying land uses is needed. Following are possible components and approaches to be incorporated into such definitions.

Livestock Raising: Federal land managers report the allowable stocking rate in Animal Unit Months (AUMs) for individual livestock allotments. In theory, the number of animal units per acre could be calculated. However, research should be done to understand how well allowable stocking rates reflect actual rates of use. In addition, we are aware of no source for consolidated information on acreage used for livestock raising on private lands.

Intensive Recreation: This category is intended to describe areas whose major purpose is recreational use, and where such use is significant enough to generate changes in the condition of the area. To adequately report on such areas, a definition needs to be devised based on factors such as levels of recreation use or number and type of recreational facilities in an area.

Energy and Mineral Development: As with recreation, adequate reporting on areas used for energy and mineral development requires a definition that accounts for the areas directly affected (e.g., drilling pad area, mine pits, tailings ponds) as well as nearby areas with visual, noise, dust, and other impacts.

Rural Residences: As with other categories, adequate reporting of this indicator component would require adoption of thresholds that identify a class of lands with low-density rural residence development. These areas are less dense than what most people would consider "suburban" but would have to be distinguished in some manner from truly rural, very low density development. The target for this component is often described as "ranchette" development.

"Protected Areas": Identifying protected areas will require adoption of a standard that distinguishes certain public or private lands based on their legal status or management practice; lands that are managed primarily in order to maintain biodiversity and natural processes should be included. Several categorization approaches have been developed including the World Conservation Union's six-category approach (see http://wcpa.iucn.org/wcpainfo/protectedareas.html) and the Gap Analysis Program's management status four-category scheme (http://www.gap.uidaho.edu/). A dataset being developed for this purpose will report the acreage of lands according to a system of categorizing management intensity developed by the U.S. Geological Survey Gap Analysis Program (see http://www.gap.uidaho.edu/handbook/Stewardship/default.htm). This database is currently under development by the Conservation Biology Institute in conjunction with the USDA Forest Service; see http://www.consbio.org/cbi/what/pad.htm.

In addition to developing definitions for these categories, mechanisms should be developed for the accounting of the acreage in each category and changes in these areas over time.

Area and Size of Grassland and Shrubland Patches

The Indicator

This measure would report the percentage of grasslands and shrublands in patches of different sizes. Patch sizes and percentages would be reported separately for grasslands and shrublands. The patch sizes for this indicator are as follows: less than 10 acres, 10–99 acres, 100–999 acres, 1000–9,999 acres, and 10,000 acres or greater.

Species and ecosystem processes are sensitive to spatial heterogeneity. Landscape diversity is an important component of species diversity, habitat conservation, and human health. While much research has been undertaken to determine these relationships in forested ecosystems, there is general agreement among grass/shrub experts that such spatial patterns are important in grasslands and shrublands as well. For example, in the sagebrush/grassland mosaic of western intermountain basins, fuel buildup after a period of minimal grazing and sufficient rainfall creates highly flammable conditions. When a fire does occur, the sagebrush is greatly reduced in abundance because it cannot sprout, unlike nearly all the grassland plants. Grassland expands, reducing the area of habitat for sagebrush-dependent species.

The Data Gap

High-quality satellite data provide an excellent baseline to assess future changes in patch sizes. Many of the indicators in this report are based on data from the National Land Cover Dataset, produced by a federal interagency consortium including the U.S. Geological Survey, the USDA Forest Service, the National Oceanographic and Atmospheric Administration, and the U.S. Environmental Protection Agency (see http://www.epa.gov/mrlc/nlcd.html and the technical note for the national extent indicator, p. 207). It is expected that satellite data will be used for this indicator.

However, the software currently available for analyzing patch size characteristics was developed for use on relatively small landscape areas. It is not designed, and cannot be used, to process datasets as large as are required for this indicator. An alternative approach involves analyzing patch characteristics for smaller landscape areas (such as 7.5 x 7.5 km), then combining the statistics on these many individual areas to describe much larger areas, such as ecoregions. However, in this approach, any patch that crosses the boundary of one of the 7.5 by 7.5 km squares is not accurately represented, because a portion is in one square and a portion is in the adjoining square. This is referred to as a right-censored distribution and will provide consistent underestimates of the number of larger patches.

In addition, the indicator should distinguish between lands that have been altered (e.g., cultivated and seeded for pastures) and more "natural" lands; only patches of the more natural grasslands and shrublands would be included.

References

Turner, M.G., and R.G. Gardner, eds. Quantitative methods in landscape ecology. Springer-Verlag Ecological Studies, Vol. 82. New York: Springer-Verlag.

Nitrate in Groundwater

The Indicator

The sources cited below provide additional information regarding the choice of nitrate as an appropriate and sensitive indicator of ecological condition (Smith et al. 1997), how vegetation composition, activity, and management affect nitrate concentrations in soil water, seeps, and streams (Ramundo et al. 1992, Tate 1990), and the relationship between soil texture and types and abundance of carbon sources (Nolan and Stoner 2000).

The Data Gap

Data on nitrate concentration in groundwater need to be collected and reported in a consistent fashion across a broad and representative set of grassland and shrubland areas. Nitrate measurement is simple, straightforward, and largely unchanged since measurements began more than 100 years ago. Because many usable wells already exist, on both public and private lands, the cost of sampling and analysis is the primary factor limiting current efforts.

In addition, careful searching of federal, state, county, municipal, and private records could produce a valuable historical archive that would serve as a baseline against which to compare current conditions.

The technical note for indicators describing nitrate concentrations in forested, farmland, and urban/suburban areas (see p. 232) provides information on the U.S. Geological Survey

National Water Quality Assessment program, which is a potential future source of data for this indicator.

References

Nolan, B.T., and J.D. Stoner. 2000. Nutrients in groundwaters of the conterminous United States 1992-1995. Environmental Science and Technology 34:1156–1165.

Ramundo, R.A., C.M. Tate, and T.R. Seastedt. 1992. Effects of tallgrass prairie vegetation on the concentration and seasonality of nitrate-nitrogen in soil water and streams, pp. 9–12. In D.D. Smith and C.A. Jacobs (eds.), Proceedings of the Twelfth North American Prairie Conference, Ames, Iowa.

Smith, R.A., G.E. Schwarz, and R.B. Alexander. 1997. Regional interpretation of water quality monitoring data. Water Resources Research 33:2781–2798.

Tate, C.M. 1990. Patterns and controls of nitrogen in tallgrass prairie streams. Ecology 71: 2007–2018.

Carbon Storage

The Indicator

This indicator seeks to track long-term changes in carbon sequestration in grasslands and shrublands. Measurements of this indicator through time can provide information on whether the ecosystem is a net source or a net sink of atmospheric carbon dioxide. An ecosystem accruing carbon is one contributing to a reduction in greenhouse gases. White et al. (2000) have estimated that grassland ecosystems worldwide store an amount of carbon that is about half of that stored by the world's forests and roughly equivalent to that stored by agricultural systems.

An ecosystem not changing in carbon content, but also not producing high inorganic nitrogen exports, is likely a late-successional, mature system possessing high biotic diversity. Systems containing high amounts of carbon are often associated with high levels of ecosystem services (i.e., responsible for clean air and clean water).

The minimum data that are required for this indicator are percentage soil organic matter (SOM) in surface soil layers and carbon stored in plant material, estimated on an area basis. Soil measurements provide an excellent index of both potential soil fertility and nitrogen storage. Soil carbon storage is the net accumulation of (mostly dead) plant matter. It represents the net accumulation of carbon inputs (plant production) minus all sources of organic carbon loss. Changes in soil carbon storage can be caused by changes in climate, changes in atmospheric chemistry, or changes in the abundance and species composition of the vegetation. Plant carbon storage varies annually while soil carbon storage changes at longer time scales.

The Data Gap

Data are not currently available to provide systematic monitoring and reporting of soil and vegetation carbon. There are, of course, many research sites at which such information is collected. Soil carbon can be found at substantial depths, although routine sampling of soils to such depths is uncommon. A variety of available models can estimate total soil carbon storage from surface measurements of SOM and estimate plant carbon from above-ground vegetation measurements. However, there is a serious concern about the use of single-point estimates to represent large areas. Some procedures for establishing the representativeness of sites will be required. Intensive, long-term data are available from the Long Term Ecological Research (LTER) sites, including those in Alaska, Michigan, Minnesota, Kansas, Colorado, and New Mexico (there are two sites in New Mexico). Such sites could provide substantial validation for more widely dispersed measurements. Relatively long-term alpine and arctic tundra SOM data are available from LTER sites as well. See http://lternet.edu/sites/ for additional information and links to the LTER network.

References

Burke, I.C., C.M. Yonker, W.J. Parton, C.V. Cole, K. Flach, and D.S. Schimel. 1989. Texture, climate and cultivation effects on soil organic matter content in U.S. grassland soils. Soil Science Society of America 53: 800–805.

Esteban, G.J., and R.B. Jackson. 2000. The vertical distribution of soil organic carbon and its relation to climate and vegetation. Ecological Applications 10: 423–436.

Jenny, H. 1941. Factors of soil formation. New York: McGraw-Hill.

National Research Council. 2000. Ecological indicators for the nation. Washington, DC: National Academy Press.

Seastedt, T.R., C.C. Coxwell, D.S. Ojima, and W.J. Parton. 1994. Controls of plant and soil carbon in a semihumid temperate grassland. Ecological Applications 4:344–353.

White, R.P., S. Hurray, and M. Rohweder. 2000. Pilot analysis of global ecosystems: Grassland ecosystems. Washington, DC: World Resources Institute.

Number and Duration of Dry Periods in Grassland and Shrubland Streams

The Indicator

This indicator has two aspects: (1) the percentage of streams with at least one day of no flow (also referred to as zero flow) in a year and (2) for sites with at least one day of zero flow, the duration of zero flow events, compared to a long-term average (50 years in this case). Together, these two variables help describe both the frequency and duration of zero-flow events. Changes in either of these could have significant effects on aquatic and riparian species.

Relatively intact/undisturbed watersheds (including their upland, riparian, and wetland components) are capable of maintaining the maximum duration of streamflow their climates will support. When soil conditions and the kinds and proportions of vegetation promote the infiltration of moisture falling in the watershed, and when evapotranspiration and groundwater recharge are in balance, rapid loss of moisture to overland flow is minimized and long-duration, frequently perennial (i.e., year-round) flow is maintained. Intact riparian areas and wetlands are capable of retaining water during high-flow periods and metering out stored moisture during periods of low flow—further supporting longer duration, or perennial flow.

Conversely, the increase of impervious surfaces through soil compaction or development and/or the loss of protective vegetation result in increased overland flow and rapid runoff events—depleting moisture storage to maintain long-duration flows. Improved management of grazing that promotes stream-side vegetation can lead to increased stream flow. In contrast, moisture loss from excessive evapotranspiration caused by plant community imbalances can also reduce the amount and duration of stream flow—this is attributed to encroachment of pinyon-juniper woodlands, Western juniper, and other species that are not actively managed.

Changes in annual weather patterns or long-term climatic changes also influence streamflow quantity and flow duration.

The Data

Data Source: Data reported here are from the U.S. Geological Survey (USGS) stream gauge network. USGS has placed stream gauges and maintained flow rate records throughout the United States since the end of the 19th century.

Data Collection Methodology: Stream gauging data are collected using standard USGS protocols.

Data Manipulation: The goal of the initial data manipulation was to identify stream gauges in watersheds where more than 50 percent of the land cover is grassland or shrubland. Each site was referenced to a watershed cataloguing unit (known as a 4-digit Hydrologic Unit Code, or HUC4) using latitude and longitude. Grassland and shrubland were defined using the National Land Cover Dataset (see http://www.epa.gov/mrlc/nlcd.html) using land cover categories 51 (shrubland), 71 (grassland/herbaceous), and 31 (bare rock, sand, clay) (see http://www.epa.gov/mrlc/classes.html and the technical note for the national extent indicator, p. 207). The HUC4s were also paired with their corresponding ecoregions (see below for description of the ecoregions used). Only sites with greater than 50% grass/shrub cover were used in the analysis.

The number of sites with at least one no-flow day in a year was determined for each water year from 1950 to 1999. The corresponding percentage value for that year was also calculated as 100 x (number of sites/total sites). The percentage values were then averaged over each decade (i.e., 1950s, 1960s, 1970s, 1980s, and 1990s). This procedure was followed for all sites with greater than 50% grassland/shrubland cover as well as for each ecoregion.

For the analysis of duration of zero-flow, only sites with at least one no-flow day in each decade between October 1, 1949, and September 31, 1999, were considered. The analysis determined whether there was a substantial increase, substantial decrease, or minimal change in the number of no-flow days, compared to the long-term (50-year) average for each stream. These categories are defined by the percent change in average zero-flow days, as compared to the long-term average, on a stream-by-stream basis. Thus, a "substantial increase" is defined as an increase of more than 100 percent in the duration of zero flow, or a change from perennial (no zero-flow) to intermittent. Likewise, a "substantial decrease" is defined as a decrease of at last 50 percent in the duration of zero flow, or a change from intermittent to perennial. "Minimal change" is defined as anything between a 100% increase and a 50% decrease.

Ecoregions: This indicator is reported using an ecoregional approach developed by the USDA Forest Service (Bailey 1995). The Bailey system has several levels into which the United States may be divided, based on dominant biological and physical attributes. The scheme has three domains, 13 divisions, and 52 provinces. We have chosen to report this indicator on the basis of divisions. We selected three major suites of Bailey's divisions:

- *Desert shrub ecoregion,* composed of the following Bailey's divisions: 320 (tropical/subtropical desert division), M320 (tropical/subtropical desert division—mountain provinces), 340 (temperate desert division), M340 (temperate desert division—mountain provinces)
- *Grassland/steppe ecoregion,* composed of the following Bailey's divisions: 250 (prairie division), 330 (temperate steppe division), M330 (temperate steppe division—mountain provinces), 310 (tropical/subtropical steppe division)
- *California/Mediterranean,* composed of the following Bailey's divisions: 260 Mediterranean division, M260 (Mediterranean division, mountain provinces)

See http://www.fs.fed.us/colorimagemap/ecoreg1_divisions.html for full definitions and a map showing the individual divisions.

Data Access: The data records used in this study are available on the Internet in the form of daily streamflow values reported as the average volume of water per second over a 24-hour period (http://water.usgs.gov/nwis/discharge).

References

Bailey, R.G. 1995. Description of the ecoregions of the United States. 2nd ed. rev. and expanded (1st ed. 1980). Misc. Publ. No. 1391 (rev). Washington, DC: USDA Forest Service.

Chaney, E., W. Elmore, and W.S. Platts. 1990. Managing change: Livestock grazing on western riparian areas. Washington, DC: U.S. Environmental Protection Agency, General Accounting Office.

General Accounting Office. 1988. Public rangelands: Some riparian areas restored but widespread improvement will be slow. RCED-88-105. Washington, DC: GAO.

Depth to Shallow Groundwater

The Indicator

Shallow aquifers, or deeper regional aquifers where shallow aquifers do not exist, are often the water source for the maintenance of riparian and wetland ecosystems (Dawson and Ehleringer 1991, Flanagan et al. 1992). Shallow groundwater is being increasingly withdrawn for agriculture, urban expansion, and mining. Reduction in stream flows, which maintain shallow alluvial aquifers, by dams or other activities also reduces the level and availability of this important water source (Shafroth et al. 2000). In addition, deep-rooted plants, such as pinyon-juniper and Western juniper, are capable of lowering shallow aquifers in the process of transpiration.

Declining groundwater has been shown to affect riparian ecosystems through a reduction of (1) the shallow water table necessary for recruitment of riparian species and (2) long-term maintenance of established woody riparian vegetation. Urban development may tap shallow groundwater associated with river basins, which can cause a gradual decline in associated riparian forests (Stromberg et al. 1992). Gravel mining may alter the natural gravel deposits along rivers, causing shallow groundwater to recede, affecting established riparian vegetation (Scott et al. 1999).

Streams in arid climates, affected by withdrawal of groundwater inputs, also show declining vigor of riparian vegetation as both the alluvial groundwater level declines and stream flow is reduced (Stromberg et al. 1996). Shallow groundwater decline is often a long-term phenomenon because it is usually caused by a gradual withdrawal of water from the shallow aquifer which may continue to be recharged, although inadequately, by stream inflows or from deeper aquifers. If the source of water replacement is affected, shallow aquifers, which are the primary water source for springs, seeps, wetlands, potholes, and riparian areas and which in some cases support declining ecosystems, will thus not be replenished.

Shallow groundwater depths are often used to determine long-term cumulative effects of groundwater withdrawal by agriculture, mining, or urban expansion. Urban expansion in the Great Basin has resulted in water claims on both shallow and deep aquifers. Modeling of this potential withdrawal shows that the shallow water table may decline by 1–3 m (Schaefer and Harrill 1995), a result that would drastically impact the isolated desert springs, the only water source for domestic livestock and wildlife in these areas. Decreasing aquifer volumes and dropping water tables also add to energy costs of water withdrawal, sufficiently so to cause decline or termination of regional agriculture in arid regions of the United States.

The technical note for Number and Duration of Stream Flow (immediately preceding this technical note) also provided relevant perspective on the interaction between groundwater, surface water, and land use.

The Data Gap

Although depth to deep groundwater or the regional aquifer is regularly measured in monitoring and functioning wells across the country and the data are reliable and maintained by appropriate agencies, these data have not been integrated either for the grassland/shrubland region or nationally (see groundwater indicator in freshwater chapter, p. 151; and USGS 1997).

Data on shallow aquifers are quite limited. Depths for shallow aquifers (e.g., groundwater under riparian communities) and deeper regional aquifers are usually treated separately. The limited shallow aquifer data from the U.S. Geological Survey and many academic and agency research projects dealing with rivers and adjacent floodplains (see citations above) may also be good sources for regional shallow groundwater data.

References

Dawson, T.E, and J.R. Ehleringer. 1991. Streamside trees that do not use stream water. Nature 350:335–227.

Flanagan, L.B., J.R. Ehleringer, and T.E. Dawson. 1992. Water sources of plants growing in woodland, desert, and riparian communities: Evidence from stable isotope analysis. US Forest Service Tech. Report INT-289:43–47.

Schaefer, D.H., and J.R. Harrill. 1995. Simulated effects of proposed ground-water pumping in 17 basins of east-central and southern Nevada. USGS Water-Resources Investigations Report 95-4173.

Scott, M.L., P.B. Shafroth, and G.T. Auble. 1999. Responses of riparian cottonwoods to alluvial water declines. Environmental Management 23:347–358.

Shafroth, P.B., J.C. Stromberg, and D.T. Patten. 2000. Woody riparian vegetation response to different alluvial water table regimes. Western North American Naturalist 60:66–76.

Stromberg, J.C., R. Tiller, and B. Richter. 1996. Effects of groundwater decline on riparian vegetation of semiarid regions: The San Pedro, Arizona. Ecological Applications 6:113–131.

Stromberg, J.C., J.A. Tress, S.D. Wilkins, and S. Clark. 1992. Response of velvet mesquite to groundwater decline. Journal of Arid Environments 23:45–58.

United States Geological Survey. 1997. Ground water atlas of the United States—Segment 1 California Nevada. Online data at http://water.wr.usgs.gov/gwatlas/ and http://water.usgs.gov/ogw/.

At-Risk Native Species

See the technical note for the core national at-risk species indicator (p. 214).

Non-native Plant Cover

The Data Gap

Data from various sources must be evaluated and synthesized to provide regional and national estimates of the area occupied by non-native plant species. There are numerous federal, state, and local government programs that collect relevant information, plus important efforts in nongovernmental organizations and academic institutions that could contribute to reporting on this indicator.

A recently established consortium includes representatives from the U.S. Geological Survey, the U.S. Fish and Wildlife Service, National Aeronautics and Space Administration, USDA Forest Service, The Nature Conservancy, Colorado Natural Heritage Program, USDA Animal Plant and Health Inspection Service, National Park Service, Bureau of Land Management, Colorado State University, the Biota of North America program (University of North Carolina), and others. This initiative, titled "One if by Land, Two if by Sea," will attempt to better coordinate and synthesize existing data on non-native species in the United States. Coordination for this initiative is being provided by Tom Stohlgren, USGS Natural Resources Ecology Laboratory, Colorado State University, Fort Collins, CO 80523, tom_stohlgren@USGS.gov.

Many agencies of the Departments of Interior and Agriculture, as well as state and local governments, nongovernmental organizations, and universities, collect important data on invasive plants in grassland and shrubland regions. Several examples of such programs are listed below.

The USDA Forest Health Monitoring program (http://www.na.fs.fed.us/spfo/fhm/), for example, collects plant cover data in forests throughout the United States, and the program is expanding to include grasslands and shrublands in some areas.

- U.S. Department of Agriculture's Center for Plant Health Science and Technology, a part of the Animal and Plant Health Inspection Service's Plant Protection and Quarantine program, maintains the Federal Noxious Weeds Database, which provides descriptive and some distributional data on many recognized invasive plants (see http://www.invasivespecies.org/fedweeds.html). The distribution data for the Federal Noxious Weeds Database (which provides data up to 1999) are from the Synthesis of the North American Flora by John Kartesz (North Carolina Botanical Garden, University of North Carolina) and Christopher Meacham (Jephson Herbarium, University of California, Berkeley). The Synthesis is available as an interactive database on CD-ROM (see http://www.bonap.org/synth.html for ordering information). It provides information at state level, although the program from which it was generated also maintains county-level data for 44 states (see www.bonap.org/summary.html).

- The University of Montana maintains the INVADERS database, which covers five northwestern states (Oregon, Wyoming, Montana, Idaho, and Washington) with information at county level. INVADERS may be accessed at http://invader.dbs.umt.edu/.
- The U.S. Geological Survey has initiated the Southwest Exotic Plant Mapping Program, or SWEMP, which is designed to develop a regional database of exotic plant distributions for the Southwest (Arizona, New Mexico, and Colorado Plateau portions of Utah and Colorado). Some data are available at http://wapiti.wr.usgs.gov/swepic/.

Standardized field techniques should be adopted to create comparable data that can be synthesized. These extensive field datasets must be linked to high-resolution maps of vegetation, soils, topography, and land use to achieve reliable national coverage.

Population Trends in Invasive and Non-Invasive Grassland/Shrubland Birds

The Indicator

This indicator reports the change in population of invasive and native, non-invasive grassland/shrubland birds. The invasive birds include both non-native birds and some native birds that spread aggressively because of a favorable change in conditions. The non-invasive birds are native birds that depend on high-quality native grasslands and shrublands.

There was some interest in separating the groups of this indicator by native/non-native; however, given the low number of birds involved, a decision was made to maximize the number of species in each group to improve the statistical reliability of the results. Thus, both natives and non-natives were included in the invasive category.

The Data

Data Source: This indicator incorporates population trend estimates for 15 invasive non-native and 35 native grassland bird species. Estimates are based on data collected for the North American Breeding Bird Survey (BBS), and were obtained from the Patuxent Wildlife Research Center (PWRC), United States Geological Survey, U.S. Department of the Interior. Trends were estimated for BBS Physiographic Strata (regions) 6–8, 32–56 and 80–91, in seven 5-year intervals from 1966 to 2000 (http://www.mbr-pwrc.usgs.gov/bbs/physio.html).

Following is a list of the invasive species included in this indicator and the reason the species is considered invasive: American crow, habitat conversion to agriculture; American robin, habitat fragmentation due to suburban development; black-billed magpie, habitat conversion and fragmentation; bronzed cowbird, forage in association with livestock; brown-headed cowbird, forage in association with livestock; cattle egret, Old World native, habitat conversion to agriculture, and forage in association with livestock; common grackle, habitat fragmentation and conversion to agriculture; European starling, Old World native; gray partridge, Old World native, habitat conversion to agriculture; great-tailed grackle, habitat conversion to agriculture; house finch, habitat fragmentation due to suburban development; house sparrow, Old World native; mourning dove, habitat conversion and fragmentation; ring-necked pheasant, Old World native, habitat conversion to agriculture; and rock dove (domestic pigeon), Old World native, habitat conversion, and fragmentation.

Native, non-invasive species, which are restricted to those native species known to be dependent upon relatively intact and high-quality native grasslands and shrublands, included Baird's sparrow, black-throated sparrow, LeConte's sparrow, bobolink, loggerhead shrike, Brewer's sparrow, long-billed curlew, burrowing owl, McCown's longspur, Cassin's sparrow, mountain plover, chestnut-collared longspur, northern harrier, common nighthawk, prairie falcon, dickcissel, sage grouse, eastern meadowlark, sage sparrow, ferruginous hawk, sage thrasher, golden eagle, savannah sparrow, grasshopper sparrow, sharp-tailed grouse, greater prairie chicken, Sprague's pipit, Henslow's sparrow, Swainson's hawk, horned lark, upland sandpiper, lark bunting, vesper sparrow, lark sparrow, and western meadowlark.

Data Collection Methodology: The BBS is jointly coordinated by the PWRC and the Canadian Wildlife Service, Environment Canada. It is conducted along randomly located routes on secondary roads throughout the contiguous United States and southern Canada. Routes are 24.5 miles long, with 50 survey points at 0.5-mile intervals. Observers survey each route annually during June (May in some southern states and desert areas). At each survey point, the observer counts all birds seen or heard within a 0.25-mile radius during a 3-minute census. The first BBS routes in 1966 were run only east of the Mississippi River. The BBS was extended to the central United States in 1967, with full coverage of the contiguous United States by 1968. The number of BBS routes has increased over time, so recent years provide more comprehensive data than early years. Summaries of the BBS methodology are provided by Peterjohn and Sauer (1993) and Sauer et al. (2000a,b), and a review of the program is provided by O'Connor, et al (2000).

Data Manipulation: W. Mark Roberts, an independent researcher, obtained trend estimates (change in population size as a percentage per year) for each species in each physiographic stratum (region) and time interval from a server program provided by John R. Sauer at PWRC (http://www.mbr-pwrc.usgs.gov/bbs/trend/tfmb.html). The program uses an "estimating equations estimator" (described in Link and Sauer 1994) to calculate each stratum's trend estimate from individual route data. Dr. Roberts performed subsequent manipulations: To reduce the influence of less reliable estimates, each stratum estimate was weighted toward the survey-wide estimate, proportionately to the variances of both estimates. Weighting used an empirical-Bayes formula (Equation 1 in Link and Sauer 1996). The mean of the variance-weighted stratum estimates was calculated for each species and time interval. The summary indicator is the proportion of species with positive (increasing) mean variance-weighted estimates. To compare native with invasive birds, Yates-corrected Chi-square statistical tests were performed on the frequencies of positive and negative mean variance-weighted estimates.

Data Quality/Caveats: Bird species differ in habits, habitat, abundance, and range, all factors that may bias trend estimates for certain species more than for others (see Droege 1990 and http://www.mbr-pwrc.usgs.gov/bbs/introbbs.html). The BBS methodology and data have been subjected to peer review, results of which are available at http://www.mp2-pwrc.usgs.gov/bbs/

bbsreview/. The trend analysis program (Sauer and Hines 2001) and manipulations performed by Dr. Roberts are based on peer-reviewed methodology. Output of these manipulations has not, however, been independently verified.

Data Access: Trend estimates are the output of a draft program (http://www.mbr-pwrc.usgs.gov/bbs/trend/tfmb.html), placed on the PWRC server but not linked to public pages. Though accessible without charge, this program should not be used without permission from John R. Sauer at PWRC.

References

Droege, S. 1990. The North American Breeding Bird Survey, pp. 1-4. In J.R. Sauer and S. Droege (eds.), Survey designs and statistical methods for the estimation of avian population trends. U.S. Fish and Wildlife Service, Biological Report 90(1).

Link, W.A., and J.R. Sauer. 1994. Estimating equations estimates of trends. Bird Populations 2:23–32.

Link, W.A., and J.R. Sauer. 1996. Extremes in ecology: Avoiding the misleading effects of sampling variation in summary analyses. Ecology 77(5):1633–1640.

O'Connor, R.J., E. Dunn, D.H. Johnson, S.L. Jones, D. Petit, K. Pollock, C.R. Smith, J.L. Trapp, and E. Welling. 2000. A programmatic review of the North American Breeding Bird Survey: Report of a peer review panel to USGS Patuxent Wildlife Research Center, Laurel, MD. http://www.mp2-pwrc.usgs.gov/bbs/bbsreview/.

Peterjohn, B.G., and J.R. Sauer. 1993. North American Breeding Bird Survey annual summary 1990-1991. Bird Populations 1:1–15.

Sauer, J.R., and J.E. Hines. 2001. Trend analysis form: Draft version. Version 2001.00, 2 April 2001 DRAFT, USGS Patuxent Wildlife Research Center, Laurel, MD. http://www.mbr-pwrc.usgs.gov/bbs/trend/tfmb.html.

Sauer, J.R., J.E. Hines, I. Thomas, J. Fallon, and G. Gough. 2000a. Details of the BBS. In The North American Breeding Bird Survey, results and analysis 1966–1999. Version 98.1, USGS Patuxent Wildlife Research Center, Laurel, MD. http://www.mbr-pwrc.usgs.gov/bbs/introbbs.html.

Sauer, J.R., J.E. Hines, I. Thomas, J. Fallon, and G. Gough. 2000b. General introduction–BBS. In The North American Breeding Bird Survey, results and analysis 1966–1999. Version 98.1, USGS Patuxent Wildlife Research Center, Laurel, MD. http://www.mbr-pwrc.usgs.gov/bbs/genintro.html.

Fire Frequency

See the technical note for the forest fire frequency index indicator, p. 243.

Riparian Condition

The Indicator

The indicator would report on the ecological integrity or health of riparian ecosystems, including both physical and biological factors.

The number and quality of streams and rivers and their associated riparian areas are a function of watershed conditions. Consequently, the condition of riparian areas may be useful as an indicator of ecological alterations of grassland and shrubland watersheds. For example, if land cover is altered, the stream flow may also be altered, changing the geomorphology of the river channel and influencing riparian dynamics. Regulation of rivers by dams and other flow-altering devices also influences downstream conditions, including streambank erosion and river meandering, sediment aggradation and seedbed development, and natural recruitment of riparian vegetation. Local land use within a floodplain, such as agriculture, grazing, and urbanization, may also greatly influence the condition of riparian ecosystems. In turn, riparian systems also influence hydrogeomorphic processes by trapping sediment and modifying flood flows and groundwater recharge.

The Data Gap

Several measures are being used nationally, but no "simple" index has received general acceptance among the research community. An appropriate "Index of Riparian Integrity" still needs to be fully developed. Several federal agencies use a combined qualitative metric called Proper Functioning Condition (PFC) when evaluating riparian systems (see Bureau of Land Management 1993). However, PFC is primarily hydrogeomorphic and includes little of the biological conditions such as species composition, age classes, understory condition, canopy condition, and successional processes. Another methodology developed in the past few years is the Hydrogeomorphic Methodology (HGM) (Brinson 1996, Smith et al. 1995). This methodology uses a complex of indices for hydrology, geomorphology, land use, biology, and other aspects to create a single index for the riparian system. It is complex, but a simplified version might be developed for broad-scale application. Yet another, simpler method is one that relies on satellite data (Iverson et al. 2001).

Aspects of the riparian condition that can be measured on a regional basis and that should be considered in any multi-metric index include hydrology (e.g., relationship to natural flow patterns), geomorphology (e.g., stream sediment transport), and biology (e.g., canopy cover condition; percentage of potential recruitment or successional measures; canopy diversity, or coverage of point bars). Many of these aspects either are being measured now or could be measured as part of a national riparian evaluation system.

Once an index is developed, it would be applied within a sampling design that would allow estimation of the conditions on all streams within a region. Thus, for example, such an approach might provide estimates of the number of miles of stream with "riparian condition index" that is "high," "medium," or "low," each of these being within a selected numerical range of the index.

References

Brinson, M.M. 1996. Assessing wetland functions using HGM. National Wetlands Newsletter 18:10–16.

Bureau of Land Management.1993. Riparian area management: Process for assessing proper functioning condition. Technical Reference 1737-9. USDOI, BLM, Denver, CO. Revised 1995, 1998.

Iverson, L.R., D.L Szafoni, S.E. Baum, and E.A. Cook. 2001. Development of a riparian wildlife habitat evaluation scheme using GIS. Environmental Management 28(5):639–654.

Smith, R.D., A. Ammann, C. Bartoldus, and M.M. Brinson. 1995. An approach for assessing wetland functions using hydrogeomorphic classification, reference wetlands and functional indices. U.S. Army Corps of Engineers, Waterways Experiment Station. Vicksburg, MS. Tech. Rep. TR-WRP-DE-9

Production of Cattle

The Indicator

This indicator reports on the U.S. cattle and calf inventory not at feedlots in July of each year. It is assumed these cattle are grazing on grasslands and shrublands (including pasture) because they are not at feedlots. Note that cattle will spend some time during the summer months in woodlands or forests if they are available; it is uncertain how this complication would affect the data reported here.

The Data

Data Source: Data presented here are from the U.S. Department of Agriculture National Agricultural Statistics Service (NASS). NASS conducts annual surveys of livestock herd sizes during January and July.

Data Manipulation: Cattle numbers on grasslands/shrublands were estimated by subtracting the number of "cattle on feed" from total cattle ("all cattle") numbers in July. Total cattle numbers include cows that have calved, bulls, heifers, steers, and calves. Most calves have not weaned by July; however, increased forage consumption by lactating cows compensates for this apparent overcounting of animals. The number of cattle on feed includes steers, heifers, cows, and bulls.

In winter, some cattle are placed on croplands to consume plant products and seeds left behind. More important, the quantity and quality (digestibility and amount of protein) of grass plants decline substantially in winter, so the forage supply on grasslands and shrublands is inadequate. Thus, in many parts of the country, ranchers must feed hay to cattle in winter.

Data Caveats/Quality: It is known that cattle will spend some time during the summer months in woodlands and forests. The effect that this caveat might have on the indicator is unknown.

Another caveat involves the fact that the indicator reports the number of cattle rather than the weight of cattle. The average weight of cattle may change over time, so the same herd size may involve more or fewer pounds of livestock. If such changes occur, this indicator may over- or under-represent the production of livestock.

Data Availability: These data are available at http://www.nass.usda.gov:81/ipedb/cattle.htm. This site allows the user to retrieve selected data for selected years from the NASS database. To obtain the total cattle July inventory, select "Inventory by Class, July 1" for years of interest. To obtain data on cattle on feed, select "cattle on feed, July 1" (data availability begins in 1994). In obtaining the data reported here, the "cattle on feed" data were subtracted from the "cattle and calves-all" column of the cattle inventory data set. These data were accessed October 25, 2001.

Data from the July inventory are available for a longer time series than is presented here, but comparable data on cattle on feed are not. In addition, data from the January inventory are available for both the total inventory and cattle on feed. However, these data are not believed to represent cattle on grasslands and shrublands, and thus are inappropriate for this indicator.

Data on the value of the U.S. cattle inventory are from NASS, 2000 Agricultural Statistics (www.usda.gov/nass/pubs/agr00/00_ch7.pdf).

Data on longer term trends in cattle inventory are from the NASS database Web site noted above. The January cattle inventory was inspected for the period from 1960 to 2000 and showed a high of 132 million in 1975. As noted, this inventory is not comparable to the July inventory, and can be used only to suggest long-term trends in cattle herd size.

Recreation on Grasslands and Shrublands

There is no technical note for this indicator.

Urban and Suburban Areas

Area of Urban/Suburban Lands

The Indicator

This indicator reports the total number of acres that are classified as "urban and suburban" and the amount of the various "undeveloped" land cover types within these areas. More detail is provided below, but "urban and suburban" is defined here as land that is substantially covered by one of the following land cover types: low-intensity residential, high-intensity residential, commercial–industrial–transportation, or urban and recreational grass. These categories are based on remote-sensing classification of land cover and are defined at http://landcover.usgs.gov/classes.html and http://www.epa.gov/mrlc/classes.html.

It was our intent that urban and suburban areas should include all major metropolises and their outlying suburbs as well as smaller settlements across the country that have a similar character even though they may not be adjacent to a metropolis. Our goal was to define those areas across the United States that should be classified as "urban and suburban"; The Heinz Center examined several possibilities before choosing the approach used here.

The use of the Census Bureau's metropolitan statistical areas (MSAs) was the coarsest approach considered. MSAs include entire counties (or cities and townships in New England) rather than only the large urban centers and those outlying areas that are connected to them in some fashion. In the West especially, vast counties are included in MSAs even though only a fraction of the county area is actually urban or suburban. MSAs account for about 20% of the land area of the lower 48 states; The Heinz Center believes this is a significant overestimate of the area covered by cities and suburbs.

Urbanized areas (UAs), also defined by the Census Bureau, offer a more refined but still incomplete solution. Metropolises and their outlying areas are included in UAs, but smaller settlements, which share many of their characteristics with suburbs, are not included. A drawback to using UAs is that they are determined in part by political/jurisdictional boundaries, in addition to the degree of development. A potentially larger confounding issue is that the rules for delineating UAs have changed significantly since their first use in 1950. The Census Bureau is well aware of this shortcoming and will be releasing newly constructed UA boundaries in early 2002. The U.S. Department of Agriculture (USDA) Economic Research Service (ERS) has estimated urban land area since 1950. ERS's estimate has incorporated the area of UAs as well as the amount of area in Census-defined "places" that have a population of at least 2,500 people. We have used ERS's estimate to gauge the change in urban land area over time in the national extent indicator (p. 40); however, due to the limitations of UAs we chose not to rely on these estimates exclusively to define urban/suburban areas.

A third and still more refined option considered would have relied totally on Block Groups (BGs), which are small regions based on political boundaries within which the Census Bureau counts the population. It would be possible to choose a density threshold—1000 people per square mile is generally accepted as "urbanized" by the Census Bureau—and define those BGs that meet or exceed this density as urban and suburban. A shortcoming of this approach is that BGs dominated by warehouses or railroad yards, for example, which are certainly urban in character, would be excluded because of their low population densities. As is discussed below, the approach chosen for this indicator does, indeed, include most BGs with densities at or above 1000 people per square mile.

A fourth option was to adopt the estimates for developed lands made by the USDA Natural Resources Conservation Service's National Resources Inventory (NRI). The definitions used by the NRI agree, in principle, with those for urban/suburban lands. However, NRI reports on any and all developed areas—including those down to about one-quarter acre. In contrast, this project focuses on those areas with sufficient density and size to qualify as "suburban" in character, as well as areas that are undeniably "urban." As noted below, this project's definition requires an area to be at least 270 acres in size before it is included within the "urban/suburban" definitions. In addition, the NRI data are derived from a statistical sampling rather than a cataloging of all developed lands. Thus, it would not have been possible to delineate individual urban/suburban areas on a map (as is done along with the national extent indicator), which would be necessary to implement several of the other indicators included in this report.

The approach adopted here (see "Data Manipulation" below) uses satellite data to classify land cover. The advantage of this method is that it includes virtually all the BGs with at least 1000 people per square mile, as well as other developed but lightly populated land, such as warehouse districts. In addition, by overlaying BGs on the urban/suburban areas, it was possible to estimate that about 75% of the 1990 population lived in these areas (note that the data used to generate urban/suburban areas came from 1992). As described in more detail below, a series of steps have been used to define the outlines of urban and suburban areas based on four different satellite land cover classifications.

A potential shortcoming of using a satellite-based approach rather than a delineation based in Census data is that it will be more challenging to correlate environmental quality trends like air and water quality with human demographic and health data. However, a geographic information system (GIS) can be generated to associate Census BGs, for example, with urban/suburban areas. This would permit such correlations to be done for studies of demographics and human health.

It may be useful in future editions of this report to consider presenting the data on urban/suburban lands based on the number of people associated with them. This would require shifting priorities for the indicator and a GIS analysis as described above. Also, as data become available, it would be good to add the proportion of native and non-native species to the graph showing the composition of the undeveloped portion of urban and suburban lands.

The Data

Data Source: Satellite data are derived from the National Land Cover Dataset (NLCD), a product of the Multi-Resolution Land Characterization (MRLC) Consortium, which is a partnership between the U.S. Geographical Survey (USGS), the U.S. Forest Service, the National Oceanographic and Atmospheric Administration (NOAA), and the Environmental Protection Agency (EPA) (see http://www.epa.gov/mrlc/ or http://landcover.usgs.gov/nationallandcover.html).

Data Collection Methodology: Please refer to the national extent indicator technical note (page 207) for a discussion of the NLCD.

Data Manipulation: The NLCD divides the lower 48 states of the United States into several billion square pixels that are about 100 feet on a side. The data presented for this indicator are based on analysis of larger pixels (1000 ft on a side), each of which contains 100 of the smaller pixels. The first step was to classify any 1000-ft pixel as urban and suburban if a majority of the 100-ft pixels within it fell into one of the four "developed" land cover types available in the NLCD: low-intensity residential, high-intensity residential, commercial–industrial–transportation, or urban and recreational grasses. Very large aggregates of the 1000-ft pixels, which were found for metropolises such as New York City, were "smoothed" to some degree; that is, small clusters of "undeveloped land" pixels that were wholly included within a metropolis were subsumed in the urban and suburban areas. Other clusters of undeveloped-land pixels within an urban and suburban area, although connected to the perimeter by one or more pixels on a diagonal, were also included in the urban and suburban area. For clusters of developed-land pixels to be counted as urban/suburban in outlying areas, at least 13 of the 1000-ft pixels had to touch at their sides or corners for a minimum size of 270 acres.

The final step for this indicator was to evaluate the proportion of different land cover types within the 1000-foot pixels. This process yielded estimates of the amount of both developed land and undeveloped land (in several categories) by region.

Data Quality/Caveats: It is important to note that the methods used to establish the NLCD relied on two different satellite images of a given area, plus ancillary data. An image taken during the "leaf-off" period in the late fall to early spring was often more important to the classification process than the fully vegetated image. This was especially true in urban settings with a good deal of tree-lined streets; the foliage of deciduous trees should not have obscured the constructed surfaces during the leaf-off period and, therefore, should not have led to an underestimate of developed lands in these regions.

Given that the method used here to establish urban/suburban areas is based on square pixels that are roughly 100 feet on a side, some detail would have been missed in a typical urban setting. Specifically, the trees on a tree-lined streets would most likely not be distinguished from the street and sidewalk. However, a large expanse of trees, such as a heavily wooded median strip or a small park, may well have been classified as forest.

Data Access: All these analyses were conducted at the Land Cover Applications Center at USGS's Earth Resources Observations Systems Data Center. The raw data from which this indicator was developed are available at no cost from the MRLC Consortium (http://edcwww.cr.usgs.gov/programs/lccp/mrlcreg.html), but vast computing power was necessary for this analysis. Note: The data available at the Web site listed here are the "raw" data from which estimates of urban/suburban area, and the size of natural areas within, were prepared. The actual data presented in this report were prepared specially for The Heinz Center for this report.

Suburban/Rural Land Use Change

There is no technical note for this indicator.

Patches of Forest, Grasslands and Shrublands, and Wetlands

The Indicator

Undeveloped land in urban and suburban areas was analyzed to identify patches of natural land. "Natural" is defined to include all lands that have been classified in the extent indicator as any of the following: forests, grasslands and shrublands, or wetlands. The indicator presents the size distribution of contiguous patches composed of any of these land cover types, or combinations of them, by region.

There is a generally understood "rule" among conservation ecologists that smaller patches of habitat generally provide lower quality habitat than larger patches. There is some debate as to whether this is true for wetlands. There is some evidence that the quality of the habitat remains fairly constant regardless of its size (see Gibbs 1993). On the other hand, there is also evidence that isolated wetlands habitats (i.e., those not surrounded by undeveloped upland vegetation) are compromised in their habitat value (see Calhoun and Klemens, 2002).

The Data

Data Source: Satellite data are derived from the National Land Cover Dataset (NLCD), a product of the Multi-Resolution Land Characterization (MRLC) Consortium, which is a partnership between the U.S. Geological Survey (USGS), the U.S. Forest Service, the National Oceanographic and Atmospheric Administration (NOAA), and the Environmental Protection Agency (see http://www.epa.gov/mrlc/ or http://landcover.usgs.gov/nationallandcover.html).

Data Collection Methodology: Please refer to the national extent indicator technical note (p. 207) for a discussion of the NLCD.

Data Manipulation: Eight of the 21 NLCD classifications were defined as "natural" for this analysis. These include three classes considered as "forest" for this report (deciduous forests, evergreen forests, mixed forests); three types considered as "grasslands/shrublands" (shrubland, grasslands/herbaceous, bare rock/sand/clay), and two wetlands types (woody wetlands, emergent herbaceous wetlands). Patches were defined as collections of 30-meter pixels in any of these eight classifications that touched one another either on their sides or at their corners. (Patches can be as few as one or as many as hundreds of pixels.) Data were processed on a state-by-state basis, and then these data were grouped based on the four regions. For a given region, the number of patches of various sizes were counted, thereby creating a distribution.

Data Quality and Caveats: Data were processed on a state-by-state basis, which means that in some cases a patch of natural land may have been broken into two segments at the state boundary by the analysis process. In addition, natural patches may well extend beyond the boundary of urban and suburban areas, meaning that the value reported here would be an underestimate of the actual size of the patch. Also, the smallest patches cannot be characterized by these methods, so estimates of the acreage (and percentage of total urban and suburban areas) in the less-than-10-acre category are an underestimate of the true value. This occurs because it is difficult to distinguish very small patches (e.g., one to a few pixels) that are mixed in with developed land cover types.

Also, the satellite data cannot be used to distinguish between a parcel of land that has always been grassland/shrubland or wooded and one that was developed but has since reverted to this apparently natural land cover (e.g., a dump or landfill). It would be misleading to label such land as "natural." It is expected that this mislabeling occurred infrequently; however, it is not possible to estimate how much of an effect this might have had on the data.

Note: Additional caveats are listed in the technical note for the Area of Urban and Suburban Lands indicator, p. 264.

Data Access: All these analyses were conducted at the Land Cover Applications Center at the USGS's Earth Resources Observations Systems Data Center. The data are available (http://edcwww.cr.usgs.gov/programs/lccp/mrlcreg.html) at no cost from the MRLC Consortium, but considerable computing power is necessary to manipulate them. Note: The data available at the Web site listed here are the "raw" data from which estimates of urban/suburban area, and the size of natural areas within, were prepared. The actual data presented in this report were prepared specially for The Heinz Center for this report.

References

Calhoun, A., and M.W. Klemens. 2002. Best development practices (BDPs) for conserving pool breeding amphibians in residential and commercial developments (MCA Tech. Paper 5).

Gibbs, J.P. 1993. Importance of small wetlands for the persistence of local populations of wetland-associated animals. Wetlands 13(1):25-31.

Total Impervious Area

The Indicator

Perhaps the single most dramatic and pervasive impact of urbanization on the functions and values of a watershed is the replacement of the natural landscape with pavement and other water-impervious (impenetrable) material such as roads, parking lots, driveways, sidewalks, and rooftops. Increased levels of impervious surfaces interrupt the hydrologic cycle, alter stream structure, and degrade the chemical profile of the water that flows through streams. These changes affect fish and wildlife in various ways, and are cumulative within watersheds. Research indicates that when total impervious area (TIA) in a watershed reaches 10%, stream ecosystems begin to show evidence of degradation. Ecological effects become severe as TIA approaches 30% (for more discussion, see Arnold and Gibbons 1996; Booth and Jackson, 1997; Schueler 1994; Schueler and Holland 2000).

Effects that have been associated with increases in impervious area include the following:

- *Increases in stream temperature,* as rain runs over heated pavement. During warmer months, water flowing over impervious surfaces is often 10–12°F warmer than water that passes through fields and forests. Higher water temperatures increase the metabolic rates of stream-dwelling plants and animals, so that an organism living in warmer water needs more oxygen than the same species in cold water. Unfortunately, warmer water cannot hold as much oxygen as cold water.
- *Changes in stream flows.* Greater stormwater volumes traveling over the surface and being delivered too rapidly to streams leads to increased stream flashiness and a reduction in summer base flows, sometimes causing perennial streams to become intermittent or to dry up completely. As a result, urbanized watersheds are prone to more frequent and bigger floods.
- *Stream channel modification.* The rapid runoff associated with increased stormwater velocity and volume quickly erodes and incises the stream channel and banks. Channels widen and straighten to accommodate higher flows. Ponds, pools, riffles, and sandbars are simplified or washed away, eliminating critical habitat for fish, waterfowl, and other species of animals and plants.
- *Increased pollutant loadings.* Concentrations of pollutants in streams increase with increases in impervious area. Common urban pollutants include pesticides, bacteria, nutrients such as phosphorus and nitrogen, and other contaminants, such as PCBs and heavy metals.

The percentage of impervious surfaces within a watershed is a good indicator of the degree of urbanization and the associated negative ecological impacts, but it can be very difficult to measure. Where such data are available, watershed urbanization is most often quantified in terms of the proportion of the basin area covered by impervious surfaces, or TIA.

The Data Gap

Existing data should be examined in order to develop a cost-effective way of estimating impervious area regionally and nationally. This may involve the use of new remote-sensing techniques; collation of existing local information; the use of surrogates, such as the amount of road surface; or other approaches.

References

Arnold, C.L., and C.J. Gibbons. 1996. Impervious surface coverage: The emergence of a key environmental indicator. Journal of the American Planning Association 62(2):243–258.

Booth, D.B., and C.R. Jackson. 1997. Urbanization of aquatic systems: Degradation thresholds, stormwater detection, and the limits of mitigation. Journal of the American Water Resources Association 35(5):1077–1090.

Schueler, T.R. 1994. The importance of imperviousness. Watershed Protection Techniques 1(3):100–111.

Schueler. T.R. and H.K. Holland. eds. 2000. The practice of watershed protection. Ellicott City, MD: Center for Watershed Protection.

Stream Bank Vegetation

The Data Gap

As discussed on the indicator page, it is not yet clear if this indicator will utilize data collected "on-the-ground" or via remote sensing. Use of satellite data would require acquisition of vegetation data, perhaps at a resolution finer than that provided by the National Land Cover Data Set (NLCD), which has 30-m resolution (see http://www.epa.gov/mrlc/ and the National Extent technical note, page 207, for more detail). It will also be necessary to decide how to characterize vegetation, which would probably be based on the ecological functioning of the cover. For example, residential lawns function differently from woods or natural grasslands in the way they shed water, passively clean stormwater runoff or provide habitat for stream-dependent animals. Secondly, the vegetation data would have to be merged with data on the location of streams (probably from the USGS National Hydrography Dataset (NHD), see http://nhd.usgs.gov/). Stream location would have to be limited to those segments that are urban/suburban in nature, which might be achieved by simply restricting the dataset to those stream and river segments that are within the urban and suburban areas defined by this project (see the Area of Urban / Suburban Lands page 181).

Nitrate in Urban/Suburban Streams

See the technical note for Nitrate in Farmland Streams, p. 232

Phosphorus in Urban/Suburban Streams

See the technical note for Nitrate in Farmland Streams, p. 232

Air Quality (High Ozone Levels)

The Indicator

The indicator reports the number of days per year when peak 8-hour average ozone concentrations exceed 0.08 parts per million (ppm). When a monitor exceeds this 8-hour average concentration four or more times per year, an area is likely to be out of compliance with the National Ambient Air Quality Standard (NAAQS) for ozone; this standard was chosen by the Environmental Protection Agency (EPA) to "protect the public health ... with an adequate margin of safety," as specified by the Clean Air Act. Note that the actual calculation to determine compliance with the NAAQS involves calculation of a 3-year average of the annual fourth-highest daily maximum 8-hour average concentration; if this value exceeds 0.08 ppm an area is in violation.

Oxides of nitrogen (NO_x), which are byproducts of fossil fuel combustion, when in the presence of sunlight in the atmosphere, will break apart and generate nitric oxide (NO) and a single atom of oxygen (O). This oxygen atom quickly combines with molecular oxygen (O_2) forming ozone (O_3). Ozone can oxidize NO back to nitrous oxide (NO_2), which allows the cycle to start over again. Volatile organic compounds (VOCs), which come from paints and solvents, unburned fuel, and industrial sources, factor into the equation because they also can oxidize NO to NO_2. Hence, with both NO_x and VOCs present, ozone accumulates in the atmosphere and ultimately poses a threat to human health, wildlife, pets, and building materials.

The Data

Data Source: Data are maintained by EPA in the Aerometric Information Retrieval System (AIRS). The Clean Air Act requires every state to establish a network of air-monitoring stations for pollutants, including ozone, using criteria set by EPA for their location and operation; there are approximately 1500 ozone monitors in this network. The states must provide EPA with an annual summary of results from each monitor.

Data Collection Methodology: Ozone monitoring instruments are intended to produce a measurement every hour, for a possible total of 8,760 hourly measurements in a year. A monitor is considered operational if it reports a measurement for more than half the hours in a year.

Data Manipulation: For each of the 1500 ozone monitors nationwide, EPA provided The Heinz Center with 10 years of data on the number of days per year that peak 8-hour average ozone concentrations exceeded 0.08 ppm. Data were not reported for years missing more than half the daily peak concentrations during the ozone season (typically May through September). From these monitors, The Heinz Center selected the 624 monitors that are located in urban and suburban areas (as defined for this report; see Area of Urban/Suburban Lands and associated technical note, pp. 181 and 264). The trend graphs include only those monitors with data for at least 8 of the 10 years between 1990 and 1999; 397 monitors meet the criteria for data completeness. For the maps, which provide the locations of monitors according to their 1999 values, 486 monitors had data.

In Hawaii, there are three ozone monitors; however, we do not have satellite data on the extent of urban and suburban areas for this state. Therefore, it was not possible to identify the urban and suburban monitors in Hawaii in the same fashion as in the lower 48 states. As is discussed in the technical note for Area of Urban/Suburban Lands (p. 264), there is reasonable overlap between the urban and suburban areas defined using satellite data and Census Bureau Block Groups having at least 1000 people per square mile. For this reason, we identified Hawaiian monitors located in Block Groups having a density of at least 1000 people per square mile. Two of the Hawaiian monitors passed this screen; ultimately one of these was dropped due to insufficient data. There is a single monitor in Alaska and following the method used for Hawaii, it was excluded from our analysis because it is not within a Block Group having at least 1000 people per square mile.

Data Quality and Caveats: The monitors that make up this national network conform to uniform criteria for monitor siting, instrumentation, and quality assurance.

Data Access: Air quality data upon which this indicator is based are collected regularly by EPA and are available at http://www.epa.gov/airs/. EPA provided the specific data used in this analysis to The Heinz Center especially for this project. However, annual summary monitoring data are available at EPA's AIRData Web site (http://www.epa.gov/air/data/index.html).

Chemical Contamination

See the technical note for the core national contaminants indicator, p. 210.

Urban Heat Island

The Indicator

Cities have modified climates based on factors such as building density and type and energy use, as well as local topography and regional weather patterns. The "urban heat island" represents the difference between urban and nearby rural air temperatures and is directly related to urban land cover and human energy use. For most cities, this difference often is negligible in the daytime but develops rapidly after sunset. Maximum difference occurs 2–3 hours after sunset and may be as great as 18°F. In general, as the population density of a city increases, the difference in minimum temperature between the urban core and rural site increases nonlinearly. Urban heat island effect for a city is calculated by comparing the temperature of a monitoring station in the urban core with a monitoring station from a neighboring rural location. This difference might be reported as the average monthly difference between urban and rural sites. Nationally, the indicator might report the number of cities with various levels of difference between urban and rural sites: 0–6°F, 6 to less than 13°F, or more than 13°F.

As constructed surfaces replace natural vegetation, an area's ability to absorb and store heat increases; the natural cooling effect mediated by trees and other vegetation is reduced (water moves from the soil into a plant via its roots, exiting ultimately by evaporation through pores in the leaves in a process called evapotranspiration—a cooling process much like when sweat evaporates from our skin). The urban heat island represents a change in the diurnal pattern of ambient temperature. Because many biological processes are temperature dependent, changes in the temperature regime may have profound effects on species and ecological processes. In fact, many of the proposed effects from elevated global temperatures occur in urban areas because of the urban heat island effect.

It is reported by the Centers for Disease Control's National Center for Environmental Health that extreme heat events, some of which may be directly attributable to the heat island effect, are responsible for greater loss of human life in the United States than hurricanes, lightning, tornadoes, floods, and earthquakes combined (http://www.cdc.gov/nceh/hsb/extremeheat/). Other effects may include physiological stress in some species, altered species composition and structure in ecological communities, modified nutrient and carbon availability, and altered home range of pathogens. For example, physiological stress results from altered phenology and modified moisture nutrient availability. The urban heat island also modifies energy use for heating and cooling buildings and vehicles.

The Data Gap

National Weather Service temperature data are available for a large number of locations in the United States and could be used to determine urban heat island effect and how this temperature differential has changed over time. Analyzing historical data would require a significant amount of time, energy, and funding to retrieve archival information, to conduct quality assurance and quality control on data, and to perform the analysis. Data problems include obtaining long-term data records for both urban and adjacent rural sites and accounting for changes in monitoring locations or instrumentation and for changes in population densities and human activities around monitoring sites. Another problem occurs for desert cities where the maximum temperature difference between urban and rural monitoring locations may occur during midday rather at sunset. Although a temperature differential exists 2–3 hours after sunset, the evaporative cooling from vegetation within the city may create cooler temperatures during the day than adjacent desert temperature.

Remote-sensing data have been used to examine temperature differences between urban and rural sites; however, these measurements record surface temperatures rather than ambient temperatures.

Species Status

The Indicator

This indicator reports the percentage of "original" vertebrate animals and vascular plants that are at risk of displacement or have been displaced from metropolitan areas (i.e., major cities and their suburbs found within the urban/suburban areas defined by this report; small, isolated cities or suburbs would be excluded because it would likely not be feasible to include them in the necessary monitoring program). "Original" is defined as existing prior to European settlement in the area that is now a metropolitan area. Using the reference point of presettlement is in some sense an arbitrary choice; its use does not necessarily mean that it would be desirable to have all original species present in urban/suburban areas. This indicator includes only vertebrate animals (not insects, worms, and the like) and vascular plants (not mosses, fungi, algae, and so on).

The Data Gap

This indicator should be reported for larger metropolitan regions, where expertise and information are likely to be available. For each of these areas, a list of plant and animal species present before settlement must be compiled. These lists can be derived from reviews of the historical literature, museum records, Natural Heritage program data, and agency files. Information on current status must be obtained through field surveys, which will need to be repeated periodically. If scientists develop standardized protocols for observation and reporting, much of the data could be collected by trained volunteers.

Many organizations collect data about the current distribution and status of species, but few of these provide information on species status or population trends within areas as small as a metropolitan area. For example, most states have Natural Heritage programs, which provide status information on a wide variety of species (http://www.natureserve.org/about_nhnoverview.htm), but generally on a statewide or larger area basis.

There are a growing number of city, county, and regional efforts to gather and use biodiversity information, and these efforts could form the basis for reporting this indicator. Two programs that exemplify this trend are the Illinois EcoWatch Network and Chicago Wilderness. EcoWatch is a series of volunteer monitoring programs coordinated through the Illinois Department of Natural Resources (http://dnr.state.il.us/orep/inrin/ecowatch). The program has an UrbanWatch component (http://www.fmnh.org/urbanwatch/splash.asp), as well as RiverWatch, ForestWatch, and PrairieWatch components. Chicago Wilderness (http://www.chiwild.org) is a partnership of more than 130 organizations working to protect, restore, and manage natural areas in the three-state Chicago metropolitan area. In addition, Robinson et al. (1994), in a study in Staten Island, New York, showed a loss of over 40% of native flora and an increase of over 33% non-native flora during the period 1879 to 1991. DeCandido (2001) found similar results for The Bronx, New York.

Finally, there must be some mechanism that will ensure adequate consistency between local and regional efforts, and that will be responsible for collating data from local sources to produce regional and national statistics.

References

DeCandido, R. 2001. Recent changes in plant species diversity in Pelham Bay Park, Bronx County, New York City, 1947–1998. Ph.D. Dissertation, The City University of New York.

Robinson, G.R., M.E. Yurlina, and S.N. Handel. 1994. A century of change in the Staten Island flora: Ecological correlates of species losses and invasions. Bull. Torrey Bot. Club 121(2):119–129.

Disruptive Species

There is no technical note for this indicator.

Status of Animal Communities in Urban and Suburban Streams

See the technical note for the freshwater indicator for status of animal communities in streams (p. 253).

Public Accessible Open Space per Resident

The Indicator

The indicator reports the amount of publicly accessible open space per resident for major urban and suburban areas in the United States. "Natural" lands include areas managed for their natural values as well as areas that are vegetated, but also relatively highly managed, such as playing fields and parks. Minor amounts of pavement or other "hard" surfaces would not preclude an area from being considered "natural."

According to the National Research Council (2000, p. 22), the natural environment provides people with a variety of ecological goods and services, including "recreation, aesthetic enjoyment, and spiritual experience." This indicator is an important measure of the capacity of urban and suburban areas to provide recreational and aesthetic enjoyment in an unbuilt environment close to home.

Definitions: "Open space" means unbuilt land or water areas dominated by naturally pervious surfaces. A grassy park or golf course would qualify as open space; a paved playground would not. A river or lake would qualify as open space, as would some cemeteries. Satellite imagery will soon provide 5-meter resolution images, but whether there should be a minimum size to qualify for inclusion—that is, whether open space or parkland loses recreational or aesthetic utility below a threshold parcel size—is a question yet to be answered.

"Publicly accessible" means publicly or privately owned open space to which the general public has legal access, with or without an entry fee. A space is not publicly accessible if access is limited to members of specific groups or organizations. For example, a public or private golf course would be considered publicly accessible unless entry was restricted to club members. A farm would not be publicly accessible, nor would a country club. A privately owned but vacant and overgrown industrial site would not be publicly accessible.

The Data Gap

There are at least two methods for calculating the amount of open space and determining whether it is publicly accessible:

Self-Reported Acreage: Cities, counties, special districts, and states can report the acreage of public parks and open spaces they administer inside metropolitan areas. Public parks and publicly owned open spaces would be assumed to be publicly accessible. Accuracy would be limited by inconsistent standards among jurisdictions in the same metropolitan area for defining parks and open spaces. Historical data from cities may be affected by boundary changes associated with annexations. Hardened playground surfaces would likely be included in the data; many water bodies would likely be excluded, as would private lands that are effectively public by virtue of the owners' access policies.

Direct Measurement: Satellite imagery can identify unbuilt open spaces with naturally pervious surfaces. Tax assessment records might be used to locate tax-exempt parcels inside the identified open spaces. The tax records normally identify the basis for each parcel's tax exemption, making it possible to infer which parcels are publicly accessible. More research is needed to determine the suitability of tax assessor records. Although tax assessment records are usually maintained by counties, in some jurisdictions cities, districts, or states may maintain the records. Some assessment records are maintained by these local jurisdictions in geographic information system (GIS) databases. GIS-based records make it easier, faster, and cheaper to derive the indicator, although it would be possible to do it with non-GIS records.

The data from both methods can be aggregated within each metropolitan area and aggregated again across all metropolitan areas for a national measure.

Before such an effort is put in place, some threshold of extent or population size would have to be developed to determine which cities, suburbs, and aggregations of the two should be included. Once this selection is completed, the per capita calculations would be carried out using population data from the Census Bureau.

References

National Research Council. 2000. Ecological indicators for the nation. Washington, DC: National Academy Press.

Natural Ecosystem Services

There is no technical note for this indicator.